Forensic Anthropology

Forensic Anthropology

CONTEMPORARY THEORY AND PRACTICE

Debra A. Komar

University of New Mexico

Jane E. Buikstra

Arizona State University

New York   Oxford

OXFORD UNIVERSITY PRESS

2008

Oxford University Press, Inc., publishes works that further Oxford University's
objective of excellence in research, scholarship, and education.

Oxford New York
Auckland Cape Town Dar es Salaam Hong Kong Karachi
Kuala Lumpur Madrid Melbourne Mexico City Nairobi
New Delhi Shanghai Taipei Toronto

With offices in
Argentina Austria Brazil Chile Czech Republic France Greece
Guatemala Hungary Italy Japan Poland Portugal Singapore
South Korea Switzerland Thailand Turkey Ukraine Vietnam

Published by Oxford University Press, Inc.
198 Madison Avenue, New York, New York 10016
http://www.oup.com

Oxford is a registered trademark of Oxford University Press

Library of Congress Cataloging-in-Publication Data

Komar, Debra A.
  Forensic anthropology: contemporary theory and practice / Debra A. Komar,
Jane E. Buikstra.
    p. cm.
  Includes bibliographical references and index.
  ISBN 978-0-19-530029-1 (hardcover)
  1. Forensic anthropology—Handbooks, manuals, etc. I. Buikstra, Jane E. II. Title.
  GN69.8.K65 2008
  614'.17—dc22   2007015839

*To the founding members of the Physical Anthropology
Section of the American Academy of Forensic Sciences,
whose vision of forensic anthropology's vital role
within the forensic sciences anchors
the ongoing success of our discipline.*

Contents

Preface

The motivation to write this book arose from the perceived need for an advanced textbook, suitable for senior undergraduates and graduate students, that addresses topics seldom synthesized in texts and case study volumes. Although numerous such works detail methodologies associated with forensic anthropology, such as aging and sex determination techniques, we felt the need for an in-depth examination of the current state of forensic anthropology from a theoretical perspective. We are committed both to the education and training of future practitioners of forensic sciences and to the development of the discipline as a recognized, professional entity.

This book is not intended as a "how-to-manual" methodology is best taught in a laboratory setting under the guidance of an experienced practitioner. Rather, the book focuses on placing forensic anthropology within the larger context of medicolegal investigations of death, including the role of the anthropologist in relation to law enforcement and the medical examiner or coroner. This focus includes integration of anthropology into the legal system, relating both to practitioners and to the theories and methods they employ. In the United States, recent revisions to the rules governing scientific and technical expert witness testimony have prompted significant changes in the admissibility of evidence, including the contributions of anthropologists. The impact of these rulings, the increased need for scientific rigor, and the evolving nature of anthropological studies are highlighted throughout this text as a means of preparing students for the demands of the judicial system that will one day evaluate their work.

Perhaps the greatest benefit derived from the recent changes in judicial guidelines is the call for increased scientific rigor in all of the forensic disciplines. In response to the revised standards, the forensic anthropology community has reevaluated many traditional techniques and theories in light of the need for testability, known error rates, and widespread acceptance. While much methodology has withstood scrutiny, some techniques have been challenged or even abandoned, prompting renewed interest in both the empirical testing of existing methods and the quest for innovative new approaches.

The impact of this research is felt far beyond forensic anthropology. Improved accuracy rates, as well as a better understanding of the external validity of methods, have had a significant effect on the study of bioarchaeology, paleoanthropology, and human variation. In a sense, forensic anthropology serves as the "canary in the mine" for physical anthropology subfields that lack the documented specimens necessary to develop or test methodology.

This book examines the vital role research plays in forensic applications of anthropology. Rather than presenting the methods as facts, we have opted for a critical review of the state of our art, particularly in light of the new judicial requirements. In so doing, we attempt to shift the focus from simply mastering techniques to gaining a greater understanding of how those techniques address our questions. What are we really saying when an unknown individual is classified as "white"? How do we define group identity in human rights investigations? Should future research focus on increasing methodological accuracy or universality? As the discipline has evolved, so too has the complexity of the questions raised.

It is this evolution of the discipline that we have aimed to capture in this book. Historically, forensic anthropologists were physical anthropologists who participated in medicolegal investigation to varying degrees. Today, students are expressly training for careers as forensic anthropologists. Forensic anthropology is no longer a mere subfield of biological anthropology but a recognized discipline in its own right. Gone are the days when physical anthropologists could simply declare themselves proficient in forensic work. Specialized training, board certification, and dedicated research programs are all welcome developments in forensic anthropology. This book is our contribution to the continuing professionalization of this dynamic and vital discipline.

ACKNOWLEDGMENTS

This book would not have been possible without the invaluable contributions of many people and institutions, to whom the authors are truly grateful. We are indebted to Chris Grivas, Heather Edgar, and Sarah Lathrop for their many contributions, from reference checking to access to collections. Claire Gordon, Eugene Giles, Linda Klepinger, and Alicia Wilbur provided comments on chapter drafts, as did the following external reviewers for Oxford University Press: Ann W. Bunch, State University of New York at Oswego; Kenneth A. R. Kennedy, Cornell University; Stephen P. Nawrocki, University of Indianapolis; Steven Ousley, Osteology Lab Director, Repatriation Office of Anthropology, Smithsonian Institution; Dawnie Wolfe Steadman, Binghamton University; Michael Warren and Laurel Freas, University of Florida; and Daniel J. Wescott, University of Missouri. Their insightful and detailed comments markedly improved the final product.

Debra Komar would like to recognize the influence of her mentor, Dr. Owen Beattie, whose commitment to human rights inspired her own and who taught her, in his own quiet way, to let the work speak for itself.

Jane Buikstra offers thanks to Dr. Charles (Chuck) Merbs, her mentor, who introduced her to forensic anthropology as a significant aspect of anthropology's mission.

This book grew from discussions between the authors at the University of New Mexico, while both were on the faculty there. They acknowledge the support of the Department of Anthropology at the University of New Mexico and the School of Human Evolution and Social Change and Center for Bioarchaeological Research at Arizona State University, the second author's new faculty home. Special thanks also to the Office of the Medical Investigator (OMI) in the University of New Mexico's School of Medicine for continued support to the first author, who holds a joint appointment between the OMI and the Department of Anthropology, as well as for the generous permission to use many of the photographs included in this book. The final revisions by the second author were completed while she was on a fellowship at the Institute of Advanced Studies at Durham University, with thanks extended to them also. Finally, this volume would not have been possible without the support and contributions of Jan Beatty and her staff at Oxford University Press.

About the Authors

Debra Komar, *PhD, University of Alberta, 1999* is an associate professor of anthropology at the University of New Mexico, a joint appointment together with her position as Forensic Anthropologist for the Office of the Medical Investigator in UNM's School of Medicine. She is a member of both the American Academy of Forensic Sciences (Physical Anthropology Section) and the Canadian Society of Forensic Science. Dr. Komar's research focuses on issues relating to identification and identity in mass death, specifically genocide, as well as taphonomic studies. She has participated in international human rights investigations in Bosnia, Kosovo, Iraq, and Darfur, working for such agencies as the United Nations Mission in Kosovo, Physicians for Human Rights, and the International Commission on Missing Persons.

Jane Buikstra, *PhD, University of Chicago, 1972* is a professor of bioarchaeology in the School of Human Evolution and Social Change and is director of the Center for Bioarchaeological Research at Arizona State University. She is a member of the National Academy of Sciences, president of the Center for American Archeology, and past president of the American Anthropological Association, the American Association of Physical Anthropologists, and the Paleopathology Association. Dr. Buikstra's research includes bioarchaeology, funerary archaeology, paleopathology, forensic anthropology, and paleodemography. She has been a member of the Physical Anthropology Section of the American Academy of Forensic Sciences since the mid-1970s and was elected Fellow in 1980. In addition, she has been a diplomate of the American Board of Forensic Anthropology since 1978, having served on the executive board and the Ethics Committee. She has been a member of the editorial board for the *Journal of Forensic Sciences* since 1989. Author or editor of 16 books and over 150 articles or chapters, Dr. Buikstra has focused her work on individuation, trauma, and the history of forensic anthropology. She is the coeditor of *Human Identification: Case Studies in Forensic Anthropology,* and *Standards for Data Collection from Human Skeletal Remains.*

Forensic Anthropology

CHAPTER 1

Introduction

Forensic anthropology, as practiced in the United States, traces its roots to the analysis of human remains from contexts as diverse as Harvard privies and Chicago's meat packing industry (Snow, 1982; Stewart, 1979a). These and other early applications of anthropological expertise to medicolegal contexts commonly focused on providing biological profiles—age-at-death, sex, stature, anomalous features—from the study of human bones, an initial step in the individual identification process. The centerpiece of such efforts was laboratory analysis, and remains were commonly studied in isolation from their contexts. Even as recently as 1979, T. Dale Stewart, a major figure in the development of forensic anthropology, noted that most bones arrived in his laboratory and that of his successor at the Smithsonian, J. Lawrence Angel, in "paper bags or cardboard cartons," well separated from their point of discovery (Stewart, 1979a: 30). At this stage in the history of our field, Stewart's influential texts—*Essentials of Forensic Anthropology, Especially as Practiced in the United States* stressed the importance of developing biological profiles and expert testimony but did not emphasize crime scene recovery, which Stewart felt might bias an objective evaluation of remains.

During the closing decades of the twentieth century, however, forensic anthropology's purview expanded to encompass the breadth of casework, from discovery to analysis to expert testimony. While neither field recovery nor testifying was previously unknown, increased anthropological engagement in these elements of forensic practice underscores the need for our students to undergo systematic and rigorous training well beyond standard osteology. In parallel, those who train the next generations of forensic anthropologists must be sensitized to the breadth of medicolegal topics that future forensic practitioners must command.

As recently as 1980, there were no recognized degree programs in forensic anthropology. Our field was alone among the sections within the

American Academy of Forensic Sciences in having only a few BA (five) and MA (four) "tracks" and nothing more. The 42 members of the Physical Anthropology Section held PhD, MA, DDS, or MD degrees, having developed their forensic specialty in human identification without formalized coursework in forensic science (Brooks, 1981).

A more recent appraisal of educational programs in forensic anthropology (Galloway and Simmons, 1997) sagely expresses concern about a number of significant issues, including the need for training that emphasizes the unique responsibilities of the *forensic* anthropologist. Despite greatly increased visibility within the forensic sciences and in the public domain, career opportunities for forensic anthropologists remain limited. It is nevertheless clear that specialized instruction in crime scene investigation and expert testimony, among other topics, must be part of a balanced curriculum to provide "superior training for an optimal number of students for whom employment in either academic or medicolegal settings is likely" (Galloway and Simmons, 1997: 801). This volume is our attempt to provide a text useful in such training programs. We also hope it will be of interest to educators and practitioners.

The primary focus of this introductory chapter is to contextualize the remainder of the volume in the current status of this vital and highly visible specialty, as practiced in the United States. We begin by placing the professionalization of American forensic anthropology within its recent historical context and then discuss forensic anthropology and forensic anthropologists today. Other historical treatments that provide deeper temporal perspectives may be found in the following references: Bass (1969, 1979), Buikstra et al. (2003), Galloway and Simmons (1997), Grisbaum and Ubelaker (2001), Kerley (1978), Reichs (1995, 1998a), Snow (1982), Stewart (1979a, 1979b), Thompson (1982), Ubelaker (1996b, 1997, 2000a–c, 2001), Wienker and Rhine (1989), and Wood and Stanley (1989). We then briefly compare U.S. forensic anthropology with that of other countries. Finally, we provide an overview of the remaining chapters of the volume

FORENSIC ANTHROPOLOGY: A BRIEF HISTORY (1972–2006)

As emphasized by Stewart (1979a, 1979b) and others, forensic anthropology was formally recognized as a field only with the establishment of a Physical Anthropology Section within the American Academy of Forensic Sciences (AAFS) in 1972. There were 14 founding members (Ubelaker, 1996b, 2000a). After more than thirty years, the membership numbers nearly 300. While anthropologists such as George Dorsey, Ales Hrdlicka, Wilton Krogman, and Ted McCown clearly contributed to medicolegal cases prior to 1972, this date marks a turning point in American forensic anthropology as a profession.

In 1977 the field was further advanced through the formation of the American Board of Forensic Anthropology (ABFA). The ABFA certifies that its diplomates (DABFA) are qualified to engage in forensic anthropological casework, though the ABFA does not certify a diplomate's performance in individual cases. In 1978 there were 22 diplomates (Ubelaker, 1996b, 2000a). By 2006, 52 additional certificates had been issued to individuals who had passed rigorous written and practical examinations. In 2006 there were 61 active DABFAs, and 2 retired members. Ten schools with DABFAs on their faculty have training programs in forensic anthropology. Diplomates are recertified annually by filing a form stating their level of engagement in forensic casework, teaching, and research. Such surveys prove valuable because they provide systematic assessments of trends in forensic activities by certified practitioners. Over time, there has been a clear increase in the presence of ABFA members in courtrooms, in international venues, and in analyses of nonskeletonized remains. Trauma analysis has assumed special prominence in expert testimony, nearly overtaking individual identification. Additional information about the ABFA can be found on their website (www.csuchico.edu/anth/ABFA), which is maintained at California State University, Chico.

To illustrate the diverse and nonforensic educational background of those engaged in the early professionalization of forensic anthropology, we list the names and dissertation titles of the founding members of the Physical Anthropology sections of the AAFS and the first DABFAs in Box 1.1. With the exception of Dan Morse, MD, all were physical anthropologists, most ($n = 17$) having focused their dissertations on the skeletal biology of archaeological samples. Three had conducted research in primatology and an additional four emphasized **human biology**. Only two, Sheilagh Brooks and Ellis Kerley, had developed dissertations with obvious and direct forensic applications.

BOX 1.1

Dissertation titles for founding members of the Physical Anthropology section of the American Academy of Forensic Sciences (AAFS) and the American Board of Forensic Anthropology (designated by DABFA number)

J. Lawrence Angel (DABFA 9)
A Preliminary Study of the Relations of Race to Culture, Based on Ancient Greek Skeletal Material, Harvard University (1942)

George J. Armelagos (AAFS)
Paleopathology of Three Archeological Populations from Sudanese Nubia, University of Colorado at Boulder (1968)

William M. Bass III (AAFS) (DABFA 6)
The Variation in Physical Types of the Prehistoric Plains Indians, University of Pennsylvania (1961)

Walter H. Birkby (AAFS) (DABFA 7)
Discontinuous Morphological Traits of the Skull as Population Markers in the Prehistoric Southwest, University of Kansas (1973)

Sheilagh Brooks (AAFS) (DBFA 10)
A Comparison of the Criteria of Age Determination of Human Skeletons by Cranial and Pelvic Morphology, University of California–Berkeley (1951)

Alice Brues (AAFS)
Sibling Resemblances as Evidence for the Genetic Determination of Traits of the Eye, Skin and Hair in Man, Radcliffe College (1940)

Jane E. Buikstra (DABFA 11)
Hopewell in the Lower Illinois Valley: A Regional Approach to the Study of Biological Variability and Mortuary Activity, University of Chicago (1972)

Michael Charney (DABFA 12)
ABO Blood Groups and Asthma—A Suspected Correlation, University of Colorado at Boulder (1969)

Michael Finnegan (DABFA 13)
Population Definition on the Northwest Coast by Analysis of Discrete Character Variation, University of Colorado at Boulder (1972)

Eugene Giles (AAFS) (DABFA 14)
A Genetic Study in the Markham Valley, Northeastern New Guinea, Harvard University (1966)

George W. Gill (DABFA 15)
Prehistoric Inhabitants of Northern Coastal Nayarit: Skeletal Analysis and Description of Burials, University of Kansas (1971)

Rodger Heglar (AAFS) (DABFA 5)
The Prehistoric Population of Cochiti Pueblo and Selected Inter-Population Biological Comparisons. University of Michigan (1974)

Richard Jantz (AAFS)
Change and Variation in Skeletal Populations of Arikara Indians, University of Kansas (1970)

Kenneth A. R. Kennedy (DABFA 16)
The Balangodese of Ceylon: Their Biological and Cultural Affinities with the Vedda, University of California–Berkeley (1962)

Ellis R. Kerley (AAFS) (DABFA 1)
The Microscopic Determination of Age in Human Bone, University of Michigan (1962)

William R. Maples (DABFA 17)
An Analysis of the Taxonomic Status of the Kenya Baboon (Papio doguera and Papio cynocephalus), University of Texas–Austin (1967)

Richard McWilliams (AAFS)
Gran Quivira Pueblo and Biological Distance in the United States Southwest, Arizona State University (1974)

Dan Morse, MD (DABFA 8)
No dissertation

Stanley J. Rhine (DABFA 18)
The Varieties of Human Prehensions: A Morphological and Functional Investigation, University of Colorado at Boulder (1969)

Louise M. Robbins (DABFA 19)
The Identification of the Prehistoric Shawnee Indians—The Description of the Population of the Fort Ancient Aspect, Indiana University (1968)

Stephen Rosen (AAFS) (DABFA 4)
A Comparative Study of the Microscopic Anatomy of Non-human Primate Head Hair, University of Kansas (1969)

Norman J. Sauer (DABFA 20)
An Analysis of the Human Skeletal Material from the Fletcher Site (20BY28), Bay City, Michigan, Michigan State University (1974)

Frank P. Saul (DABFA 21)
Disease and Death in an Ancient Maya Community: An Osteobiographic Analysis, Harvard University (1972)

Clyde C. Snow (AAFS) (DABFA 3)
The Physical Growth and Development of the Open-Land Baboon, Papio doguera, University of Arizona (1967)

Richard Gerald Snyder (AAFS) (DABFA 2)
The Dental Morphology of the Point of Pines Indians, University of Arizona (1959)

Audrey Sublett (AAFS)
Seneca Physical Type and Changes Through Time, State University of New York–Buffalo (1966)

Judy M. Suchey (DABFA 22)
Biological Distance of Prehistoric Central California Populations Derived from Non-Metric Traits of the Cranium, University of California–Riverside (1975)

We further explore the histories and contributions of five forensic anthropologists in Boxes 1.2 to 1.6. These examples reflect the diverse experiences and career paths associated with the architects of forensic anthropology as a profession. Most arrived at their forensic casework following broad training in **physical anthropology**, with specialties ranging from primatology to **bioarchaeology**. Few dissertations contributed directly

BOX 1.2
William M. Bass III (b. 1928)

Although Bass majored in psychology at the University of Virginia (BA 1951), coursework in anthropology also caught his attention, particularly archaeology and physical anthropology. Following service in the Korean conflict, his intended graduate program in psychology at the University of Kentucky was diverted by Prof. Charles E. Snow (Bass, 1968), whose human osteological research and forensic anthropology work influenced Bass's decision to study with Wilton M. Krogman at the University of Pennsylvania. While his dissertation research focused on human skeletal variation in Plains Indians (PhD 1961), Bass also participated in forensic casework, tutored by Krogman. The Smithsonian's T. Dale Stewart served as a member of his dissertation committee, reflecting the beginning of a professional relationship that would span the decades until Stewart's death in 1997. Following a brief tenure at the University of Nebraska, Bass joined the faculty at the University of Kansas (1960–1971), subsequently moving to the University of Tennessee (UT) as professor and department head (1971–1992). during this period, he founded the Forensic Anthropology Center (FAC) at UT and remained its director after his headship ended in 1992. Following retirement (1994) from the faculty, he continued to head the FAC until 1999.

Bass's many achievements include the breadth of his research and written communications in forensic anthropology. His total publications number around 200, including a volume for nonspecialists that recounts his forensic research and casework, *Death's Acre* (Bass and Jefferson, 2003). A pioneering achievement was the Anthropological Research Facility at the University of Tennessee, created in 1971, where human cadavers are used to systematically study postmortem changes. His caseload increased from an average of 7.2 in the early 1970s to 45 per year in the 1990s, reflecting Bass's growing engagement with forensic anthropology and paralleling law enforcement's recognition of anthropology's important role in forensic science. A founding member of both the Physical Anthropology Section of the AAFS and the ABFA, Bass was active in shaping the profession through both direct action and his student legacy. A 1992 survey found that among practicing forensic anthropologists, Bass was the most frequently listed adviser. Ever popular as teacher and mentor, he supervised 20 PhD dissertations. For his service to the profession, Bass was awarded the Distinguished Fellow Award from the AAFS in 1994, the same year he was honored by a symposium at the annual AAFS meeting.

(Ubelaker and Hunt, 1995)

BOX 1.3
Sheilagh T. Brooks (b. 1923)

Sheilagh Brooks' BA degree was awarded (with honors) from the University of California—Berkeley in 1944. Following an MA in paleontology and anthropology, she concentrated on human skeletal biology and obtained the first PhD in physical anthropology to be awarded to a woman by UC–Berkeley (1951). Her adviser, Theodore C. McCown, had developed an interest in forensic anthropology through the identification of WWII military deaths, and he influenced Brooks's interest in forensic subjects (Brooks, 1970). Having spent brief periods at various universities in the western United States, Brooks joined the faculty at the University of Colorado–Boulder (1963–1966) and then in 1966 became the first full-time anthropologist on the faculty of the University of Nevada–Las Vegas' Department of Sociology and Anthropology. She remained at UNLV throughout her career, during which she played a key role in establishing a separate Department of Anthropology and Ethnic Studies (1972).

Although skeletal aging criteria were an enduring interest (Brooks, 1955; Brooks and Suchey, 1990), Brooks also contributed to other forensic topics, such as human and nonhuman bone differentiation, curriculum, archaeological recovery, stature estimation from incomplete remains, and descriptions of soft tissues in wound analysis. Her most visible research products include the standards for pubic symphysis aging known as the Suchey-Brooks method. She developed these in collaboration with Judy M. Suchey, based on the observation of 1,225 pubic symphyses of men (739) and women (486) of known age, autopsied in the Department of the Coroner, County of Los Angeles between 1977 and 1979. The technique has gained broad acceptance in the United States and elsewhere.

Brooks was among the founders of the Physical Anthropology Section of the AAFS and the ABFA. She and her husband, Richard Brooks, along with Stan Rhine and Walter Birkby, were a driving force behind the creation and maintenance of the annual regional meetings of forensic anthropologists from the Southwest. Known as "Mountain, Desert and Coastal," this regional group began its existence in a log cabin near Cedar City, Utah, and continues to meet annually at Lake Mead, in Nevada. For her many contributions to forensic anthropology, Brooks received the T. Dale Stewart Award from the AAFS in 1993.

(Powell et al., 2006)

BOX 1.4
Ellis R. Kerley (1924–1998)

Following service in World War II, Kerley completed a BA degree at the University of Kentucky (1950). His emphasis on physical anthropology was influenced by the research and teaching of Charles E. Snow. Intensely involved with identifying the dead from WWII, Snow had been appointed the first head of a central identification laboratory established in Hawaii during 1947 (Bass, 1968; Grisbaum and Ubelaker, 2001). The year Kerley graduated, he requested permission from T. Dale Stewart to study the dentition of archaeological remains from Indian Knoll, Kentucky, thus starting a professional relationship that would extend throughout their respective careers. He next served as staff anthropologist at the Bowman Gray School of Medicine at Wake Forest University, North Carolina, participating in a human biology project (1950–1953). Following this experience, he joined the Department of the Army, Graves Registration, Central Identification Laboratory in Kokura, Japan, where those killed in the Korean conflict were being identified.

Upon return to the United States, Kerley enrolled in graduate school at the University of Michigan, where he received both the MS (1956) and PhD (1962) degrees. His dissertation focused on the estimation of age through the microscopic study of cortical bone from documented individuals, which were made available by the Armed Forces Institute of Pathology. The "Kerley method" became widely used, and served as a model for other workers who extended and refined it.

Kerley joined the faculty of the Department of Anthropology at the University of Kentucky in 1965, subsequently moving to the University of Kansas (1966–1971) and the University of Maryland (1971–1987). After leaving his professorial position at the University of Maryland, he returned to the identification of military remains at the U.S. Army Central Identification laboratory in Hawaii (CILHI), serving first as a consultant (1987–1988) and then director (1988–1991). During his long career, Kerley participated in a number of high-profile cases, including the JFK assassination and the Josef Mengele identification. Later in his career, he participated in war crimes investigations in Bosnia (1997, 1998).

Along with Clyde Snow and William Bass, Kerley was one of the principal architects of the Physical Anthropology Section of the AAFS. By all accounts, Kerley was the prime mover. Previously, only he (Pathology & Biology Section) and Wilton Krogman (Retired Fellow, Criminalistics Section) had been Fellows of the AAFS. Bill Bass and Clyde Snow had become provisional members of the Pathology & Biology Section prior to the formation of the Physical Anthropology Section. Kerley served an unprecedented three terms as chairman of the section and was also the first president of the ABFA. Following numerous other positions within the administrative structure of the Academy, he became the first physical anthropologist to serve as AAFS president (1990–1991). Deservedly, he was the first recipient of the Physical Anthropology Section Award (later the T. Dale Stewart Award).

(Ubelaker, 2001)

BOX 1.5
William R. Maples (1937–1997)

William Maples received his BA degree from the University of Texas (UT–Austin) in 1958. An English major until his final semester, Maples' shift to anthropology was influenced by the arrival of Thomas McKern at UT after the termination of Operation Glory, the recovery, exchange, and identification of those killed in the Korean conflict. McKern invited Maples to enter the graduate program, which he did, completing an MA thesis on archaeological materials and then shifting his focus to Africa, where he was employed by the Southwest Foundation for Research and Education, trapping baboons at a field site at Darajani, Kenya. After three years in Kenya, interrupted by a year's return to UT to complete degree requirements, Maples had collected sufficient osteological data to complete a dissertation on baboon taxonomy (1967).

Following two years as an assistant professor at Western Michigan University (1966–1968), Maples joined the faculty at the University of Florida (UFL). Although his early research focused on primate behavior, in 1972 he began to engage in forensic casework. Between 1972 and 1996, Maples authored or coauthored 1,230 forensic reports. Prior to 1987 (1972–1986), his average caseload was 10.1/year. This jumped to 84.8/year during the next decade, as the C. A. Pound Human Identification Laboratory was being established with support from UFL and a private benefactor solicited by Maples. The laboratory opened in 1991, a state-of-the-art facility. It continues today as a permanent establishment for research, training, and casework, thus institutionalizing forensic anthropology at UFL. Maples always emphasized research within his forensic program, conducting primary research himself and encouraging students to address topics ranging from dental and skeletal aging to facial superimposition. Maples' high professional standards, coupled with his supportive yet demanding mentorship, influenced many loyal students who benefited from his guidance.

Maples joined the AAFS in 1974 and became a DABFA in 1978. He subsequently served the Physical Anthropology Section and the Academy in numerous elected roles. For his many contributions to the profession, Maples was honored with the T. Dale Stewart Award, presented by the Physical Anthropology Section of the AAFS in 1996.

The CILHI also benefited from Maples' forensic expertise. From 1985 until his death, he served as a member of a consultant team that regularly reviewed CILHI's procedures and casework. Dedicated to improving and then maintaining the scientific integrity of this facility, he was extremely influential in establishing the professional base from which JPAC-CIL (Box 1.7) operates today.

Beginning in 1984, Maples became involved in biohistorical cases, including Francisco Pizzaro, the Romanovs, Joseph Merrick (the Elephant Man), and Zachary Taylor. These became the basis for his popular book, *Dead Men Do Tell Tales* (Maples and Browning, 1994). The book is engagingly written, but also clearly reflects the respect Maples accorded the subjects of his inquiry, whether high visible personages or not.

(Buikstra and Maples, 1999; Falsetti, 1999; Hoshower, 1999)

BOX 1.6

Clyde Collins Snow (b. 1928–)

While Snow's initial encounters with higher education produced uneven results, he ultimately received his BS degree from Eastern New Mexico University (1951), followed by an MS in zoology from Texas Technical University (1955). After three years in the U.S. Air Force Medical Service Corps, as administrative officer in charge of the Histopathology Center at Lackland Air Force Base, he enrolled at the University of Arizona (1958), seeking an archaeology degree. He soon switched to physical anthropology and ultimately received his PhD based on a study of growth and development in Savannah baboons (1967).

By the time he completed his PhD, Snow had been employed by the Civil Aeromedical Institute (CAMI) for six years. He was part of a team focused on developing better safety procedures in air transportation. Among his responsibilities was investigating airplane crashes on site, an activity that included identifying human remains. This experience anchored his forensic consultations in various U.S. and foreign locations. His high-profile cases included the assassination of John F. Kennedy and the identification of victims of John Wayne Gacy and the Oklahoma City bombing. Biohistorical initiatives ranged from Josef Mengele to Tutankhamen. His contributions are recounted in a popular volume, *Witnesses from the Grave: The Stories Bones Tell* (Joyce and Stover, 1991).

One of three principal architects of the Physical Anthropology Section of the AAFS, Snow engaged in human rights initiatives, sponsored by groups such as Americas Watch. In 1984 Snow was part of a delegation of forensic experts from the American Association for the Advancement of Science who traveled to Argentina to advise human rights groups on procedures for the identification of those who had "disappeared" during the 1976–1983 period of rule by military juntas. Over the ensuing five years, Snow directed forensic recovery and analysis initiatives and helped form and train the Argentine Forensic Anthropology Team (EAAF), a highly visible team working today in human rights initiatives around the globe. Subsequently, in 1992, he directed training during the period the Guatemala Forensic Anthropology Foundation (FAFG) was formed.

Snow has investigated human rights abuses in many other international contexts, including Bosnia, Congo, Chile, El Salvador, Ethiopia, Iraq, and Zimbabwe. For his many contributions to the profession of forensic anthropology, Snow received the T. D. Stewart Award from the AAFS in 1988.

(Schick, 1997)

to advancing forensic anthropology, though several were influenced by the forensic experiences of their advisers and some participated in casework while in graduate school. Their varied careers focused on local cases, human rights initiatives, mass disasters, and military decedents. Some maintained research and teaching agendas throughout their careers, while others seldom taught or left professorships to accept other forensic responsibilities. Some few frequently served as expert witnesses and engaged in high-profile cases; most spent their time creating biological profiles, distinguishing animal from human bone, and identifying prehistoric materials. Despite such diverse experiences, however, they were united in their concern for advancing the profession of forensic anthropology.

WHAT IS FORENSIC ANTHROPOLOGY TODAY?

In 1979 T. Dale Stewart defined forensic anthropology as "the applied branch of physical anthropology that deals with the identification of more or less skeletonized human remains for legal purposes" (Stewart, 1979a: 169). In the same year, he also described the tasks that forensic anthropologists performed. "Beyond the elimination of nonhuman elements, the identification process undertakes to provide opinions regarding sex, age, race, stature, and such other characteristics of each individual involved as may lead to his or her recognition" (Stewart, 1979b: ix).

The current definition of forensic anthropology, as published on the web page of the American Board of Forensic Anthropology, is the following:

Forensic anthropology is the application of the science of physical anthropology to the legal process. The identification of skeletal, badly decomposed, or otherwise unidentified human remains is important for both legal and humanitarian reasons. Forensic anthropologists apply standard scientific techniques developed in physical anthropology to identify human remains, and to assist in the detection of crime. Forensic anthropologists frequently work in conjunction with forensic pathologists, odontologists, and homicide investigators to identify a decedent, discover evidence of foul play, and/or the postmortem interval. In addition to assisting in locating and recovering suspicious remains, forensic anthropologists work to suggest the age, sex, ancestry, stature, and unique features of a decedent from the skeleton.

There are four key distinctions between these two definitions. First, the 1979 formulation focused primarily on the *identification* of remains. Stewart's

1979 handbook of forensic anthropology did include a chapter on judging time and cause of death, but it was clear that individual identification was the central theme (Stewart, 1979b). A generation late, we continue to stress identification in addition to underscoring that forensic anthropologists also "assist in the detection of crime" and "discover evidence of foul play, and/or postmortem interval." Today, for example, forensic anthropologists work alongside other medical professionals in the recognition of child abuse, through the identification of fracture patterns by location and by providing osteological evidence of repetitive trauma (Dirkmaat et al., 2002; Walker et al., 1997). The mandate for today's forensic anthropologists has clearly broadened since the time of Stewart's landmark volume.

Second, humanitarian initiatives such as the investigation of human rights violations are recognized in the 2006 definition. As we shall see in Chapter 9, these have become highly visible, especially within the past decade. They are a major source of employment outside the traditional university and museum settings.

A third point is the shift from Stewart's "more or less skeletonized remains" to "skeletal, badly decomposed, or otherwise unidentified human remains." Recent surveys of DABFA casework reveal increased involvement with decomposed and fresh remains. In 1992, for example, most DABFA cases involved skeletons/skulls (63% of cases) rather than decomposed (11%) and fresh (11%) remains (Reichs, 1995). In 1996 these figures had shifted to 38, 7, and 23%, respectively (Ubelaker, 2000b). While this trend may not be heartening to those with queasy digestive systems, it does reflect the increased engagement of forensic anthropologists in a broad array of medicolegal contexts.

Finally, Stewart's volume (1979a) did not envision forensic anthropologists as part of recovery teams. Stewart acknowledged that in burial contexts, archaeological techniques are important, but he expressed skepticism about the development of forensic archaeology as a field. Many of his contemporaries were also concerned that contextual knowledge or other information developed in the course of an investigation might compromise the anthropologist's objectivity in characterizing the remains. The issue of objectivity is real, in that forensic anthropologists are obligated to report precisely what they observe rather than generating results distorted to be comparable with other lines of evidence. Often a seemingly anomalous observation proves crucial later in the investigation, and the forensic anthropologist must maintain scientific objectivity. That said, the forensic anthropologist's presence in certain recovery contexts (as indicated in Chapter 4) is invaluable and far outweighs the concerns expressed by earlier Stewart and others.

WHERE DO FORENSIC ANTHROPOLOGISTS WORK?

Many forensic anthropologists continue to be employed in university or museum settings, where their casework is considered to be "public service."

While their contributions are truly significant and frequently quite visible, the sad truth is that this form of public service is commonly less important in promotion and **tenure** decisions than teaching, research grants, and publications. High-quality research, of course, is the cornerstone of forensic anthropology, but a high caseload can limit the amount of time available for research. Time management thus can become a difficult balancing act for forensic anthropologists working in university settings.

Employment patterns are clearly changing, however, with many more options open to individuals at both junior and senior levels. In 1972, when the Physical Anthropology Section of the AAFS was formed, it is doubtful that any of the founding members spent a majority of their time on forensic casework and research. Similarly, the 22 founding DABFAs were primarily based in traditional settings. By contrast, in 2006, of the 61 practicing diplomates, fully a quarter are employed outside "the **academy**" in a variety of private, state, or federal agencies where they practice forensic anthropology.

The primary employer of full-time forensic anthropologists today is the Joint POW/MIA Accounting Command's Central Identification Laboratory (JPAC-CIL). Calling itself "the world's largest forensic anthropology laboratory," the CIL focuses on searching for, recovering, and identifying personnel unaccounted for from hostilities in Southeast Asia, Korea, World War II, and the Cold War (www.jpac.pacom.mil/Operations.htm). The CIL website reports that more that 1,800 persons are still missing from the Vietnam War, 120 from the Cold War, 8,100 from the Korean conflict, and more than 78,000 from WWII, an indication the ongoing need for the CIL's engagement.

As noted in Box 1.7, the JPAC CIL are also engaged in humanitarian efforts across the globe, many times engaging with the victims of mass disasters. Other governmental and nongovernmental agencies involved in humanitarian initiatives employ either full-time or part-time forensic anthropologists. In 2001, Steadman and Haglund conducted a preliminary survey concerning the involvement of forensic anthropologists in humanitarian efforts during the period from 1990 to 1999. The following groups were reviewed: Physicians for Human Rights (PHR), the Argentine Forensic Anthropology Team (EAAF), the Guatemalan Forensic Anthropology Team (EAFG), and the International Criminal Tribunal for the Former Yugoslavia (ICTY). For security reasons, data from the ICTY were limited. Steadman and Haglund reported that a minimum of 97 anthropologists had participated in human rights investigations during this period. Twenty different nationalities were represented, with deployment to 32 nations.

Steadman and Haglund (2001) caution that the number of full-time positions for forensic anthropologists identified in their survey was relatively low. While incidents of human rights abuses are unfortunately not decreasing, it is unlikely that large numbers of forensic anthropologists will be employed full time in the near term

There obviously continues to be, however, an important role for both part-time and permanent staff, including forensic anthropologists, in these

> **BOX 1.7**
>
> ## JPAC-CIL (Joint POW/MIA Accounting Command–Central Identification Laboratory)
>
> The CIL employs 30 forensic anthropologists, approximately half holding doctoral degrees and including six diplomates of the ABFA. Several are postgraduate fellows supported by the Oak Ridge Institute for Science and Education (ORISE). CIL forensic anthropologists typically have both field recovery and laboratory casework responsibilities, as well as expectations for presenting and publishing research results.
>
> A typical year sees 75 to 100 recovery missions directed toward missing U.S. servicemen. There are 18 recovery teams, which deploy for 14 to 60 days at a time, most commonly 30 to 45 days. During recent years CIL has identified over 100 individuals a year. An average of two sets of remains are returned home each week.
>
> CIL uses a variety of methods for establishing identity, integrating forensic anthropological and odontological evaluations with assessments of mitochondrial DNA. An extensive odontological database has also been generated (ODONTOSEARCH) and is available through the CIL website, www.jpac.pacom.mil/Operations.htm. This database contains information on missing, filled, and unrestored teeth from a large sample of the U.S. population. Thus, it is possible to determine the frequency of certain features and clusters of features as one develops an identification.
>
> CIL teams are also deployed around the world on humanitarian missions. During recent years, CIL has sent missions to North Korea, South Korea, Vietnam, Laos, Cambodia, China, Iran, Kuwait, Burma, Tibet, Russia, Panama, Nicaragua, Papua New Guinea, Solomon Islands, Marshall Islands, Indonesia, Gilbert Islands, Guam, France, Germany, Holland, Belgium, Italy, Albania, Turkey, England, Norway, Latvia, and the United States.

important humanitarian efforts. As Steadman and Haglund (2001) emphasize: "the need for the objective collection and evaluation of physical evidence by forensic scientists for use in international courts has never been greater. When mass graves are found, anthropologists and archaeologists become essential components of the scientific investigation."

IS FORENSIC ANTHROPOLOGY EXPANDING IN THE UNITED STATES?

We have already noted that the Physical Anthropology Section of the AAFS has expanded from 14 founding members in 1972 to nearly 300 in 2006. In

BOX 1.8

Forensic Anthropology and the Media

Forensic anthropologists are frequently queried about the relationship between fictionalized accounts and "real" casework. "Who are forensic anthropologists" they are asked. Is Gideon Oliver, an invention of anthropologically trained mystery writer Aaron Elkins, a good example? What about Temperance Brennan, a fictionalized forensic anthropologist created by Kathy Reichs, who is herself a "real" forensic anthropologist?

As emphasized in this volume, during the 15 years between the publication of Elkins' first Gideon Oliver mystery (*Fellowship of Fear*, 1982) and Reichs' initial contribution (*Déja Dead*, 1997), the field of forensic anthropology changed remarkably, a trend that continues today. Through most of the twentieth century, Gideon Oliver would have been considered a typical forensic anthropologist. His full-time job was teaching physical anthropology at a university (or a museum), which had a laboratory where he could study both ancient remains and bones from the recently deceased. Skills learned in human anatomy classes were readily transferred to remains of medicolegal significance. Forensic anthropologists of this era rarely visited crime scenes or engaged in search and recovery efforts mounted by law enforcement personnel. Many forensic cases at this time would have involved discriminating chicken, dog, and bear bones from human remains or assuring law enforcement personnel that materials they had encountered were prehistoric and not of medicolegal significance. Seldom would the practitioner appear in court and even more rarely in a medical examiner's autopsy suite. Standard forensic osteological analyses generated biological profiles (age-at death, sex, stature, inherited features, pathology) similar to those developed for ancient archaeological materials.

By contrast, Temperance Brennan's primary employment is as a forensic anthropologist in the Laboratoire de Médecine Légale in Montreal, which would have been a rare and bold career move until the closing years of the twentieth century. Now, however, the "applied" field of forensic anthropology is developing rapidly and Temperance's job security is assured. In fact, Dr. Brennan has taken her message to television, where she appears weekly as the star of *Bones*.

1978 the number of ABFA diplomates numbered 22. The total number of certificates issued currently stands at 74.

Increased activity is also clearly reflected in the pages of the *Journal of Forensic Sciences*, the flagship journal of the AAFS, which was founded in 1956. For example, the number of articles in forensic anthropology nearly tripled between the 1970s and the 1980s, almost doubling again during the 1990s (Buikstra, 2002). Another measure of vigor and visibility is the engagement of students in professional venues. Of 82 abstracts accepted for presentation at the 2006 annual meeting of the AAFS, nearly half (45%) the first authors held BA or MA degrees. While this may be only an approximate measure of student engagement, as the PhD degree is normally required for employment as a forensic anthropologist and to attain DABFA status, it does suggest a remarkable level of youthful activity in forensic anthropology.

A different dimension of forensic anthropology's visibility is its popu-larity beyond the academy—on the printed page, in movies, and on televi-sion. While television shows such as *Bones* and *CSI* have markedly increased the visibility of forensic anthropology, the downside is that jurors and lawyers may have expectations that extend well beyond our science.

In Box 1.8 we offer a comparison of two forensic anthropologists in the media whose fictional characters reflect the changing character of forensic anthropology over recent decades. We note that attempts to move Aaron Elkins' creation onto television lasted only a few episodes during the 1980s. By contrast, Kathy Reich's Temperance Brennan appears to have made a more successful transition to twenty-first-century television. This reflects the media and the American public's current fascination with forensic science, especially forensic anthropology. To paraphrase Carl Sagan (1993: 3–4), it indeed appears that people are hungry for forensic science explained in nonintimidating ways with some of the grandeur and wonder left in.

FORENSIC ANTHROPOLOGY OUTSIDE THE UNITED STATES

In Latin America, non-U.S. teams that include forensic anthropologists are also regularly deployed, both nationally and internationally. The most well known of these is the Argentine Forensic Anthropology Team (EAAF), which was founded as a **nongovernmental organization (NGO)** during 1984 to investigate those "disappeared" between 1976 and 1983 during the "Dirty War." American forensic anthropologist Clyde Snow (Box 1.6) was crucial in the development of this organization and the training of its members. During 2004, the 13 members of the EAAF, along with consultants and volunteers, worked within Argentina and traveled to Bolivia, Brazil, El Salvador, the Ivory Coast, Mexico, the Republic of Georgia, Spain, and Sudan (EAAF Annual Report 2005). Since 1984, EAAF has worked in 26 countries (Doretti and Snow, 2003). Influenced by the EAAF example, similar organizations have developed in Chile, Guatemala, and Peru. A Latin American forensic anthropology association (ALAF) has also been created.

Forensic anthropology as a professional discipline in Canada is often viewed as a hybrid of the U.S. and British systems because the criminal courts in Canada follow English traditions, while the development of the forensic science mirrors that of the United States. The Canadian Society of

Forensic Sciences (CSFS) was formed in 1953 with 17 members. Today, the society has over 450 members, including 12 regular and 4 student/provisional members in the Anthropology Section. The society does not offer board certification to anthropologists, although a few members have opted to seek diplomate status in the ABFA. The society's flagship publication, the *Journal of the CSFS*, began as a newsletter in 1963 and now has a circulation of over 700. Another important milestone was the creation of the Centre of Forensic Sciences in Toronto in 1966. The center has a mandate to assist law enforcement through the analysis and interpretation of evidence, the provision of educational programs and materials, and research to expand forensic science services. There are no full-time forensic anthropologists employed in the Canadian medicolegal system, although numerous members consult with local authorities and law enforcement. Although no formal forensic anthropology graduate training programs exist in Canada, both the University of Toronto at Mississauga and the University of Alberta advertise forensic anthropology graduate studies. Bachelor degrees in forensic science can be pursued at Erindale College (University of Toronto), Laurentian University, and Mount Royal College.

In Europe, the professionalization of forensic anthropology is more recent than in the United States (Black, 2003a, 2003b; Ubelaker, 2000b). While one can trace forensic anthropological studies to much earlier examples from eighteenth- and nineteenth-century France, Spain, and Portugal (Brickley and Ferllini, 2007b; Cunha and Catteneo, 2006; Iscan and Quatrehomme, 1999), as well as exquisitely detailed twentieth-century case studies in the United Kingdom (Glaister and Brash, 1937; S. A. Smith, 1939, 1959), the first use of the term in Europe occurred in 1954 and referenced paternity identification (Schwidetzky, 1954; Stewart, 1984). Although previous important book-length treatments have dealt with forensic subjects (e.g., Fazekas and Kosá, 1978), the tempo appears to be increasing (Brickley and Ferllini, 2007a; Cox and Mays, 2000; Hunter et al., 1996; Schmitt et al., 2006).

European forensic anthropological practitioners are today found primarily at medicolegal institutions and academic departments within the social and natural sciences. As of April 1, 2006, there were 47 members of the Forensic Anthropology Society of Europe, which was organized in 2004. Only three of the European members list anthropology units as their primary professional affiliation. According to Brickley and Ferllini (2007b), disciplinary affiliation does not reflect the degree to which forensic anthropologists are engaged in casework. Medicolegal expertise is sought rarely in the United Kingdom, for example, and frequently in Hungary, even though practitioners in both countries typically have similar relevant training. The varied legal systems across Europe also influence variation in the use of forensic anthropologists in court proceedings. Consultations by the prosecution are more frequent than by the defense, though the latter appear to be rising (Brickley and Ferllini, 2007b).

Black (2003a, 2003b) emphasizes that forensic anthropology's history in the United Kingdom is relatively brief, partly because of disciplinary separation of archaeology and physical anthropology. Forensic anthropology is now, however, in an expansionist state, driven by a number of factors, including popular media and increased global awareness of mass disasters, human rights abuses, and war crimes. Training programs have increased, leading to the need for professional standards and regulation. As Brickley and Ferllini note (2007b), the United Kingdom is the only country presently registering (accrediting) forensic anthropologists, which occurs within the Council for the Registration of Forensic Practitioners. Forensic anthropology, first listed in 2003, includes a total of 9 registrants apportioned across the following categories: general forensic anthropology (8); osteology (1, individual also listed in general forensic anthropology); facial reconstruction (modeling) (1). Archaeology is considered a separate specialty, and as of April 2006, there were five individuals registered.

RESEARCH IN FORENSIC ANTHROPOLOGY TODAY

Recent research in forensic anthropology covers a variety of topics. Some of these are specialized, relevant primarily to contemporary forensic casework. Others, including age, sex, stature, and ancestry, present important new evidence about subjects of traditional concern to both forensic anthropologists and skeletal biologists/bioarchaeologists. In a survey of papers and posters presented by forensic anthropologists at annual meetings of the American Academy of Forensic Sciences between 1979 and 2000, Benedix and Belcher (2006) report that case studies appear to be decreasing, "standard" biological profile research is holding steady, and specialized forensic subjects, including taphonomy, trauma, and forensic archaeology, appear to be increasing. While case studies continue to be edifying, increasing emphasis on research that can be generalized to establish analytical principles and foundations signals the maturation of forensic anthropology as a field.

The practice of forensic anthropology, including the form taken by research, is increasingly focused on developing theories and methods that will be effective during expert testimony. As we shall emphasize in Chapter 3, in the United States there are three rules of law that generally influence a judge's decision to permit an expert to present evidence: the Federal rules of Evidence 702; the *Frye* test, and the *Daubert* ruling. The 1923 *Frye* ruling specified that expert testimony must be both beyond the general knowledge of the jurors and based on methods considered generally reliable among scientists. The 1993 *Daubert* ruling, which superseded *Frye*, places emphasis on testing through scientific methodology, peer review, error rates, and acceptance within the scientific community (Christensen, 2004a). Strictly speaking, while *Daubert* applies only in federal courts, it has been adopted by the majority of states, and its impact on expert testimony is more widespread.

For research in forensic anthropology this means increased emphasis on quantification (Hefner and Ousley, 2006; Konigsberg and Jantz, 2002; Ousley and Hefner, 2005; Steadman et al., 2002a). We must explore, for example, the statistical bases for inferring identity between antemortem records and post-mortem observations. "How probable is it that a given fracture pattern might be drawn from the population at large?" now becomes a significant question. Simply identifying multiple points of similarity between the two sets of records is not sufficient for a positive identification. Scientific, statistically valid rigor should enter each inferential argument. Chapter 5, which considers subjects long held to be fundamental in developing biological profiles—age-at-death, sex, stature, and ancestry—will emphasize the status of each in relationship to the *Daubert* ruling criteria.

In our critical review of standard methods, we also note in Chapters 5 and 8 that certain methods currently fail to meet the *Daubert* criteria. These methods include facial reconstructions and photographic superimposition to establish positive identity. Similarly, occupational marker and handedness inferences are neither methodologically rigorous nor widely accepted. While such methods may be useful aspects in certain contexts, including bio-historical treatments (Chapter 10), they are not admissible in court.

Generating the data required to meet the *Daubert* criteria requires that methods be tested on appropriately documented "knowns." For skeletal attributes, this means recently living, documented individuals. Autopsy series, with clinical records, are commonly used to test osteological methods for estimating ancestry, age-at-death, and sex diagnosis (Box 1.9); see also Usher (2002) and www2.potsdam.edu/usherbm/reference. Another database of "knowns" useful in forensic research is the Forensic Data Bank (FDB) maintained by the University of Tennessee's Forensic Anthropology Center (FAC). The FDB, which contains records for 2,100 cases from across the country, with nearly 1,600 documented for sex and ancestry, is constantly growing through contributions by forensic anthropologists (web.utk.edu/~anthrop/index.htm). Twelve hundred of the FDB cases also carry positive identifications. Extensive documentation for many individuals include place of birth, medical history, occupation, stature, and weight. Skeletal data include cranial and postcranial measurements, along with information on age indicators, nonmetric cranial attributes, and dental features.

The importance of using reference series that are contemporary with the case or population about which forensic inferences are to be made cannot be overemphasized. Recent research compared modern forensic cases in the Forensic Data Bank (Ousley and Jantz, 1996, 1998, 2005) with earlier autopsy collections such as the Terry and the Hammann-Todd, made in the twentieth century but largely representing individuals born during the 1800s.

BOX 1.9

Documented skeletal collections in the United States and Canada

Collection Name and Current Location	Sample Size	Age Range	Sex M/F	Samples Collected
Bass Donated Skeletal Collection University of Tennessee Forensic Center (FAC)	669[a]	Fetal–101 y	491/170	1981–present
W. Montague Cobb Human Skeletal Collection and Biological Laboratory at Howard University	965	17–106 y	684/287	1850–1935
Grant Anatomical Collection Department of Anthropology University of Toronto	282	17–90 y	176/26	1930–1955
Hammann-Todd Osteological Collection Cleveland Museum of Natural History	~3100	Fetal–>90 y	2M/1F	1830s–1930s
George S. Huntington Collection U.S. National Museum of Natural History	4054	5–96 y	75% male	1893–1921
Maxwell Museum Documented Collection Maxwell Museum University of New Mexico	254[a]	Fetal–100 y	158/89	1978–present
Maxwell Museum Forensic Collection Maxwell Museum University of New Mexico	204[a]	Fetal–adult	132/52	1977–present
NMNH Fetal Collection U.S. National Museum of Natural History	320	Fetal	152/129	1904–1917
Robert J. Terry Anatomical Collection U.S. National Museum of Natural History	1728	14–102 y	949/658	1919–1960s
Trotter Fetal Bone Collection Washington University (St. Louis) Medical School	144	Fetal	78/59	1950–1969
Stanford–Meyer Collection San Diego Museum of Man	~3500	Unknown	Unknown	1920s–1950s
Stanford Collection University of Iowa Office of State Archaeologist	~1100	27–96y[b]	4M/1F[b]	1920s–1950s

[a]—As of June 2006.
[b]—Estimated.

Differences have been discovered in morphologies of those classified as "white" and "black," for example, in the temporally distinct samples. Similarly, relationships between limb length and stature apparently differ in the Terry Collection and in the more recent FDB cases. We may be seeing secular trends, the result of sampling biases, or distinctions between those who were cast into socially constructed racial categories during different portions of the twentieth century. While all three factors may be implicated, the most clearly documented appear to be secular trends, which are confirmed by historical sources (Jantz, 2001; Jantz and Meadows Jantz, 2000; Meadows and Jantz, 1995; Meadows Jantz and Jantz, 1999; Ousley and Jantz, 1998). These research results underscore the importance of contemporary reference data sets, whether collections of remains or data banks, in forensic anthropological research.

FORENSIC ANTHROPOLOGY IN PERSPECTIVE

Forensic anthropology is obviously a vital and rapidly developing applied subfield of physical anthropology. Career paths and training programs are proliferating. Training for future careers in forensic anthropology should include the biomedical and physical sciences, as well as biological anthropology. Broader anthropological concerns, especially those relating to human rights and ethics, are also issues of importance. Research is becoming increasingly sophisticated, including topics of specialized interest to forensic anthropologists and others of more general anthropological significance.

The organization of this volume reflects the need for twenty-first-century forensic anthropologists to be trained beyond skeletal biology. Knowledge of medical (Chapter 2) and legal terminology and procedures (Chapter 3) is essential to the anthropologist, who increasingly must communicate with other team members in the course of recovery, in the autopsy suite, and in the courtroom.

Having introduced key terms and their definitions, we then turn to crime scene investigation (Chapter 4). This chapter applies concepts of jurisdiction and authority to differentiating medicolegal and nonmedicolegal remains, scene work, protocols, scene evaluation, search strategies, evidence collection, documentation, and report writing.

Following scene investigation, we begin a series of chapters (Chapters 5–8) that reflect the process of individual identification. In Chapter 5, we evaluate methods for developing descriptive biological profiles, concentrating on sex diagnosis, age and stature estimation, and ancestry. As noted in the preceding section, we emphasize the degree to which these methods satisfy the *Daubert* criteria. Specialized aspects of forensic analysis are treated in Chapter 6, Pathology and Trauma Assessment; Chapter 7, Forensic Taphonomy, and Chapter 8, Personal Identification.

We then close the volume with chapter–length treatments of two subjects that account for part of the recent global visibility of forensic anthropology. In Chapter 9, Mass Death and International Investigations of Human Rights Violations, we consider topics such as mass disasters, war crimes, genocide, establishing individual identity, and establishing medicolegal jurisdiction in international contexts. In our final chapter, Biohistory: Historical Questions, Methods, and Ethics, we sketch a brief history of bio-historical applications, outline various biohistorical approaches, and provide an extended case study of Billy the Kid. The significant ethical issues that must be considered in biohistorical research are also raised.

CHAPTER 2

The Medicolegal System

As emphasized in Chapter 1, forensic anthropology is rapidly changing from an applied aspect of physical anthropology to a profession with distinctive training, requirements, and practitioner responsibilities. Increasingly, forensic anthropologists are appearing in courtrooms and in autopsy suites. To ensure effective communication with members of the medical and legal professions, certain concepts and terminology must be shared. Forensic anthropologists should therefore develop a thorough understanding of the forensic sciences, the medicolegal community, and the judicial system. Once this foundation is in place, forensic anthropologists can transition easily through interactions with law enforcement at crime scenes, pathologists at autopsy, and lawyers at trial.

This chapter is designed to present an overview of the medicolegal concepts and terminology necessary for the forensic anthropologist to function effectively in casework and within the courtroom. We begin by examining the medicolegal system and its legal boundaries, known as jurisdiction. The primary goals of the medicolegal authority—determining cause and manner of death—are then discussed, as well as the related concepts of motive, intent, and volition. Finally, we introduce the principal medicolegal investigative tool, the autopsy or postmortem examination.

THE MEDICOLEGAL SYSTEM

The term **medicolegal** (ML) refers to the application of medical science to law. The **medicolegal system** is the formal mechanism for death investigation and certification within a specific geographic area. The system's head is either a coroner or a medical examiner. A **coroner** is an elected or appointed public official. A coroner need not be a physician and does not necessarily perform or attend autopsies. Unless the coroner is a licensed pathologist, autopsies must be performed by a trained and certified pathologist, who will then report the autopsy findings to the coroner. The training and educational background of coroners vary; in addition to physicians, coroners are also former law enforcement officers or funeral directors. A **medical examiner** (ME) is an appointed public official and must be a licensed, board-certified forensic pathologist. Medical examiners perform autopsies. The professional organization that oversees medical examiners is the National Association of Medical Examiners (NAME). The NAME website (www.thename.org) provides information on the organization and on training programs for those interested in joining the profession as well as a listing of current members.

Both medical examiners and coroners can convene inquests, sometimes called inquiries or fatality inquiry hearings. Similar to a trial, an **inquest** is an inquiry by the medicolegal authority, in the presence of a judge and sometimes with the aid of a jury, into the death of an individual who died under suspicious circumstances or in prison, or whose identity is in question. The proceedings may be open to the public or closed. An inquest differs from a trial in that no one stands accused of a crime, and the findings do not result in penalties. Inquests are used to gather information that may be used at trial and are often held in the interest of public safety. For example, the results of medicolegal inquests led to a change in how manufacturers configured the drawstrings on venetian blinds after a number of small children accidentally choked on the drawstrings.

One of the primary roles of the medicolegal system is death certification. Using independent investigations, inquests, and autopsies, coroners and medical examiners are responsible for determining the cause and manner of deaths that fall under their jurisdiction.

JURISDICTION

One of the most important yet misunderstood aspects of forensic work is the notion of jurisdiction. **Jurisdiction** is the legal power, authority, and responsibility of an agency or investigator to perform an act. Jurisdictional authority is derived from three sources: a government's general power to exercise authority over all persons and things within its territory, a court's power to decide a case or issue a decree, and a geographic area within which political or judicial authority may be exercised (Garner, 2001).

Some of the most common mistakes made by forensic anthropologists relate to jurisdictional issues. It is not an academic degree or becoming a board-certified diplomate that gives an anthropologist the authority to participate in casework or to attend a crime scene or autopsy. It is a formal and express extension to an anthropologist of the existing jurisdictional authority of an agency, such as a medical examiner or a coroner, or law enforcement. This may be granted on a case-by-case basis, called **consultancy**, or as an ongoing formal **appointment**. In other words, you must be invited by those having jurisdiction and the power to extend their jurisdictional authority to you.

Any investigator having jurisdictional authority must be fully aware of the geographic boundaries of his or her authority. For example, the state of New Mexico is one of 19 states in which jurisdiction is centralized (for a list of the 18 other states, see Box 2.1). This means that the geographic boundaries of the medical examiner encompass the entire state, excluding federal land such as military bases or autonomous Native American lands (tribal lands legally considered to be sovereign nations). The state medical examiner may participate in death investigations on Native or federal land if invited by the authority with jurisdiction over such lands. In states with very large populations, such as New York, or large geographic areas, such as Texas or California, jurisdictions follow county or municipal lines. Each jurisdiction has a set of laws, state statutes called medicolegal acts, which clearly outline the medicolegal investigators' rights, responsibilities, and duties.

The location of death is critical in determining jurisdiction; so too are the circumstances of death. Expressly stated within medicolegal acts are the circumstances and types of deaths that fall under the jurisdiction of

---

BOX 2.1

U.S. states having centralized medical examiner systems

Alaska
Connecticut
Delaware
District of Columbia
Iowa
Maine
Maryland
Massachusetts
New Hampshire
New Jersey
New Mexico
Oklahoma
Oregon
Rhode Island
Tennessee
Utah
Vermont
Virginia
West Virginia

the local medical examiner or coroner. Although details may vary nominally by region, the following circumstances normally warrant a medicolegal investigation:

1. All non-natural deaths (homicides, suicides and accidents)
2. All "suspicious" death, such as those occurring outside a doctor's care or sudden deaths of those less than 50 years of age
3. All deaths occurring in nursing homes, state-run facilities, or prisoners in custody
4. All deaths of children less than 1 year of age and women who have given birth within the last year

When a death occurs that meets any one of these criteria and is within the geographic boundary as defined, then the medical examiner or the coroner is obligated to investigate that death.

Death Investigation

Death investigation requires establishing a time frame for the decedent. This time frame represents three distinct periods: antemortem, perimortem, and postmortem. Although each may initially appear to be self-explanatory, defining the perimortem period can be problematic. Since defining both the ante- and postmortem periods relies on first establishing the perimortem period, we initially review this concept.

The **perimortem period** encompasses the events surrounding death. An individual's perimortem period begins with the interaction between the individual and his or her cause of death. Since the cause of death is any disease, trauma, or action that initiates the sequence of events leading to death, this interaction may be brief, such as a fatal gunshot wound, or lengthy, as in the case of a prolonged illness or sequelae from a remote injury. For example, if death resulted from exposure to a toxic chemical during an industrial accident eight years prior, the perimortem period was eight years.

Defining the perimortem period is further complicated in cases of homicide where the interaction between assailant and victim extends beyond the death of the victim. For example, should a killer choose to dismember and dispose of a victim, the dismemberment and disposition of the body occurs within the perimortem period. The perimortem period ends at the conclusion of the interaction between the individual and the causative agent. Thus, the perimortem period does not automatically end with death.

The **antemortem period** (AM) begins at birth and terminates at the beginning of the perimortem period. The **postmortem period** (PM) begins at the end of the perimortem period and continues until the body is discovered.

There remains considerable debate within the forensic community over how these temporal periods are defined. Discrepancies exist between how pathologists and anthropologists differentiate ante-, peri-, and postmortem periods, specifically in terms of what can be scientifically established. For example, dismemberment of the body after death can be identified by pathologists by the absence of hemorrhage in soft tissues. In skeletonized remains, however, the timing of the injury cannot be determined solely from cutmarks on bone. Further, in cases of natural disease such as cancers that create bone lesions, most anthropologists would describe the lesions as antemortem, while the pathologist who determines the cancer to be the cause of death would consider the disease and all of its manifestations to be part of the perimortem period. The purpose of this discussion is not to terminate the debate but rather to highlight for students and practitioners the ambiguities present in current thinking. While the definition of the periods is widely known, its level of acceptance or application is not. For a subject of such import, clearer definitions and a more widely accepted understanding is crucial. Clearly, additional consideration and consensus is urgently needed.

Although individuals may vary in their approaches to investigation, the flow or progression of a case typically follows the sequence illustrated in Figure 2.1. The process begins with the pronouncement of death. The legal definition of death and how it is determined is provided in Box 2.2.

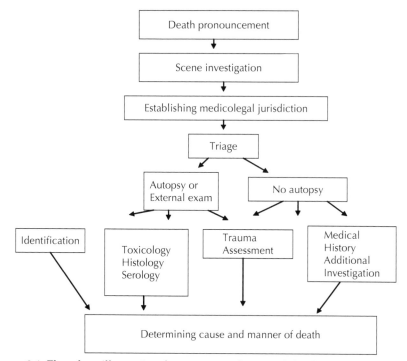

Figure 2.1 Flowchart illustrating the sequence of a typical death investigation.

BOX 2. 2

Definitions of death

Cultural and religious attitudes toward death are highly variable. Differing perceptions of death and the concept of an "afterlife" result in little agreement among scholars or theologians as to what constitutes death. Even from a medical perceptive, with the advent of artificial respirators and technology capable of sustaining life, it is sometimes problematic to define clearly when an individual is legally dead.

To remove any ambiguity, the Western medicolegal definition of death relies on a series of tests and signs to determine when death has occurred. As medicolegal investigators are called on to pronounce death at a scene, recognizing and documenting the following criteria is paramount.

1. Cerebral death: all brain function has ceased, and the condition is irreversible. In a clinical setting, this is evident through a flat brain wave reading. However, in the field, investigators must rely on the following four criteria as evidence of a lack of brain activity.
2. Bilateral dilation and fixation of the pupils.
3. Absence of all reflexes. Tests for pain, such as applying pressure to the sternum or supraorbital region, are standard.
4. Cessation of respiration without mechanical assistance.
5. Cessation of spontaneous cardiac activity (Spitz and Fisher, 1980).

When all the preceding criteria have been met, the individual is pronounced dead and the time and location are noted, since this information will be needed for the death certificate. Only those with the proper authority, such as licensed physicians, medical investigators, or (in some jurisdictions) emergency medical technicians, can legally declare an individual to be dead. The recorded time-of-death reflects the time of pronouncement, not the actual time of death. For a decedent who has clearly been dead for some time, showing advanced signs of decomposition, the time-of-death is declared as the time of pronouncement following discovery, rather than attempting to estimate when the actual death may have occurred.

When a death is pronounced, responsibility for determining whether the death falls under the jurisdiction of the local medicolegal authority rests with the declaring individual or physician. All unnatural deaths (homicides, suicides, and accidents) are medicolegal cases and must be investigated. This remains true in the case of a wounded individual who dies in a hospital under a doctor's care, even if there is a significant time lapse between the injury and death. When a natural death occurs under a doctor's care and the treating physician is prepared to sign a death certificate attesting to the cause of the death, there is no need to involve the medical examiner or the coroner. When a natural death is unexpected or occur outside a doctor's care, the death will be investigated by the medicolegal authority. If, during the course of the investigation, a treating physician is found who can attest to a preexisting condition that caused the death, the ME or coroner will terminate jurisdiction and allow the death certificate to be signed by the treating physician. If no personal physician is found, or if the physician is unwilling to sign the death certificate, the medical examiner will sign the death certificate attesting to the cause of death. To determine the cause of death, the medicolegal authority relies on investigation, interviews, prior medical histories, and autopsies.

Death investigation culminates in the issuance of a death certificate. Only licensed physicians may sign a death certificate in the case of a natural death, and only the medicolegal authority can sign a death certificate in the case of a nonnatural death. Clarity on this issue is paramount: while an anthropologist may document injury, offer interpretation about how the injury occurred (i.e., tool matching), or address other aspects relating to the cause of death, only a licensed medical doctor can determine and certify the cause and manner of death. The death certificate, like a birth certificate, is a social tracking mechanism that allows the Bureau of Vital Records to compile and maintain information on the population as a whole. The death certificate is a legal document in that it is a necessary component in the resolution of the deceased's estate or other civil matters. The coroner or ME's ruling of manner on the death certificate, however, does not constitute a legal charge or accusation. Should a death result from a criminal act, such as homicide, it falls to the district attorney in that jurisdiction to file charges against an accused person based on evidence provided by local law enforcement. The ruling by the medicolegal authority may initiate or support the investigation or, at times, contradict the findings of other agencies such as law enforcement. The death certificate is a statement of opinion by the certifying physician, not a legal absolute. Death certificates and the rulings they contain are often changed with the introduction of new evidence or information. Although the final determination of cause and manner remains the duty of the medical examiner, forensic anthropologists involved in the case are expected to recognize and document any evidence that contributes to making that determination.

CAUSE AND MANNER OF DEATH

The primary responsibility of the medicolegal system is to determine the cause and manner of death. The **cause of death** is the disease or injury responsible for initiating the sequence of events, brief or prolonged, that

BOX 2.3

Homicidal violence

Although the cause of death in most cases is identifiable, there exists a special classification—homicidal violence. **Homicidal violence** may be invoked when the specific cause of death cannot be determined and there is adequate evidence of violence or postmortem modifications of the remains. The postmortem modifications must be sufficiently violent in nature that the decedent could not have survived, regardless of whether he or she was still alive when the modifications began. Examples include decapitation and dismemberment. Although illicit burial or disposal of remains often indicates homicide, this alone does not constitute sufficient evidence for a finding of homicidal violence. When the cause of death is certified as homicidal violence, the manner of death must be certified as homicide.

(DiMaio and DiMaio, 2001)

result in death (the exception is detailed in Box 2.3). The **manner of death** is the fashion in which the cause of death comes into being (Spitz and Fisher, 1980). While there are a large number of possible causes of death, including cancer, gunshot wound, drug overdose, and heart failure, there are only five recognized manners of death: natural, accidental, suicide, homicide, and undetermined.

Natural deaths are those resulting exclusively from disease or advanced age. Emphysema, heart disease, and stroke are examples of natural causes of death. **Accidental deaths** occur when trauma or elevated toxicity cause or contribute to the death, yet the harm inflicted was not intentional. Car accidents, falls, drowning, and drug overdoses are ruled as accidental, provided there is no evidence of intent (a detailed discussion of intent is provided in an upcoming section). **Suicides** are deaths in which one intentionally kills oneself. **Homicides** involve deaths at the hands of another. The legal definitions of homicide will be examined in Chapter 3. **Undetermined** is the designation used when the manner of death is equivocal or elusive.

To aid investigators in establishing the manner of death, many employ the "but for" rule. The rule is simple: **But for** what person, action, or circumstance would the individual still be alive? But for the actions of self typically indicate suicide or accident. But for the actions of others argues for homicide or accident.

Although NAME has proposed guidelines for death certification, determining the manner of death is solely at the discretion of the certifying medicolegal authority. Some rulings may initially appear contradictory. For example, a death resulting from an illicit or prescription drug overdose is ruled nonnatural—accident, homicide, suicide, or undetermined. However, a death resulting from smoking or chronic drinking (excluding deaths resulting from impaired driving or alcohol toxicity) is ruled natural. Given that alcohol and tobacco are drugs, albeit legal ones, the rulings may seem hypocritical. It is important, however, to note that the manner reflects the cause of death resulting from these different drugs. In illicit or prescriptive drug overdoses, the cause of death is drug toxicity, a nonnatural death. In contrast, prolonged use of alcohol or tobacco leads to natural diseases such as liver cirrhosis and lung cancer. As these causes are natural, the manner of death is natural.

Unique situations continue to challenge medical examiners and coroners. The first is "suicide by cop." In such cases, suicidal individuals orchestrate circumstances, often involving armed standoffs, with the goal of provoking law enforcement officers to shoot and kill them. How should such deaths be ruled—homicide or suicide? Based on the actions of the law enforcement officer, who shoots with deadly force, the manner is homicide. However, based on the intent of the victim, whose provocation compelled the officer to shoot, the manner should be suicide. As such deaths must be evaluated on a case-by-case basis, NAME offers no express standards for ruling on these types of deaths.

Similar ambiguity was encountered in the tragic example of the people who jumped from the World Trade Center buildings during the terrorist attacks of September 11, 2001. Based solely on their actions, the manner of death for these individuals would be suicide. The medicolegal authority, however, is permitted to consider the intent of the terrorists, which was to kill those aboard the planes and occupying the buildings. Furthermore, there was no evidence of prior suicidal intent on the part of those who jumped. None were believed to have arrived at the towers that morning with the intention of committing suicide. But for the actions of the terrorists and the circumstances they initiated, none of these individuals would have jumped. Therefore, the manner of death was justifiably ruled homicide.

Motive, Intent, and Volition

All crimes consist of two elements: the physical element or *actus reus* (prohibited act) and the mental element or *mens rea* (guilty mind). It is the responsibility of the prosecutor to demonstrate both elements during trial. It is the responsibility of all forensic investigators and expert witnesses, including anthropologists, to recognize, document, analyze, interpret, and present any and all evidence that addresses these two elements.

Determining the manner of death often requires understanding the state of mind of the decedent in the case of suicide or the assailant in the case of homicide. Motive, intent, and volition are similar concepts but with important legal distinctions. **Motive** is the personal justification for action or the reason or willful desire that leads one to act. It answers the question

"Why?" From a prosecutor's perspective, establishing in the minds of the jury members an accused person's motive for committing a crime can be an important part of successful prosecution. **Intent** is the state of mind accompanying an act. While motive may be the reason to commit the act, intent is the mental resolution to do it. It is important to note, however, that when the intent to commit an act that violates the law exists, motive becomes immaterial (Garner, 2001). Under the law, there are no "good" reasons to commit criminal acts. Whether a murder was committed for revenge, jealousy, or money is immaterial. It is the intentional act that carries the penalty, not the perceived reason for committing it. An excellent discussion of intent and volition, relevant to death investigation, is found in *A Guide for Manner of Death Classification*, prepared by the National Association of Medical Examiners in 2002.

Different forms or types of intent have been defined under the law. The ability to recognize examples of each type is crucial for all medicolegal investigators, including anthropologists, tasked with understanding manner of death. **Constructive intent** exists when the outcome of an act can be reasonably expected. For example, playing "Russian roulette" with a loaded weapon has a reasonably predictable outcome, and therefore the intent to harm ones self can be presumed for those who choose to engage in such behavior. **Implied intent** infers a person's state of mind from speech, conduct, or written word. A suicide note is an example, as is an uttered threat to kill. **General intent** normally takes the form of recklessness or negligence. The accused person's conduct suggests either awareness of risk or indifference. Examples include drunk or dangerous driving or child abandonment. **Manifest intent** can be more difficult to discern, in that the intent is apparent only on the basis of indirect or circumstantial evidence. For example, repeated exposure to toxins or poison can be interpreted as manifest intent. A single exposure could be accidental, whereas repetition of the act reveals the intent of the poisoner. **Transferred intent** shifts the intent from the original intended act to the actual committed act. The killing of an innocent bystander during a drive-by shooting illustrates transferred intent (Garner, 2001).

Like intent, **volition** is the act of making a choice. It is an act of free will, representing an understanding or acceptance of the consequences of that act. For example, smoking and drinking alcohol are acts of volition. Users of tobacco or alcohol do so of their own free will and must accept the consequences for such use. Volition is also an important legal component of activities that carry an elevated risk, such as skydiving or surgery. Those who choose to engage in such activities typically sign a waiver beforehand, acknowledging the dangers inherent in their actions and accepting any consequences that may result.

To illustrate how to differentiate intent and volition in practice, a case study involving an unclear manner of death is presented. Although dramatic, deaths resulting from the use of power tools such as chain saws are not as uncommon as once believed (see, e.g., Betz and Eisenmenger, 1995; Campman et al., 2000; Clark et al., 1989; De Letter and Piette, 2001; Rainov and Burket, 1994; Reuhl and Bratzke, 1999; Segerberg-Konttinen, 1984).

CASE STUDY 2.1

The decedent was a 34-year-old white male with a history of alcohol use. His family reported he had no known history of depression or prior suicide attempts. He was divorced and had been employed as a general contractor. He lived alone in a rural area in a trailer with a detached work shed. He was found by a neighbor in the work shed. The decedent had last been seen alive by the neighbor three days before. At that time, the neighbor described the decedent as being "in good spirits."

The decedent was found lying on his right side surrounded by a large pool of blood. A gas-powered chain saw lay on the floor adjacent to the man's head. A large gaping wound was noted on the left side of the man's neck. There was no evidence of a struggle, and no suicide note was found. An autopsy documented a 5-inch incised wound to the neck, and the cause of death was listed as exsanguination (blood loss) as a consequence of sharp force trauma.

Discussion: Because of the unique circumstances, forensic investigators initially had to consider the potential manners of death as homicide, suicide, or accidental. At autopsy, the angle of the wound was determined to be consistent with a self-inflicted wound. The only fingerprints recovered from the chain saw belonged to the decedent, and the footprints evident at the scene matched the decedent and the neighbor who had discovered the body. Analysis of the blood spatter at the scene indicated that the decedent was standing when the fatal wound was inflicted, there were no voids or disruptions in the spray pattern to suggest the presence of an assailant. Based on this information, investigators rule out homicide as a possible manner of death, leaving accident and suicide.

The operation of a chain saw is an act of volition that carries obvious risks. It is the responsibility of the operator to use all recommended safety precautions and to obtain adequate training. The possibility of serious or fatal injury as a result of mishandling or mechanical failure is high; such deaths would be ruled accidental.

For a death to be ruled a suicide, the intent of the decedent to take his or her life must be shown. In this case, investigators must search for evidence of the various types of intent. Traditional examples of implied intent include suicide notes, past suicide attempts, and uttered threats to harm self. Investigators in this case had already established that no suicide note was recovered at the scene or from among the victim's possessions, the deceased had no known previous suicide attempts or history of depression and, when last seen alive, the decedent was described as being "in good spirits." Investigators were now obliged to look to the scene to determine whether evidence of intent exists.

Manifest intent does not require direct evidence but can be inferred from circumstance. In this case, the location of death and the surrounding environment provide some clues. The death occurred in a work shed. Despite manufacturer's warnings not to operate a chain saw indoors, it is possible that the decedent chose to ignore the warning. However, investigators at the scene found no wood, sawdust, or any indication of materials requiring the use of a chain saw. Investigators also considered the possibility that the decedent was cleaning or repairing the saw in the shed but they found no tools, cleaning materials, or gasoline.

It was the chain saw itself that provided unequivocal evidence of constructive intent. The chain saw was equipped with a safety mechanism that stops the blade when pressure on the hand grip is released. The decedent had taped the safety switch down to override the mechanism, ensuring that the saw would continue to operate even if the hand grip was released. This purposeful action on the part of the decedent served as evidence of his intent to commit suicide. Despite the absence of a history of depression or a suicide note, the physical evidence of manifest and constructive intent was sufficient for the medical examiner to rule the manner of death as suicide.

The concepts of volition and intent often raise as many questions as they answer. For example, at what age can an individual be said to have sufficient capacity to demonstrate volition or intent? This question is foremost in the mind of the medical examiner in cases involving possible suicide in children. It is also relevant to prosecutors involved in homicide trials with underage assailants. What substances, ailments, or events could reasonably negate an individual's volition or intent? In other words, is being drunk a legal excuse for illegal behavior? What about Alzheimer's, senility, or posttraumatic stress? Do these mind-altering conditions render a person inca-pable of forming intent? These questions have been and will continue to be argued in courts of law. As forensic investigators, we are required to interpret all evidence available in the hopes of addressing these complex questions. In addition to evidence recovered through scene investigations, the medicole-gal system relies on the process of autopsy.

AUTOPSY AND POSTMORTEM EXAMINATIONS

Autopsy means "to see for one self." Although medicolegal statutes define the circumstances of death that warrant an investigation, the medical exam-iner or forensic pathologist in charge of the case determines what type of postmortem examination an individual will receive: full autopsy, partial or limited autopsy, external examination, and anthropological examinations involving skeletal remains. The National Association of Medical Examiners (NAME) has developed Performance Standards for Forensic Autopsy, avail-able at http://www.thename.org/.

The training of forensic pathologists follows a time-honored tradition in which autopsy techniques are learned from a senior pathologist in the autopsy suite. The same is true for students of forensic anthropology, who must learn applied methods from a practicing anthropologist. Therefore, while textbooks and other teaching aids defining autopsy procedures can provide useful overviews, it is the tutorial aspect that is of paramount importance. With this in mind, we present certain basic principles intended to familiarize anthropologists with the types of autopsy in preparation for defining the role of the anthropologist within the autopsy process.

Most pathologists distinguish between full autopsies and limited autop-sies. Full or **typical autopsies** follow a predictable sequence. These autopsies begin with an external examination and description of the body, including weight and height, after which radiographs are taken. This is followed by the primary incision and removal of the thoracic and abdominal contents for examination and dissection. Techniques vary depending on the training of the pathologist but may include *en bloc* removal of complete systems, *en masse* removal of the entire contents, *in situ* dissection of selected organs, or use of the **Virchow technique**, which is the removal of individual organs. The individual organs are weighed, examined grossly, and then sectioned to reveal any morphological signs of pathology or trauma. Samples are col-lected for histological preparation and examination. The cranial vault is exposed and opened, allowing examination of the brain in a similar manner. Additional samples, such as blood, hair, and fingerprints, may be collected at the discretion of the pathologist. The contents of each cavity are returned and all incisions are closed.

Limited autopsies, also known as partial autopsies, may be substi-tuted for full examinations. Partial autopsies can be performed when the next-of-kin object to full autopsy. Pathologists limit invasive procedures to those necessary for evidence collection. Needle or endoscopic autopsies can be conducted to procure samples when more invasive procedures are not possible. Limited next-of-kin permission may also result in the restriction of skin incisions. In these cases, autopsy is limited to the reopening of existing surgical incisions or the examination of only one cavity, such as the abdomen or thorax (V.I. Adams and Ludwig, 2002).

A medicolegal autopsy provides evidence that forms the basis for opinions rendered by pathologists in court. Depending on circumstances, the certifying pathologist may be comfortable forming such opinions without a formal autopsy. Medicolegal authorities can appoint deputies or field investigators to perform external examinations at death scenes or in hospitals (for more information on deputy investigators, see the website of the American Board of Medicolegal Death Investigators: http://www.slu.edu/organizations/abdmi/index.phtml). Such exams focus solely on inspection of the body's external surface, as well as easily visualized cavities such as the mouth, and limited sample collection such as blood draws or fingerprinting. Investigators look for signs of trauma or illicit drug use, such as track marks, that indicate the death was not natural. Based on the findings of such external examinations, as well as the decedent's medical history and information from the scene, the pathologist may decide to forgo a full autopsy and certify the cause and manner of death.

Advances in noninvasive imaging techniques have created the possibility of a **virtual autopsy**. The use of multislice **computed tomography** (MSCT) and **magnetic resonance imaging** (MRI) allows pathologists to visualize internal structures, pathological conditions, and trauma without making an incision. Such technology is, however, not widely available in medicolegal offices and its use is costly. Virtual methods have been presented as an alternative to invasive autopsy (Box 2.4).

BOX 2.4

Virtual autopsy

Certain death scenarios challenge death investigators. For example, fire-related changes to a body can obscure or destroy indicators of identity, cause of death, injury, and toxic agents. Thali et al. (2002a) used multislice computed tomography (MSCT) and magnetic resonance imaging (MRI) to conduct a postmortem examination of a badly burned body. Following both the MSCT and MRI imaging examinations, the team conducted a traditional autopsy (with standard radiography) to compare the results. The team found MSCT to be superior to both radiography and MRI for documenting bone injury and gas embolism. MRI was best suited to soft tissue injury investigation. Neither MSCT nor MRI contributed significantly in the identification of the victim.

Triage and Mechanisms of Death

The primary goal of an autopsy or postmortem exam is to establish the cause and manner of death. However, not every decedent requires a full autopsy to determine cause and manner. This introduces the practice of triage. **Triage** is the medical screening of individuals to assess their needs, priority, and sequence of treatment. Similar to hospital emergency rooms, bodies entering a morgue or medicolegal facility must be evaluated. Based on information gathered in the field, as well as the circumstances of the scene and morphological appearance of the body, the attending or supervising pathologist will determine whether an individual requires an autopsy.

It might surprise students of forensic science that an autopsy is not necessary in most deaths. For example, a full autopsy in cases involving poisoning or drug overdose will reveal little. As circumstances surrounding death determine which deaths fall under medicolegal jurisdiction, a suspected or presumptive cause of death exists for the majority of decedents entering the medicolegal system. Whether an autopsy will contribute to the investigation depends on the suspected mechanism of death.

Causes of death fall into three categories known collectively as **the mechanisms of death**: physiological; toxicological, and morphological/anatomical. **Physiological deaths** stem from a disruption of function, either on a systemic or a cellular level. Examples of physiological causes of death include hypothermia, electrocution, asthma, and diabetes. **Toxicological deaths** occur as the result of an overabundance of an exogenous agent, such as a poison, toxin, or drug. Examples include drug overdose and lead or carbon monoxide poisoning. Anatomical or **morphological deaths** result from a lethal physical change in the body, such as a gunshot wound. For an expanded discussion of mechanisms of death, see DiMaio and DiMaio (2001).

Upon death, systemic function in the body ceases. As such, assessing function after death can be problematic. Autopsy primarily allows for the assessment of morphology. As evaluating toxicological and physiological deaths is largely a matter of evaluating functioning, the information derived from autopsy may be limited. Whether decedents with suspected toxicological or physiological causes of death receive full autopsies is at the discretion of the pathologist. Often, such cases can be resolved through external autopsies as well as toxicological or serological tests. Investigations involving decedents with suspected morphological causes of death benefit from an autopsy. Autopsy not only provides an opportunity to collect evidence such as projectiles but allows the pathologist to visualize the extent of morphological change.

Five potential outcomes involving morphological findings at autopsy are recognized. The first outcome represents gross evidence of a cause of death. The observed morphological change is incompatible with life. Examples of such findings include lethal gunshot wounds, decapitation, diffuse cancer, and **myocardial infarctions** (heart attacks). Many forensic pathologists consider such findings optimal or preferable, as they provide independent physical evidence regarding the cause of death.

The second outcome includes a condition with lethal potential but no evidence of an immediate cause of death. For example, **stenosis** (narrowing) of the arteries of the heart without complete blockage indicates a potentially life-threatening condition but is not necessarily incompatible with life. With such chronic and progressive disorders, it is often a matter of a "threshold." If the individual was alive with the same condition hours before, what change occurred that caused the condition to become lethal? For the forensic pathologist to certify a condition as the cause of death, he or she must be satisfied that a condition had reached its lethal potential.

The third possible outcome involves morphological evidence of a pathological state but without immediate lethal potential. Examples include many chronic but controllable diseases such as asthma or diabetes. Again, this outcome invokes the notion of thresholds and requires careful review of medical histories as well as detailed interviews with family members. Even minor changes in treatment regimens or lifestyle can explain the lethal change in a once manageable disease.

The fourth outcome is associated with individuals having a history of a pathological condition but with no morphological evidence at autopsy. The example most often cited is epilepsy, a condition with no immediate lethal potential but the possibility of fatal complications. Many pathologists feel this outcome is the most difficult to address: the pathological condition offers a potential explanation for the death but without the objective evidence to support it.

Sometimes the final outcome results in no findings. There are no significant morphological changes or pathological conditions evident at autopsy. No morphological cause of death can be identified, and the pathologist must now look to toxicological tests or histological studies in the hopes of finding the cause of death. Forensic anthropologists are well acquainted with such an outcome, as many cases involving skeletal remains fall into this category. Frequently, this outcome results in a ruling of the cause and manner of death as "undetermined."

The categories of mechanisms of death are mutually exclusive relative to causes of death. Each specific cause is contained in a specific category. However, the three categories are not mutually exclusive in terms of the types of evidence found at autopsy. For example, physiological causes of death such as electrocution can produce gross morphological changes, such as burns, that are observable at autopsy. It is important not to misinterpret such findings and be led to assume that all morphological changes evident at autopsy equate to a morphological mechanism of death.

Anthropology and Autopsy

To what extent anthropologists fit into the medicolegal process of autopsy depends on a variety of factors. First, the type of autopsy may limit the need for anthropological consultations. External exams or limited autopsies seldom require the services of an anthropologist beyond consultation on radiographs or assistance in establishing personal identification. Second, the nature of the death may not warrant involvement from an anthropologist. Finally, the personal preferences and prior experiences of both pathologists and anthropologists often dictate the timing or extent of an anthropology consultation. Some pathologists prefer to conduct an autopsy and then provide the remains to the anthropologist for additional testing. Other pathologists request the presence and involvement of the anthropologist during autopsy. Some anthropologists begin their exams by removing any retained soft tissue from the remains, while others prefer to examine the remains as found. Most practitioners adopt styles based on their experiences during training.

Anthropologists participating in autopsy or conducting separate exams should know the rules governing sample collection and retention. Since the authority for anthropologists to conduct such examinations is an extension of medicolegal jurisdiction, laws regulating tissue retention apply to the consulting anthropologist, even if not expressly named within the statute. Temporary removal of tissue for examination is an accepted part of autopsy (Box 2.5). Following autopsy, limited samples may be retained by the medicolegal authority only if deemed in the public interest. For example, a skull fragment with a gunshot wound could be needed as evidence in court

---

BOX 2.5

Internal morgue tracking systems

Critical to the effective functioning of any morgue or medicolegal office is an internal tracking mechanism, sometimes referred to as the "green board" because information was historically recorded on a green chalkboard. Bodies and remains entering a morgue are evidence for which the medicolegal authority is responsible. The internal tracking system meets the chain of evidence requirement to be described in Chapter 3. The green board allows the tracking (often with the use of bar coding) of large numbers of individuals at various staging of processing, with the status and location of the remains known at all times. Removal of any portion of the remains, such as brains or skeletal elements, needs to be documented within the tracking system. Should body parts be separated for cleaning, photography, or specific testing, the body must be placed on hold to prevent its release, and the location of all elements must be recorded. Anthropologists must communicate any element removal to the pathologist or morgue technician to ensure that a body is not released to a family incomplete or prematurely. Transportation of remains or elements to labs or locations outside the morgue facility must be authorized and tracked to maintain chain of custody.

and can therefore be retained. Such specimens are evidence and must be treated and maintained as such. Anthropologists seeking to retain any portion of the body for teaching or research purposes must obtain express permission from the next-of-kin. Permanent retention of nonevidentiary tissues without express next-of-kin consent is an actionable offense (V.I. Adams and Ludwig, 2002). This guideline was established in response to ethical as well as legal issues and should be adhered to regardless of circumstance or geographic location and in the absence of express laws or regulations.

Such policies also cover the use of photographs taken during autopsy or at a crime scene. Anthropologists should obtain written permission from the medicolegal authority prior to publishing or disseminating any photograph obtained in the course of a medicolegal investigation. Scientists wishing to engage in research or data collection during autopsy or to use materials generated at autopsy must also obtain express permission. In most cases, the research review committee overseeing the medicolegal institution where the research is to be conducted must grant prior permission. In regions where no formal review system is in place, anthropologists seeking to do invasive research (in which samples are removed solely for research purposes) must seek express approval from the medicolegal authority with jurisdiction.

Public Perception of Autopsy

Family members often wonder why an autopsy is necessary in cases where the cause of death is obvious, such as a gunshot wound to the head in a presumed suicide. Two governing principles explain the need for autopsy in such cases: "Never bury evidence" and "Never trust the obvious." The type of examination required is often determined by the need to recover any associated evidence, such as the bullet from a gunshot wound or partially digested pills in the stomach of a suspected suicide. To certify a non-natural death, the medical examiner must observe and collect evidence to support the finding.

"Never trust the obvious" refers to the rare cases that present as accident, suicide, or even natural deaths but may in fact represent homicides. All medicolegal acts retain for the coroner or the ME the right to investigate any "suspicious" deaths and leave the responsibility of defining "suspicious" to those authorities. As the certifying agent, the ME or the coroner has the right and obligation to investigate the circumstances of death until satisfied and prepared to declare the cause and manner of death. Normally, thorough investigations, reviews of medical records, and interviews with family members are sufficient to answer any lingering uncertainties. In cases when the ME's suspicions persist, however, an autopsy may be ordered.

Medicolegal investigators require signed next-of-kin consent prior to performing optional or **hospital autopsies**. These occur when the death falls outside the jurisdiction of the medicolegal authority but, at the request of the family, the ME performs an autopsy for a fee. When the death falls under medicolegal jurisdiction and there are family objections to autopsy, the attending pathologist can exercise his or her discretion regarding the need for a full autopsy, opting instead for an external exam. However, if the pathologist believes that an autopsy is necessary, the autopsy will be conducted, despite family concerns, unless the family obtains a court order to prevent it.

Many religious denominations oppose the practice of autopsy, and all medicolegal authorities are aware of and sensitive to religious attitudes toward autopsy. Most medicolegal offices work with families to resolve these issues. Even so, next-of-kin who refuse to permit an autopsy on religious grounds may encounter difficulties, depending on the circumstances of the death.

Justification can be found in the three major monotheistic faiths—Judaism, Christianity, and Islam—for prohibiting or objecting to autopsy, although such objections are not based on specific religious law or text (Davis and Peterson, 1996). Individual or family objections to autopsy generally reflect personal beliefs or text interpretations given by religious figures centuries ago. In these instances, the ME or coroner, often with cooperation from law enforcement personnel, will work with the family to minimize objectionable procedures or to accommodate specific religious practices associated with death. For example, a rabbi may attend an autopsy at the request of a Jewish family, or a Catholic priest may administer a form of last rites prior to autopsy in cases of sudden, unexpected death. Similarly, Islam has forbidden the "disfigurement" of the body (although autopsy is not expressly mentioned). However, in 1982 a fatwa (legal opinion) committee found that the benefits of autopsy outweighed the prohibition (Davis and Peterson, 1996), providing greater acceptance of autopsies in the Muslim community. Sensitivity and willingness to answer questions on the part of the medicolegal authority contributes greatly to addressing a family's objections to autopsy.

Attitudes toward autopsy also vary by ethnic or cultural group (Orlowski and Vinicky, 1993). Sanner (1994), in a survey conducted in Sweden, found that 70% of respondents felt uncomfortable with the thought of autopsy for themselves or their loved ones. Older women were the most hesitant or negative toward autopsy. Denninghoff (2000) reported an imbalance in the recruitment success of various racial and ethnic groups in autopsy-based research protocols (Box 2.6). Denninghoff attributed the underrepresentation of blacks in recent studies to an earlier history of forced participation, unethical research, and unauthorized dissections of African Americans. Overcoming negative public perceptions of autopsy is the responsibility of all professionals engaged in the process. Communication and education are keys to successfully changing the public's attitudes toward autopsy.

BOX 2.6

Therapeutic versus nontherapeutic human tissue donation

After death, tissues and organs can be donated for therapeutic purposes, including life-saving transplants. Although personal preference and participation regarding such donations varies among religious (Campbell, 1998) and ethnic groups (Perkins et al., 2005), many individuals feel that the use of the body is justified for therapeutic reasons.

More contentious, and less well understood, is nontherapeutic use of human tissue. The donation of select tissues or entire bodies solely for the purposes of teaching or research is often perceived in a less positive light than therapeutic donation. Despite the important role of autopsy and body donation in the training of medical personnel and the contributions of autopsy data to medical research, there continue to be misconceptions about the use of human remains in research and education. In a recent study that actively solicited brain donations for research purposes, Azizi et al. (2006) found the response (58% of families gave permission) more encouraging than prior studies had predicted. The researchers concluded that greater education and awareness regarding research donation is needed.

Not everyone is reticent regarding body donation. Sanner (1997) asked registered bone marrow donors for their views on therapeutic and nontherapeutic tissue donation. Compared to a prior study of the general public involving similar questions, marrow donors were much more positive about all forms of body donation. Sanner concluded "that if one is prepared to give from the body in life, one is also prepared to give after death" (1997: 67). Botega et al. (1997) surveyed medical students at various stages of their training regarding autopsy and donation. First-year students had more difficulty in both attending autopsy and approaching families of the deceased regarding donation. These objections decreased in the student's fourth and sixth years of training, as students began to view autopsy as a vital tool in medical research.

Forensic anthropologists interested in conducting research requiring body or tissue donation should note the example of successful nontherapeutic body donation seen at the forensic anthropology decomposition facility at the University of Tennessee at Knoxville. Commonly known as the Body Farm, the facility has hosted numerous studies involving donated human remains since its inception by Dr. William Bass in the early 1980s. This facility, as well as countless anatomical donation programs associated with medical schools, demonstrates that individuals are willing to donate bodies for research, provided the benefits of the donation can be clearly understood.

CONCLUSION

Like all forensic scientists participating in an investigation, the anthropologist is part of an integrated medicolegal team and, as such, must understand not only his or her role and responsibilities but also how the entire team functions. Although the medicolegal authority alone rules on cause and manner of death, the forensic anthropologist, who provides supporting evidence, is a vital member of the investigative team. A well-informed anthropologist, knowledgeable in all aspects of the medicolegal system, makes significant contributions to death investigations without inadvertently overstepping his or her authority or responsibilities. The use of shared terminology allows forensic anthropologists to communicate effectively with other team members and removes any ambiguity from written reports. Anthropologists who respect and understand the structure and operation of the medicolegal system inevitably function well within it.

CHAPTER 3

Evidence and the Judicial System

One of the most misused words in the forensic sciences is "forensics" itself. The media have redefined the adjective as a verb and thus misrepresent "forensics" as activities done by a forensic team as part of a forensic investigation. In reality, **forensic** simply describes anything used in a court of law. Methods and theories found in any recognized discipline become forensic when they are employed to address issues before the court. In this way, traditional methods of anthropology, odontology, and even accounting become forensic methods when used within a legal system. In other words, it is not that a skeleton is modern or that it is analyzed by a forensic anthropologist that makes a case "forensic"—it is the use of the results of that analysis in a court of law that makes the case forensic. Alternately, the term "medicolegal" should be used to describe cases that fall under medicolegal jurisdiction but are not the subject of court proceedings.

To successfully participate in forensic investigations, practicing forensic anthropologists must have a thorough understanding of the judicial system, and their role in it. The key to integrating into the system is communication. The disciplines of law, medicine, and forensic science all require highly specific terminology that allows practitioners to express subtle yet crucial differences in legal concepts, morphology, or investigative techniques. Mastering legal terminology and concepts allows forensic anthropologists to both understand and be understood in their interactions with the legal process.

The foundation of the legal system is the presentation and interpretation of evidence. As forensic investigators, anthropologists must recognize and collect all available evidence in each assigned case. In their role as expert witnesses, forensic anthropologists must analyze, interpret, and present this evidence to a court. In this chapter, we will examine issues relating to evidence and evidence interpretation, and offer an introduction to the judicial system, expert witness testimony, and the investigation of homicide.

EVIDENCE

Evidence is testimony, documents, or physical evidence presented in court that proves or disproves an alleged fact. In legal terms, a **fact** is evidence that has been stipulated to by both parties. **Stipulation** is the voluntary agreement between opposing parties concerning a piece of evidence. For example, if the identity of the victim has been established to the satisfaction of both the prosecutor and the defense team, then both parties will stipulate to the victim's identity. By this stipulation, the victim's identity becomes a fact of the case and no further evidence need be presented to the court regarding this point. In effect, the two sides agree that they are not going to argue about who the victim was; rather, they are agreeing to let the court's attention focus on other matters relevant to the case.

Evidence can be classified in a number of ways. Evidence is either objective or subjective. Three types of evidence are recognized: circumstantial, material, and physical. Evidence can be grouped by type or categorized individually. At the level of individual pieces of evidence, each piece can be categorized into class evidence or individual evidence. Each piece of evidence can also be evaluated relative to a specific defendant at trial and categorized as either inculpatory or exculpatory. These systems of classification are described as follows.

Systems of Classification of Evidence

Objective versus Subjective Evidence

Evidence exists in two realms: the objective and the subjective. In court, that which is **objective** exists independent of a single mind. It is observable and verifiable. Knowledge of objective evidence is derived from sense perception—it can be seen, touched, smelt, heard, or tasted. A fingerprint, a bullet, and drops of blood are examples of objective evidence. That which is **subjective** belongs only to reality as it is perceived. It exists solely in the mind and is lacking in reality or substance. A motive is an example of subjective evidence.

There are three types of evidence: circumstantial, material, and physical. **Circumstantial evidence** is based on inference or speculation, rather than knowledge or observation. It is inherently subjective. It is not the action itself but rather the perceived motive ascribed to the action—say, a suspect fleeing the country after a crime—that makes it both circumstantial and subjective. **Material evidence** is witness testimony or other forms of evidence based on personal knowledge or observation and having some logical connection with the issues of the case. Material evidence can be both subjective and objective. Eyewitness testimony and a video tape from a surveillance camera placing the accused at the crime scene are two examples of material evidence. **Physical evidence**, such as projectiles, fingerprints, or autopsy findings, is obtained by scientific means. Physical evidence is inherently objective evidence.

Class versus Individual Evidence

Each piece of evidence can be assigned into one of two categories: class evidence or individual evidence. **Class evidence** is group affiliation based on common characteristics. Class characteristics create subsets within a larger population or sample, excluding that which does not share the trait used to define the class. Examples include classes such as males 25 to 35 years of age, high-top sneakers, and .38 caliber handguns.

Class categories become more specific depending on the number of traits introduced; for example, all circular saws versus a Black & Decker 10-inch Skilsaw. The strength of class evidence is contextually specific, depending on the number of different classes. Interpretation of isolated class evidence is often subjective. However, evidence encompassing a large number of different classes strengthens the argument. This is expressed through statements of probability, such as "what is the probability that the accused would match the description given by a witness AND have the same make and model of car reported at the scene AND have clothing that produces fibers identical to those recovered from the scene?"

Individual evidence is material traced to a single source. It is equivalent to a positive identification of a specific tool, weapon, or source. Individual evidence relies on the presence of unique characteristics or identifiers that allow investigators to exclude similar items of the same class. It is important for all forensic scientists and investigators to distinguish the potential to individualize evidence from the ability to actually do so. Repetitive use may produce wear patterns that result in unique features in some objects, such as shoe soles. Most objects, however, can never have traits that are sufficiently unique to allow for individualization.

Inculpatory and Exculpatory Evidence

Both individual and class evidence may be introduced in a court of law as part of the case against an accused person. **Inculpatory evidence** supports the argument against the accused person. It is evidence that suggests the accused is guilty of the crime. **Exculpatory evidence** excludes the accused. It is evidence that indicates the accused is innocent. Inculpatory and exculpatory categories of evidence are relative only to a specific accused individual. Exculpatory evidence supporting the innocence of one individual may prove the guilt of another, thereby becoming inculpatory for the second individual.

While the types of evidence are readily categorized, recognizing what is evidence at a scene or autopsy can prove more difficult. Protocols and guidelines for the collection of evidence at a scene will be addressed in Chapter 4; however, theories addressing evidence recognition are best discussed here.

Recognizing Evidence

The foundation of forensic science is Locard's principle. **Locard's principle** states that any two objects that come in contact with each other leave traces one upon the other. This simple principle provides the rationale behind DNA, fingerprints, and all forms of **transfer evidence**. Finding these traces of contact is the goal and responsibility of all forensic investigators.

There are two theories governing physical evidence recognition: association and matter out of place. **Association** implies a direct link between the crime and a piece of physical evidence. The knife protruding from the victim's chest is an obvious example of associated evidence, as are the fingerprints on the knife handle. Associated evidence follows Locard's principle; physical evidence such as fingerprints, hair, DNA, or blood literally connects the object to the crime. Associated evidence is recognizable at scene and is collected as part of a routine, thorough analysis of a crime scene.

More difficult to discern is evidence based on matter out of place. With **matter out of place**, there is no immediate link between a physical finding and the crime, but the unusual nature of the item or its location warrants further investigation. A gas can in a bedroom is an example of matter out of place. Even with no fingerprints on the object or bloody footprints leading to it, the mere presence of the gas can in a room where gas cans are not normally found suggests that it has some evidentiary value. If the gas can were found in the bedroom at the scene of a homicide or robbery, it is possible the assailant planned to start a fire to destroy the evidence of the other crime. Even if the gas can ultimately plays no significant role in the crime, it is important to recognize its potential evidentiary value during the initial crime scene analysis.

The scientific theory of matter out of place extends beyond finding unusual objects in atypical locations. An everyday item, such as a chair, takes on evidentiary importance simply by being knocked over. In this case, it is not the chair's presence that is unusual, but rather its position. A displaced chair at a crime scene such as an assault or homicide is usually interpreted as evidence of a struggle. The evidentiary value of the chair is based on the assumption that, in the course of normal daily living, if a chair were to be knocked over it would typically be righted. While it is not necessary to collect the chair as evidence, it is crucial to document its position as found through still and video photography and written notation.

The value of evidence in court depends largely on the interpretation of that evidence by the investigators and prosecutors responsible for presenting it to the judge or jury. Associated evidence is considered to be very strong or compelling evidence because it may place the accused at the scene or connect the murder weapon with the assailant. Matter-out-of-place evidence is often used to reconstruct the events surrounding the crime. Evidence based on matter out of place can prove more difficult to present to the court because its subjective nature renders it open to multiple interpretations. An overturned chair or curtains pulled from their rod might be argued to be signs of a struggle or simply poor housekeeping. Ultimately, all evidence is subject to interpretation—by anthropologists, investigators, lawyers, judges, and juries. Establishing the mental and physical elements of

---

BOX 3.1

Bias

In a legal sense, **bias** is an inclination or prejudice. Bias can manifest as a judge who favors the prosecution by giving unbalanced instructions to the jury or who requires the accused to wear prison garb instead of street clothes at trial. It is favoritism toward one of the opposing sides. In a scientific sense, bias is a skewing of results in one direction or another. Selecting data to produce a specific outcome or failing to include in a sample specimens that would alter the results are examples of such bias. In forensic contexts, bias is the loss of objectivity. True **objectivity** in forensic investigations requires disinterest in the outcome. This does not mean apathy in one's job but rather allowing the evidence to lead where it will and not personally investing in a specific result. It is for this reason that investigators are discouraged or forbidden from working on cases involving relatives or friends, as a personal relationship with a decedent inevitably results in a loss of objectivity.

The opposite of objectivity is advocacy. **Advocacy** is assisting, defending, or pleading on behalf of another. An advocate is no longer an objective observer but an active, invested participant. Where there is advocacy, there is a strong potential for bias. An example of advocacy familiar to anthropologists is the renowned primatologist, Dianne Fossey. Fossey's passion and personal investment in the conservation of the gorillas she studied led her to become a powerful advocate for their welfare, often at the cost of scientific inquiry.

a crime requires not only thorough evidence collection, but unbiased interpretation (Box 3.1).

EVIDENCE INTERPRETATION: RATIONALITY VERSUS PARSIMONY

To fully participate in criminal investigations, anthropologists must have a command of evidence collection and interpretation as well as the ability to recognize and avoid potential sources of bias. Regardless of evidence type, there are two modes of investigation: parsimony and rationality. An argument that arrives at an understanding of events based on the minimum number of steps needed to explain all available data or observations represents **parsimony**. An argument in which events are expected to follow a logical course exemplifies **rationality**. While both modes have scientific merit, employing a strictly rational argument in forensic investigations may prove problematic. For example, deaths involving homicide or suicide cannot be approached from a purely rational perspective. Such acts, and the mindset of those committing them, are inherently not rational. To expect the actions and events associated with these acts to follow a logical course is not reasonable.

The following case study demonstrates the difference in structuring forensic investigations parsimoniously versus rationally.

CASE STUDY 3.1

Fire fighters are dispatched to a burning vehicle on a rural bridge. After extinguishing the blaze, they discover charred human remains in the driver's seat. The driver's-side window is partially open and the driver's seat belt is fastened. Fragments of the decedent's left hand, including several burned finger bones and a gold ring, are embedded in the plastic cover of the steering wheel. Similar fragments of the right hand are noted in the melted plastic of the stick shift. A search of the car reveals an open bottle of vodka and a cigarette lighter on the floor of the front seat and a loaded .32 caliber handgun in the glove compartment. At autopsy, the decedent is identified as the 24-year-old Hispanic female who was the vehicle's registered owner. There is no evidence of trauma or gunshot wounds. Soot deposits in the trachea and lungs, as well as an elevated carbon monoxide level, indicate that the woman was alive when the fire began. Toxicology tests reveal the decedent's blood alcohol level to be almost double the legal limit.

Police interview two witnesses who were traveling in the area the night of the fire. Both witnesses report seeing the decedent's car parked on the side of the bridge at two specific times. The first sighting was at 9:20 P.M., as the witnesses were traveling westbound on the bridge. At that time, the victim was seen talking through the car window with a man who was leaning against the driver's door. A second car (presumably the man's) was parked in front of the decedent's car on the bridge. Both witnesses describe the interaction between the decedent and the man as "angry" and "argumentative." The witnesses then report that, while traveling eastbound on the bridge at 9:35 P.M., they again passed the decedent's car parked in the same location, but with flames and smoke billowing from the driver's-side window. The witnesses called for police assistance. The witnesses noted that the man and the second vehicle were gone at the time of their second observation.

Discussion: Law enforcement and medicolegal investigators are tasked with reconstructing the events of this scene. Although the cause of death remains the same, three possible scenarios can be reconstructed, each representing a different manner of death. In the first scenario, the woman's death is a homicide. The man witnesses place at the scene is responsible for starting the fire and the woman is unable to escape the car owing to panic, intoxication, or incapacitation. The second scenario suggests the woman's death is a suicide. Despondent following the argument, the woman douses herself with the alcohol and sets herself on fire. The third scenario indicates the death is accidental. The couple argues and the man leaves; the woman continues drinking. Intoxicated, the woman passes out, dropping both the vodka bottle and her lit cigarette on the floor mat of the car, igniting the fire.

Parsimony argues that all three scenarios are possible, if not equally probable. In the absence of any additional evidence, such as testimony from the man involved, all proposed scenarios parsimoniously explain the events as reconstructed. Although less parsimonious scenarios cannot be excluded, such as the possibility of a second assailant or a freak mechanical flaw that ignites the fire, investigators are correct in initially focusing on the more parsimonious scenarios.

Invoking a *prima facie* rational argument in this case would limit the number of possible alternatives prematurely and thus could introduce a potential source of bias. By insisting that the decedent behaved in a rational manner, plausible or parsimonious scenarios would be excluded solely on the basis of the investigators' need for logical explanations. For example, while other more common or effective methods of committing suicide were open to the decedent (such as the gun), such courses of action may not have been apparent to her while in an agitated state. Individuals contemplating suicide are not necessarily rational. For investigators to dismiss the possibility of suicide because they feel there were more rational ways to commit the act would introduce a bias and would be inappropriate in this example.

Having examined evidence in death investigations, we will now focus on the court system in which such evidence is presented. This overview is intended to introduce anthropologists to the major components of the judicial system, particularly those issues relevant to providing expert witness testimony.

THE JUDICIAL SYSTEM: AN OVERVIEW

A **crime** is a social harm that the law makes punishable. Examples range from vandalism, petty theft, and trespassing to rape, kidnapping, and murder. The definition of crime also extends to a breach of a legal duty treated as the subject matter of a criminal proceeding—medical malpractice trial, police misconduct hearing, or military court martial, for example. There are two classes of crime: felony and misdemeanor. A **misdemeanor** is a crime punishable by fine, penalty, or confinement in a place other than **prison**, such as a county **jail** or adult detention facility. Parking violations, minor drug possession, and vandalism are examples of misdemeanors. A **felony** is a more serious crime punishable by imprisonment for more than one year or possibly by death. Examples of felonies include murder and rape. All crimes are defined by laws and statutes. **Laws** are the aggregate of legislation, judicial precedents, and accepted legal principles within a prescribed geographic region. **Statutes** are acts of legislation declaring, commanding, or prohibiting something.

The formal written accusation of a crime, made by a grand jury and presented to the court for prosecution against the accused, is an **indictment**. Contained within the indictment is a statement of *res gestae*. *Res gestae* means "things done." Such a statement describes the physical element of the crime, encompassing events relevant or contemporaneous to the crime. *Res gestae* is a statement detailing what the defendant is accused of doing, to be established through evidence and facts presented to the court (Box 3.2).

BOX 3.2

Double jeopardy and "lesser included offense"

The concept of **double jeopardy** stems from the Fifth Amendment provision that no person can be charged twice for the same offense. The notion of lesser included offense conveys many of the same components but with an important distinction. While the double-jeopardy clause prevents charging an individual with the same crime twice, "lesser included offense" prohibits filing charges for a crime that is composed of some, but not all, of the elements of a more serious crime and that was necessarily committed in carrying out the greater crime (Garner, 2001). Conviction or acquittal of either offense prohibits a separate trial for the other. For example, if the accused stood trial for first-degree murder for shooting her boyfriend and was acquitted, she cannot then be charged with illegal possession of a firearm for the same offense. This prevents prosecutors from repeatedly indicting offenders for progressively less serious crimes if unable to secure convictions on the more heinous offenses. Once the *res gestae* of a crime is established in the indictment, all "lesser included offenses" are automatically attached, and the prosecution is afforded only one opportunity to secure a conviction on the body of the crime as a whole.

The court system itself is divided into two parts—civil and criminal. **Civil court** has jurisdiction over noncriminal cases such as divorce or insurance claims. Should you choose to sue someone, the case will be heard in civil court. The court's decisions cannot result in jail time; only punitive and compensatory damages can be awarded. **Criminal court** has jurisdiction over cases involving felonies and misdemeanors. Guilty verdicts result in jail time and/or fines or possibly the death penalty. Civil courts have a specific local jurisdiction, typically at the city or county level; jurisdictions of criminal courts are either local (county or state) or federal.

All courts have the power to compel participation. To **compel** is to cause or bring about by force or overwhelming pressure. With respect to courts, judges can compel witnesses, defendants, or petitioners to appear, under the threat of fine or imprisonment if the individual fails to comply. The power to compel manifests in warrants and subpoenas. A **warrant** is a court order or writ authorizing a law enforcement agency to make an arrest, a search, or a seizure. Although warrants are drafted by the law enforcement agency seeking the authority to act or by a prosecutor acting on their behalf, the warrant is not valid until reviewed and signed by a judge. Once signed, the warrant compels those named to cooperate with law enforcement or incur imprisonment or fines as a result. A **subpoena** is an order to produce documents, testimony, or evidence for the court's review. A subpoena can be issued by the prosecuting and defense lawyers in a case as well as by appointed medicolegal investigators. A subpoena compels the person named to cooperate or risk penalties determined and enacted by the presiding judge for failing to comply.

TESTIFYING AS A WITNESS

Testimony is the providing of witness under oath or affirmation at trial. Testimony can also be given through written affidavit or spoken deposition outside the courtroom. There are three types of witness testimony: material, fact, and expert. Whether a judge can compel the witness to testify depends on the type of witness. A **material witness** is a member of the general public whose testimony contributes to establishing *res gestae*. A material witness is someone who saw the crime (an eyewitness) or is called to establish some aspect of the defendant's personality (a character witness). Eyewitnesses can be compelled to testify, provided such testimony does not constitute a hardship. A **fact witness** is drawn from within the law enforcement or medicolegal systems. Fact witnesses include the police officers or medicolegal investigators who attended the crime scene or crime lab employees who processed specific pieces of evidence. The obligation to provide testimony in criminal proceedings is a condition of employment for these individuals and, as such, their testimony rarely needs to be compelled, although it can be if circumstances warrant. An **expert witness** is an individual with specific qualifications who offers testimony focusing solely on an area of expertise. To be declared an expert, the witness must supply a résumé or certification attesting to his or her qualifications (Box 3.3). Opposing counsel has the right

---

BOX 3.3

Detailing your work history (curriculum vitae/résumés)

A complete work history is mandatory for qualifying as an expert witness at trial, as well as a necessary step in securing employment in the first place. Employment in the forensic and medicolegal communities inevitably entails background checks of candidates as a condition of hire. These reviews are more extensive than those in academia or other disciplines. **Background checks** are comparisons of employment records, transcripts, and other forms of social data to confirm the qualifications and experience claimed by applicants in their curricula vitae (CV), résumés, and letters of application. As such, accuracy and honesty in these statements is paramount. For students wishing to reflect their case experience in a CV, translating activities at crime scenes and in the autopsy suite into a formal written document may be especially challenging. We offer a few guidelines

- Titles such as anthropologist, investigator, instructor, and intern carry specific meanings and responsibilities. Do not create job titles to describe your student experiences. If you receive compensation for your work, your title appears on your pay stub or contract. This is the title you must use on your CV, as supporting documents will identify you only as having held that title. Teaching assistants cannot change titles to instructor, even if the duties performed are best described by the higher title. Ask the writers of your letters of recommendation to highlight instances in which your responsibilities exceeded your title.

- For unpaid experience, ask your supervisor or mentor how the experience should be noted on your CV. Better yet, ask for an open letter of recommendation that outlines your experience and responsibilities.

- Recognize that observing is not the same thing as participating, which is not the same as being responsible. The members of the hiring committee to which you are applying are well aware of the level of responsibility students and trainees can have in forensic contexts. Avoid the temptation to embellish or overstate. All experience is valuable and relevant but must be presented appropriately.

Any discrepancies between your CV and your background check may be viewed as deception. It would be tragic to be identified as an excellent job candidate, only to lose the opportunity because of an embellishment or inaccuracy on your résumé or to have such discrepancies discredit you on the witness stand.

to question the witness regarding training and qualifications; however, the decision to qualify the witness rests with the judge. While material and fact witnesses can testify only with respect to their direct knowledge and observations, an expert witness is allowed to offer opinion based on experience as well as direct observation. Expert witnesses cannot be compelled to testify unless they had some direct participation in a case and would therefore be categorized as fact witnesses as well.

In criminal proceedings, opinions must be stated to a *reasonable degree of certainty*. There can be no other reasonable possibilities, and speculation by any witness, fact or expert, is not appropriate or permitted. In civil court, expert opinion must meet a *degree of probable certainty*. Under this standard, there is no need to eliminate competing reasonable possibilities. The law requires only that such possibilities be less likely than the proposed possibility (V. I. Adams, 2002). Although speculation is also not permitted in civil court, the expert witness is often granted more latitude to explain the basis for his or her opinions.

An anthropologist preparing to testify as an expert witness should consider two related and important legal concepts: disclosure and work product. **Disclosure** is the act of making known something that was previously unknown or the revealing of facts or evidence. All witnesses, including experts, are required to disclose information throughout the course of a legal proceeding. The process of discovery controls what information is disclosed, when it is disclosed, and to whom. **Discovery** is compelled disclosure. On discovery, all relevant information, reports, evidence, and findings must be presented to the opposing counsel. The exception to the discovery rule is work product.

To avoid inadvertently or inappropriately disclosing information, it is important for practicing forensic anthropologists to understand the notion of work product. **Work product** is any tangible material or its intangible equivalent that is prepared by or for a lawyer or an act of litigation. Materials deemed to be work product are exempt from discovery or compelled disclosure. All work produced at a crime scene, in reports, or during autopsy must be produced on request by subpoena (Box 3.4). However, any written or oral communication between counsel and an expert or fact witness need not be disclosed to the court, even under subpoena. For example, a forensic anthropologist taking notes at a crime scene may note that the decedent is a female. Upon closer examination at autopsy, the anthropologist may revise that opinion and conclude the individual is male. Before going to the grand jury, the prosecutor receives copies of the original scene notes and the final autopsy report. Noting the discrepancy, the prosecutor contacts the anthropologist, who explains the reasons for the change in a letter. Once the accused has been indicted, the defense lawyer begins the process of discovery and subpoenas all relevant documents from the forensic anthropologist. The anthropologist is obligated to send the original scene notes and the final report, but is not obligated to send a copy of the letter sent to the prosector,

BOX 3.4
Work product in the workplace

A variation of the work product rule also exists among different agencies. For example, a medicolegal investigative file includes internally generated materials such as the autopsy report and the field investigators' notes, as well as externally generated materials such as police reports and photos. If the file is subpoenaed, the medicolegal agency is required to produce only documents generated within the agency and may not release any externally created documents such as police reports. Although lawyers can obtain all such materials from the respective issuing agencies, it is crucial that such rules be respected and followed. Consultants such as anthropologists or odontologists must carefully review files submitted in response to subpoenas so that they do not inadvertently or unknowingly release inappropriate documents.

since it is work product. Should the defense lawyer notice the same discrepancy, he or she is free to question the forensic anthropologist about it, either prior to or during trial.

Qualifying as an Expert Witness

The role of the expert witness was, is, and will likely continue to be assisting lawyers, judges, and jurors in understanding complex scientific and technical issues (Auxier and Prichard, 2001). The rules governing expert witness qualification and acceptance of scientific and technical methodologies have evolved over time and are likely to be rewritten again as technology advances. Which rule applies to a scientist in a specific case is determined by jurisdiction and is often the source of considerable confusion and misunderstanding.

Regardless of the expert witness ruling used in a jurisdiction, the final determination of who qualifies as an expert and what methods or theories that expert may introduce in testimony rests solely with the presiding judge. All expert testimony rules identify the judge as the "gatekeeper" or final arbiter of who qualifies as an expert witness and what scientific or technical issues may be presented to the court. While either counsel can challenge the qualifications, training, or expertise of a proposed expert, the final decision is rendered by the judge. To begin to comprehend the complex nature of expert testimony, it is important to separate the two entities—the experts themselves and the methods, theories, and opinions expressed in their testimony.

Unlike material or fact witnesses, expert witnesses are permitted to offer opinion based on experience or experimental findings rather than direct observation. As such, the qualifications required to be declared an expert are more stringent. Federal Rule of Evidence 702.1 outlines the general qualifications of an expert witness as an individual having sufficient knowledge, skill, experience, training, or education to provide the knowledge that will assist the trier of fact to understand the evidence or to determine a fact in issue (Graham, 2003). What constitutes "sufficient" knowledge, training, or education is left to the judge's discretion. For certain professionals, such as medical examiners, the minimum requirements for qualifying as an expert in court are the same as the minimum qualifications for employment—a valid medical license and appropriate certification from a licensing board in forensic pathology. For other forensic scientists, including anthropologists, the minimum credentials required to qualify as an expert have yet to be established and vary by jurisdiction. A terminal degree, such as a PhD, DDS, or MD, is often considered a minimal qualification for many specialties. As noted in Chapter 1, a PhD is required for DABFA certification.

Defining an expert's precise area of expertise is an integral part of the qualification process. For example, not all practitioners who identify themselves as forensic anthropologists have received the same training or possess equal knowledge of different aspects of anthropology. While all forensic anthropologists should, in theory, be able to reliably determine the sex of an individual, not all can provide an accurate estimate of the time-since-death or testify regarding the class of weapon that produced the sharp force wound. An expert's areas of expertise are normally established through review of work history, publications, and other scholarly works, as well as any certifications obtained from accredited licensing boards. Once such areas have been established, it is the responsibility of the judge as well as the expert to ensure that the testimony of the expert falls within the scope of his or her expertise. Aggressive attorneys may attempt to lead an expert witness into offering opinions on evidence outside his or her areas of expertise. Should this occur, the judge will require the question to be reworded or withdrawn. However, if the judge fails to do so, it falls to the experts to protect themselves and confine their comments to their recognized fields of expertise.

Once an individual has been qualified by a judge as an expert, and areas of expertise have been established, it again falls to the judge to determine which scientific methods and theories may be presented to the court. In the United States, three different rules of law exist that guide or assist the judge in making this determination: Federal Rule of Evidence 702, the *Frye* test, and the *Daubert* ruling. Federal Rules of Evidence 403, 702, and 703 concern testimony by experts and provide five criteria for the qualification of expert witnesses and the admissibility of scientific evidence (Box 3.5). These rules apply only to federal courts using the Federal Rules of Evidence (FRE).

For most of the twentieth century, courts applied a standard first outlined in *Frye v. United States* (293 F.2s 1013, 1923). Known as the *Frye* test, the ruling requires scientific expert testimony to be "generally accepted" within the particular field to which it belongs (Box 3.5). In 1993 the Supreme Court handed down a decision that set forth criteria federal courts must follow in admitting scientific evidence. In its ruling in *Daubert v. Merrell Dow*

BOX 3.5
Rules of evidence

FEDERAL RULE OF EVIDENCE 702 CRITERIA

1. Requires the validation of the expertise of the witness, through demonstrated skill, experience, training, or education.
2. The responsibility for testing the admissibility and reliability of experts falls to the judge.
3. The testimony is based on sufficient facts or data.
4. The testimony is the product of reliable principles and methods.
5. The witness has applied the principles and methods reliably to the facts of the case.

(Graham, 2003)

THE FRYE RULING CRITERIA

1. The expert qualifications of the presenting witness.
2. That proposed testimony be generally accepted within the particular field in which it belongs.

(Masten and Strzelczyk, 2001)

THE DAUBERT RULING CRITERIA

1. Whether the theory or technique used by the expert can be or has been tested.
2. Whether the theory or technique has been subjected to peer review and publication.
3. The known or potential rate of error of the method used.
4. The degree of the method's or conclusion's acceptance within the relevant scientific community.

(Brautbar, 1999)

THE CANADIAN STANDARD—THE MOHAN RULING

1. Relevance
2. Necessity in assisting the trier of fact
3. The absence of any exclusionary rule
4. Qualifications of the expert

(Bruce, 2006; Rogers and Allard, 2004)

Pharmaceuticals (No. 92–102 509 US 579, 1993), commonly known as *Daubert*, the Supreme Court clarified some ambiguity in Federal Rule of Evidence 702. The ruling removed the "general acceptability" criterion required under the *Frye* test and focused on a method's validity, replicability, and error rate (Box 3.5). Anthropologists and other scientists facing a court challenge under

---

BOX 3.6
Preparing for a challenge under the Daubert ruling

The judge, who serves as the "gatekeeper" of all court proceedings, is solely responsible for determining which methods or theories are admissible. While it is technically possible for an expert to be challenged on any evidence presented, experts should be prepared to be challenged on all new, controversial, or previously inadmissible methodologies. In pretrial preparations with prosecuting and defense attorneys, anthropologists and other experts will review the substance and content of their expected testimony. Should questionable methods be identified, the anthropologist will be expected to prepare the necessary supporting documentation to meet a challenge under the *Daubert* ruling. Challenges may be raised by the judge or by the opposing counsel.

Challenges to exclude testimony may be heard as pretrial motions or can be raised prior to the swearing in of the expert witness who intents to introduce the evidence. In either instance, the jury (if present) will be excused from the courtroom. The judge will hear preliminary statements from both sides addressing the issue, or will voice his or her concerns regarding the admissibility of the evidence. The challenge may relate to the inadmissibility of the method due to inadequate scientific rigor, or it may focus on the expert's lack of qualifications or expertise in applying the method. In either case, the anthropologist or expert must present to the court a full or annotated curriculum vitae, detailing his or her experience, training, and scholarly works expressly related to the issue at hand. If the judge is satisfied that the witness has the necessary background to be declared expert, attention then shifts to establishing the scientific credibility of the method. The anthropologist is responsible for presenting to the court the necessary documentation (a full bibliography relating to the method, as well as copies of all cited publications or other scholarly works) to address the four *Daubert* criteria (testing, publication and peer review, known error rate, and acceptance). After the judge has reviewed the documentation, he or she will rule on the admissibility of the evidence and the jury will be returned. If admitted, the expert may introduce results or opinions based on the method. If excluded, neither the expert nor counsel can discuss or even allude to the method or its results.

the *Daubert* ruling regarding a specific method or theory must be prepared to present supporting evidence to the court (Box 3.6).

Response to the *Daubert* ruling in the scientific and legal communities has been prolific and mixed (see, e.g., Auxier and Prichard, 2001; Bertin and Henifen, 1994; Bohan and Heels, 1995; Brautbar, 1999; Gatowski et al., 2001; Gold et al., 1993; Kaufman, 2001; Sturner et al., 2000). In response to the *Daubert* criterion that calls for techniques to be subject to peer review and publication, anthropologists have been attempting to refine probabilistic estimates for the application of standard osteological method (see, e.g., Christensen, 2004a, 2005; Koot et al., 2005; Rogers, 2005a; Rogers and Allard, 2004).

Any expert preparing testimony should be mindful that the *Daubert* ruling is binding only in federal courts applying the Federal Rules of Evidence.

BOX 3.7

"Scientific" versus "technical" expert testimony

Two U.S. Supreme Court decisions have ruled on the interpretation of "scientific, technical or other specialized knowledge" contained in Federal Rule of Evidence 702. In *Daubert v. Merrell Dow Pharmaceuticals, Inc.* ([1993] 509 US 579), the court identified four criteria for determining scientific expert testimony (Box 3.5). The issue in dispute in the *Daubert* case was whether the drug Bendectin, when taken by pregnant women, caused birth defects. However, the *Daubert* criteria are not applicable to issues involving technical expertise, such as that offered by engineers. In *Kumho Tire Co. v. Carmichael* ([1999] 131 F.3d 1433), a number of passengers in the plaintiff's vehicle had been injured when a tire failed. An expert in tire failure relied on his own experience to conclude that the blowout was the result of a defect and not misuse on the part of the plaintiff. The Supreme Court ruled that technical testimony was not subject to the same criteria as "scientific knowledge." Testimony regarding technical matters can be considered expert if it focuses on specialized observations, a specialized theory, or the application of such theory, in a particular case. Further, the ruling requires a technical expert to employ the same level of intellectual rigor, whether based on professional studies or personal experiences. This ruling impacts archaeologists, anthropologists, death investigators, police, and scene reconstructionists, who can be declared technical experts on the basis of specialized knowledge acquired through personal experience. For example, an anthropologist with years of experience working in a particular geographic region would be permitted to offer opinions on the scavenging patterns of animals common to the area that frequently interact with human remains. Such testimony could rely on observations and experience, rather than experimentation.

Many states use variations of the FRE and have either adopted *Daubert* or are moving toward doing so. The remaining states apply a local variation of the *Frye* ruling (Masten and Strzelczyk, 2001). It is also important to note that *Daubert* rules apply only to scientific knowledge, not technical expertise (Box 3.7). Experts, including forensic anthropologists, are strongly advised to determine which rule is applied within their jurisdiction prior to presenting testimony. If in doubt, a discussion with the district attorney will indicate which rule applies.

In Canada, the guidelines for admissibility of expert testimony are less complicated. In *Regina v. Mohan* (1994, 25, CR.9), the Canadian Supreme Court held that the following criteria must be met: relevance, necessity of assisting the trier of fact, the absence of any exclusionary rule, and the qualifications of the expert (Boxes 3.5 and 3.8).

BOX 3.8
Expert witness testimony in Canada

In *Kelliher (Village of) v. Smith* ([1931] S.C.R. 672, p. 684), the Supreme Court of Canada declared that for testimony to be considered expert "[t]he subject matter of the inquiry must be such that ordinary people are unlikely to form a correct judgment about it, if unassisted by persons with special knowledge." In *Regina v. Mohan* ([1994] 2 S.C.R. 9), the Supreme Court of Canada expanded on this requirement, stating that expert evidence must be both *necessary in assisting the trier of fact* and *relevant*. Regarding the necessity in assisting the trier of fact, the Court contended that expert evidence was not to be admitted if the testimony concerned a matter of common knowledge. "An expert's opinion is admissible to furnish the court with scientific information that is likely to be outside the experience and knowledge of the judge or jury. If on the proven facts a judge or jury can form their own conclusions without help, then the opinion of an expert is unnecessary" (*R. v Mohan* [1994]: 24). An example pertinent to anthropologists would be an attempt to introduce testimony describing an unmarked grave as an indication of unnatural death, as it does not take an expert to realize the dead cannot bury themselves.

In terms of expert witness testimony, one of the principal differences between the U.S. and Canadian courts is the Canadian criterion of relevance. The Canadian Supreme Court ruled that expert testimony was relevant if it "so related to a fact in issue that it tends to establish it" (*R. v. Mohan* [1994]: 20). Relevancy is evaluated *prima facie* (on first appearance, subject to further evidence or information). Furthermore, relevance must also include a test to determine whether the evidence is essential and reliable. In evaluating essentialness, evidence that is otherwise logically relevant may be excluded "if it involves an inordinate amount of time which is not commensurate with its value or if it is misleading in the sense that its effect on the trier of fact, particularly a jury, is out of proportion to its reliability" (*R. v. Mohan* [1994]: 21). An example of excluding evidence under the essential rule would be the decision to forego testimony on anthropological methods of determining the sex of an individual who has been positively identified or who retained external genitalia.

Reliability is one of a number of terms that the Canadian Supreme Court did not adequately define, providing the lower courts with little direction concerning the characteristics of expert testimony. The introduction of new methodologies was addressed: "[e]xpert evidence which advances a novel scientific theory or technique is subjected to special scrutiny to determine whether it meets a basic threshold of reliability and whether it is essential in the sense that the trier of fact will be unable to come to a satisfactory conclusion without the assistance of the expert" (*R. v. Mohan* [1994]: 25).

Many Canadian professional organizations have issued position statements concerning ethics in courtroom testimony. These statements establish guidelines for expert witness testimony for their respective membership. For example, the Canadian Psychiatric Association issued a code of ethics pertaining to psychiatric expert testimony (Mellor, 1980). A later amendment included the following stipulation: "a psychiatrist should testify in a court of law as to the mental state of a particular person only if he has examined that person" (Canadian Psychiatric Association, 1986, Annotation #9). Enforcement of such regulations is restricted to terminated membership within the professional organization and possible professional censure.

(Bruce, 2006)

The preceding section was intended to serve as an introduction to the judicial system. Given an understanding of the fundamental structure of the system, as well as the rules governing expert witness testimony, anthropologists can successfully navigate the complex world of jurisprudence. With a foundation of legal terminology established, we will now focus on applications of law that are of interest to the medicolegal community, specifically those relating to homicide.

HOMICIDE

Of the five manners of death, suicide elicits the most emotion while homicide evokes the most interest. Homicide is the killing of one person by another. Under the law, homicide is a "neutral" charge. It implies only responsibility, not intent. While all homicides imply responsibility, only one particular type of homicide—murder—implies intent. Terminology becomes critical: although the popular media use the terms interchangeably, it is imperative that practitioners recognize the distinctions between homicide and murder. All murders are homicides; not all homicides are murders. Homicide is a manner of death certification, while murder is a legal charge. Murder is only one of several types of homicide including negligence, misadventure, murder, and manslaughter.

Negligence is any conduct that falls below the legal standard established to protect others against unreasonable risk of harm (Garner, 2001). Negligence excludes conduct that is willful, intentional, or wanton. Like other forms of homicide, negligence is a neutral charge. Acts such as a property owner failing to remove ice or snow from a public walkway can result in a finding of negligence should a pedestrian be injured as a result.

Another neutral form of homicide is misadventure. **Misadventure** is committed accidentally by a person performing a lawful act and with no intent to kill. An individual driving a car in a completely lawful manner who kills a pedestrian or another driver is one example of misadventure. Misadventure is an accident in which the lawful action of another person is the causative agent. While the concept is embraced by the British legal system (Garner, 2001), many jurisdictions in North America do not recognize misadventure. In such places, deaths resulting from acts of misadventure are called accidental.

While negligence and misadventure are neutral forms of homicide that may not result in any criminal penalties, murder and manslaughter are felonies that carry harsh penalties. **Murder** is the killing of a human being with malice aforethought. **Malice aforethought** is the requisite mental state of the accused for the finding of murder. To demonstrate malice aforethought in a court of law, the act in question must encompass any of the following: the intent to kill, the intent to inflict grievous bodily harm, reckless disregard for human life, or the intent to commit a dangerous felony.

Demonstrating malice aforethought in a criminal trial requires evidence of premeditation, or *mens rea*. **Premeditation** is any evidence of conscious consideration and planning that precedes the commission of a crime. Examples of premeditation include obtaining the murder weapon or creating an alibi. *Mens rea* means "a guilty mind," The accused shows an understanding of the unlawful nature of the act through his or her behavior before or after the crime. Examples of *mens rea* include lying to the police and destruction of evidence.

Although terminology and associated penalties vary by jurisdiction, most legal systems acknowledge the following types of murder: felony; first, second, and third degree; serial; and mass. **Felony murder** is committed during the course of another dangerous felony, such as armed robbery, rape, or kidnapping. **First-degree murder** is willful, deliberate, and premeditated. Many jurisdictions do not differentiate between felony murder and first-degree murder, opting for the hybrid first-degree/felony murder charge that covers both circumstances and carries the same penalty. **Second-degree murder** is an act not aggravated by any of the circumstances of first-degree murder. The significant difference between first- and second-degree murder is that second degree murder lacks the requirement for evidence of premeditation. **Third-degree murder** is statutorily defined murder that is considered less heinous than first- or second-degree murder. The death results from an act that does not constitute murder under common law. For example, in some jurisdictions, prosecutors may bring such a charge against parents having strong religious beliefs who refused medical treatment that might have prevented the death of their child.

Both serial and mass murder require the killing of a minimum of three individuals but under different circumstances. **Serial murder** involves the killing of at least three people at different times and locations. The elapsed time between killings can range from hours to years and is referred to as a "cooling off" period. Serial murders are linked through similar methods of operation or by victims who share similar characteristics. Ted Bundy, John Wayne Gacy, and Jack the Ripper were serial killers. **Mass murder** involves the killing of three or more people at the same approximate time and place. Although the deaths may occur over a matter of minutes or hours, they represent a single homicidal event. The killings at Columbine High School in 1999 and Virginia Tech in 2007 examplify mass murder.

Manslaughter is the unlawful killing of a human being without malice aforethought. All jurisdictions recognize two types of manslaughter: involuntary and voluntary. **Involuntary manslaughter** is homicide in which no intent to kill or harm is demonstrated, although the act was committed with criminal negligence or during the commission of a crime not included within the felony murder rule (Garner, 2001). **Voluntary manslaughter** is an act of murder reduced to manslaughter because of extenuating circumstances or diminished capacity. These acts are considered to be "crimes of passion." The primary distinction between charges of murder and manslaughter is that with manslaughter, although the act resulting in death was unlawful, the intent or malice that is the essence of murder is not evident.

CONCLUSION

Ignorance of the law is not a valid defense for an accused person and is unacceptable in any practicing member of the medicolegal profession. It is the responsibility of all forensic scientists to know and understand the relevant laws. Forensic anthropologists should, at a minimum, have clear knowledge of the rules of evidence, as well as a thorough understanding of the structure of the criminal justice system.

As in all disciplines in medicine and law, forensic science requires a specificity of language that is crucial in effective communication. There are subtle but important distinctions in terms such as "homicide" and "murder." The correct use of terminology is vital, particularly in a discipline so prone to misrepresentation in the media.

This chapter was intended as an introduction to the structure, concepts, and terminology of the legal system. It is by no means definitive or exhaustive. Students, infrequent practitioners, or scientists transferring to new locations are strongly advised to review all the statutes, guidelines, and protocols in their jurisdictions. To advance and professionalize forensic anthropology, practicing anthropologists must acknowledge and embrace the legal system that makes their work "forensic."

CHAPTER 4

Crime Scene Investigation

Identifying Medicolegal Remains, Search Strategies,
and Scene Documentation

Having introduced the medicolegal system in Chapter 2 and discussed its interaction with the judicial system in Chapter 3, we now focus on the role of forensic anthropologists in the death investigation process. This process begins at the death scene. Depending on circumstances, the location of death can be of significant evidentiary value to the investigation, providing clues to the cause and manner of death and to the victim's identity, as well as an understanding of perimortem events. Understanding how scenes are processed allows anthropologists to become fully integrated into the investigation.

The participation of forensic anthropologists can be invaluable to the proper handling of a death scene. Forensic anthropologists are uniquely qualified to answer specific questions, such as how many individuals are represented or how much of the body is present. Forensic anthropologists are also specially trained to locate clandestine graves and recover remains from a variety of depositional environments, including burials, fire scenes, and mass death scenarios, such as mass and natural disasters. Depending on the geographic region and environment, cases involving skeletal remains may not be a common occurrence. For example, in Edmonton, Alberta, skele-tonization occurs within months regardless of the season (Komar, 1998). In contrast, in Arizona and New Mexico, bodies may be skeletonized within days (Galloway et al., 1989). Because skeletal cases are relatively rare, law enforcement officers and medicolegal investigators may have only minimal experience in dealing with such cases. Therefore, the opportunity exists for anthropologists to play a vital role in field recovery, providing continuity of approach and much-needed expertise.

As with fleshed bodies, the primary question facing investigators handling skeletal material is whether the remains fall under the jurisdiction of the medicolegal authority. This requires anthropologists to evaluate whether the remains are of medicolegal significance, an issue so significant that we address it before turning to issues of jurisdiction, authority, and protocols specific to scene work, as well as scene evaluation, search strategies, evidence collection, documentation, scenes involving burials, and report writing.

An experienced anthropologist on scene can address issues as they arise, assist in the recovery of the remains and develop appropriate search strategies. Anthropologists are also uniquely qualified to answer the following questions:

- *Is it bone or tissue?*
- *Is it human?* These initial questions will be examined shortly in greater detail.
- *How many people are represented?* Establishing the **minimum number of individuals (MNI)** may be necessary in cases of mass or natural disaster (see Chapter 9), fire scenes, or any circumstance in which the bodies are no longer intact. Establishing MNI relies on recognizing the duplication of specific elements, such as recovering two right femora, as well as differences in the age, sex, or size of elements.
- *How long has the body been there?* Establishing the time-since-death may be crucial in many investigations. Methods of estimating postmortem interval will be discussed in Chapter 7.
- *Have the remains been modified postmortem?* Has the body been disturbed, moved, or altered? Recognizing scavenging, intentional dispersal, or modification, as well as identifying the possibility of a second scene, is vital to proper scene analysis and interpretation. This subject will be examined in Chapter 7.
- *Has the entire body been recovered?* When dealing with disarticulated, fragmentary, or partial remains, anthropologists can identify and inventory recovered elements. This allows investigators at the scene to know when the search has been successfully completed.
- *Who is it?* Anthropologists on scene may alert law enforcement personnel to any inconsistencies between the evidence at the scene, such as a wallet or car registration, and the remains themselves. The process continues at autopsy with the development of the biological profile (Chapter 5) and culminates in a positive identification, which will be examined in detail in Chapter 8.
- *Are the remains of medicolegal significance?*

IDENTIFYING REMAINS OF MEDICOLEGAL SIGNIFICANCE

Medicolegal remains are defined as those individuals for whom cause and manner of death are determined and a death certificate is generated.

Frequently, medicolegal investigators encounter other types of human remains. These remains may be of legal interest or may even represent criminal activity. However, such cases do not fall under the jurisdiction of the medicolegal authority. The involvement of the medicolegal authority or anthropologist may continue in these cases, but only at the request of law enforcement or the district attorney. Medical examiners or anthropologists can assist in evaluating or identifying the nature of the remains. Some cases even result in a trial, making the remains of forensic interest. Nevertheless, such cases are not expressly of medicolegal significance, no death certificate will be issued, and the remains are therefore designated "nonmedicolegal." Types of **nonmedicolegal remains** include anatomical specimens, trophy and souvenir skulls, "shrunken heads," and historic or prehistoric remains, as well as improperly disposed of mortuary remains.

Provenience is a record of the source or origin of artifacts or materials such as skeletal remains. Skeletal remains for which there is no provenience are classified as unidentified remains. All medicolegal statutes grant the ME or coroner the right to evaluate unidentified remains and to establish jurisdiction over those remains believed to be of medicolegal significance. Medicolegal authorities who initially claim jurisdiction over material that ultimately proves to be nonmedicolegal are responsible for safeguarding and housing such material and forwarding it to the appropriate agency or authority for disposition. This responsibility can prove time-consuming or costly, and most medicolegal authorities prefer that such remains be correctly identified in the field before jurisdiction is established. A well-trained anthropologist who is capable of rendering such identifications is invaluable in preventing the needless commitment of resources on the part of the ME or coroner.

Types of Nonmedicolegal Remains

Anatomical specimens are cleaned dry bone prepared for teaching or research purposes. In 1985 a U.S. embargo on the importation of human remains made anatomical specimens a "regulated" material. Sale by licensed medical supply companies of existing specimens is restricted to medical professionals and educational institutions. Possession of anatomical specimens is not illegal, although resale by or to unlicensed private citizens is a trade violation. Despite this, the sale of anatomical specimens on the Internet continues largely unchecked.

Typically, anatomical specimens are easily identified. Most show evidence of mounting or wire articulation, as well as autopsy skull sectioning (Figure 4.1). Some specimens bear serial numbers or specimen codes, while others have anatomical terms or muscle insertions painted on the bone. Remains that have been buried or embalmed, as evidenced by color changes or condition, cannot be considered anatomical specimens and may represent illegally procured remains. Anatomical specimens received by the

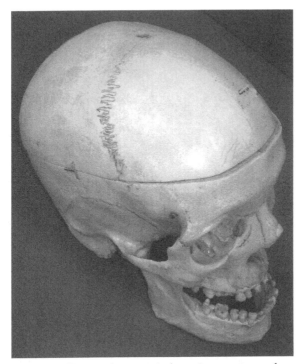

Figure 4.1 Prepared anatomical specimen, a type of nonmedicolegal human remains frequently encountered by law enforcement or the ML authority. Note the autopsy cut, drilled hole at bregma to accommodate mounting, wire articulation hardware, and cement repair to the alveolar processes and distal nasal bones. (*Photo by Debra Komar*)

medicolegal authority can be donated to medical schools or universities or may be retained for teaching purposes, if local statutes permit.

Trophy or **souvenir skulls** are crania obtained from foreign sources. Trophy skulls include remains brought to the United States as battle souvenirs by returning soldiers. Souvenir skulls are historic or archaeological skulls purchased by U.S. tourists overseas from flea markets or antique shops (Figure 4.2). In-depth examinations of trophy skulls and diagnostic criteria are provided in Bass (1983), Sledzik and Ousley (1991), and Taylor et al. (1984). Shrunken heads are mummified skulls (often those of monkeys or rodents) with retained scalps. It is important to note that, under the law, it is immaterial whether the shrunken skulls are "real"; any biological material representing a shrunken head is subject to similar restrictions in the United States.

Possession, import, or trade of trophy and souvenir skulls is restricted. These items cannot be bought or sold and are subject to seizure by customs

(a)

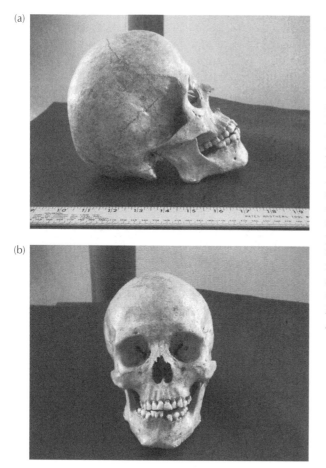

(b)

Figure 4.2 (a) and (b) Two views of a trophy skull, illegally imported to the United States from Guam by a tourist. This type of specimen represents nonmedicolegal remains in that it does not fall under the jurisdiction of the medicolegal authority. Although many trophy skulls are painted or decorated, this skull is an example of a cleaned, prepared skull that might potentially be mistaken for an anatomical specimen. (*Photos courtesy of the New Mexico Office of the Medical Investigator*)

officials or law enforcement, with the holder subject to penalties and prosecution. A number of laws or statutes regulate trophy and souvenir skulls, including the following:

- U.S. Code Title 18 Sections 2314 and 2315, National Stolen Property Act
- Public Law 97–446 96 Stat 2329, Implementation of Convention on Cultural Property Act
- USPHS 42 CFR, Part 71, Importation of Etiologic Agents
- Customs Prohibited and Restricted Items List
- 1970 UNESCO Convention

Determining which laws have been violated by possession of a particular specimen depends on its origin, its date of importation, and other factors. Because of the illicit nature of these materials, documentation is usually nonexistent and prosecution often depends on catching someone in the act of importing or selling the remains. Medicolegal offices in possession of trophy or souvenir skulls should contact local customs law enforcement to begin the process of repatriation.

Historic remains are skeletal, fleshed, or preserved remains of individuals who died during the historical period of a given region. The beginning of the historical period varies by geographic region but represents the beginning of written documentation or governmental structure within the area. The end of the historical period reflects the point at which the death falls under the jurisdiction of the medicolegal authority. A number of variables determine the temporal sectioning point within a region or jurisdiction, including the availability of social data providing a means of personal identification, the statute of limitations on crimes, and the possibility of unresolved missing persons reports or outstanding criminal investigations. Ultimately, determining whether a set of remains falls under medicolegal jurisdiction rests solely with the ML authority. In cases involving remains recovered following a significant postmortem interval, there is often little chance of identifying the individual and these remains are designated historic or nonmedicolegal. Responsibility for historic remains lies with the state Office of Historic Preservation if the remains were recovered on state or private land, and with the U.S. Department of the Interior for remains discovered on federal lands.

Prehistoric or **ancient remains** involve individuals whose deaths predate the historic period in a given region. In 1990 the **Native American Graves Protection and Repatriation Act** (Public Law 101–601; 25 U.S.C. 3001) was created to protect Native American remains. Commonly known as NAGPRA, the act renders the excavation or disturbance of remains on federal or tribal lands illegal, with U.S. district courts having jurisdiction over violators. Complete information regarding **NAGPRA**, including its impact on teaching and research, is available at the governmental website: www.cr.nps.gov/ nagpra. Chapter 53 of 18 U.S.C., Section 1170, also makes punishable the sale, purchase, use, or transport of the human remains of a Native American, with penalties including imprisonment and fines. Penalties under this section are not restricted to violations involving remains recovered from federal lands. Legislation in other countries concerning repatriation varies. While neither Canada nor the United Kingdom, for example, has a national law such as NAGPRA, repatriation remains a matter of concern, as outlined in Box 4.1.

In the United States, the lawful excavation of prehistoric remains requires participating anthropologists and archaeologists to possess a valid permit. Regulations regarding the issuance of permits are covered in Section 4 of the Archaeological Resources Protection Act of 1979 (16 U.S.C. 470cc). Archaeologists working within the boundaries of any Native American Reservation require a permit from the Bureau of Indian Affairs, as specified in the Code of Federal Regulations (25 CFR 262). Those excavating

BOX 4.1

Canadian and European laws concerning repatriation

Canada does not have federally regulated repatriation laws similar to those outlined in NAGPRA in the United States. However, Canadian anthropologists have worked to establish and maintain a respectful, cooperative relationship with First Nations groups, fueled in part through the ongoing repatriation of both cultural artifacts and human remains. Excellent reviews of Canadian perspectives on repatriation issues can be found in Conaty and Janes (1997) and Reynolds (1980). Case studies involving the Kwakiutl Indians of British Columbia (Jacknis, 1996) and the reinterment of Thule Inuit remains (McAleese, 1998) also provide valuable insight into cultural property repatriation in Canada.

In 2001 the British government established the Working Group on Human Remains to review the legal status of osteological material from indigenous communities held in publicly funded museums. The group recommended that the laws governing national museums be changed to allow for repatriation of human remains, as well as transparent procedures for addressing repatriation claims, and the creation of a governmental licensing authority to monitor the treatment and return of human remains (L. Smith, 2004). The response to the report from the anthropological community was mixed, reflecting two disparate viewpoints concerning repatriation. While some felt it a positive move toward correcting past wrongs and offering respect to the dead and their relatives (Payne, 2004), others decried the recommendations as unworkable, overly bureaucratic, and threatening to collections and research on human remains by giving indigenous rights primacy over scientific values (L. Smith, 2004). Although, to date, no formal action has been taken, the report is seen by many as an important first step in addressing the current lack of regulation of curated prehistoric remains in the United Kingdom.

on state or private land require permits from the state Office of Historic Preservation. Several types of permits are issued, included those specially designated for the excavation of human remains. It is the responsibility of the archaeologist to ensure possession of the appropriate permit. Such permits are typically issued for one year and must list the specific sites where permission has been granted. Permits require all archaeologists to contact the local medicolegal authority and law enforcement whenever any human remains are encountered in the course of the excavation. Only the medicolegal authority can legally determine whether the remains are of medicolegal significance.

Forensic anthropologists working in areas where prehistoric remains are frequently encountered should possess valid excavation permits.

Anthropologists working under the jurisdiction of the medicolegal authority are permitted to participate in the initial assessment of remains to determine their significance. However, once the remains have been designated as nonmedicolegal and the ME or the coroner terminates ML jurisdiction, continued interaction with the remains by the anthropologist is only lawful if the anthropologist possesses the necessary permit. Affiliation with the ME or the coroner does not grant an anthropologist the authority to disturb ancient remains, nor does it automatically protect the anthropologist from any legal ramifications resulting from such actions after jurisdiction has been terminated.

Construction and public works crews frequently encounter prehistoric remains in the course of their excavations. Such inadvertent discovery is also regulated on federal lands by NAGPRA and by the Archaeological Resources Protection Act on state or private land. All relevant laws require that companies encountering human remains immediately cease activity and contact local authorities. Failure to do so carries significant penalties. Individuals who inadvertently discover prehistoric remains are also required to report such remains to law enforcement or local authorities. Disturbance of *in situ* remains is a fourth-degree felony (resulting in penalties including imprisonment for up to 18 months). Such inadvertent discoveries will initiate an assessment by the medicolegal authority to determine the nature of the remains.

Medicolegal offices often encounter human remains that are out of context. Excavated remains are recovered from flea markets, antique stores, private homes, or for sale on the Internet (Huxley and Finnegan, 2004). While simple possession of prehistoric remains is not illegal, there are very few circumstances under which anyone could have lawfully obtained the remains. Despite this, successful prosecution of those violating the laws relating to prehistoric remains is rare. Forensic anthropologists who possess the appropriate permits can analyze these remains and initiate the repatriation process.

Distinguishing Recent from Archaeological Remains

Anthropologists are often called upon to determine the temporal origin of human remains. In her review of ABFA diplomate casework from 1984 to 1992, Reichs (1995) reported that 26% of all cases represented historic and prehistoric remains. Despite laws enacted to protect them, historic and prehistoric graves are increasingly being exposed by construction, vandalism, and natural forces. The prompt recognition of such remains prevents further medicolegal investigation, saving time and resources (Berryman et al., 1997).

Distinguishing recent from historic or prehistoric remains is a common responsibility for forensic anthropologists. Of primary importance in a forensic investigation is identifying remains that are of medicolegal importance.

Grisbaum and Ubelaker (2001) use the term "archaeological cases" to describe remains that are no longer of medicolegal significance.

Aiding in the recognition of cemetery remains encountered in a forensic investigation is an understanding of basic funerary customs (both modern and historic), as well as body preparation techniques and grave artifacts, (Berryman et al., 1997). Body position and orientation may provide clues. Clandestine graves could contain individuals deposited face down or with erratically arranged limbs, while careful, purposeful placement of the body in a grave is consistent with formal burial (Figure 4.3). Depending on religious or funerary customs prevalent in the region, the orientation of the grave (i.e., heads positioned to a specific cardinal direction), as well as its depth, stratigraphic context, and associated materials, may further contribute to determining the grave's temporal origin.

Additional indicators of postmortem interval may be found in the skeletal elements recovered. Biological indicators of cultural practices such as cranial deformation (Figure 4.4) or extensive dental wear (Figure 4.5) are often associated with specific time periods in a geographic region. Trauma assessment, in particular identifying wounds resulting from modern

Figure 4.3 Illustration of body position in a grave. The position of the bodies indicates that the grave was created with care and respect for the decedents, two prehistoric individuals: an adult female (left) cradling a juvenile (right). Such mortuary practice is inconsistent with clandestine graves. (*Photo courtesy of the New Mexico Office of the Medical Investigator*)

(a)   (b)

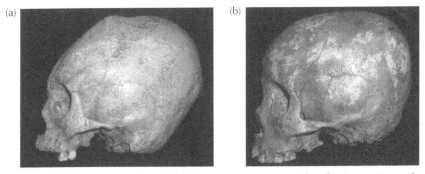

Figure 4.4 Illustrations of cranial deformation, an example of interpreting cultural affiliation from the skeletal remains. Deformation can be intentional, such as cranial binding (**a**) or inadvertent, as in cradle-boarding (**b**). (*Photos courtesy of the New Mexico Office of the Medical Investigator*)

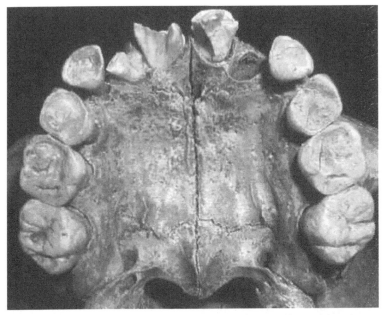

Figure 4.5 Dental attrition or wear, commonly seen in prehistoric remains of the American Southwest. Note the exposed dentine. This view is recommended when one is relying on photographs as part of a distance consultation. (*Photo courtesy of the New Mexico Office of the Medical Investigator*)

weaponry, can also differentiate recent remains from those of greater antiquity. Finally, bone quality, color, and condition may provide some evidence, although variability in decomposition and preservation rates prohibits forming stringent guidelines. Typically, bones that are yellow or pink (indicating the presence of lipids or blood) are of recent origin. The presence of a strong odor or insect activity also suggests a relatively short postmortem interval. The retention of desiccated soft tissue, however, should not be used in isolation as a means of differentiating recent and archaeological material. A variety of external factors, such as a cold or arid environment, and internal conditions, such as embalming, can result in the prolonged retention of soft tissue postmortem (Berryman et al., 1997; Komar, 1998; Manheim, 1997; Micozzi, 1997; Sledzik and Micozzi, 1997).

In cases involving equivocal or highly fragmentary remains, or when remains are recovered out of context, additional invasive testing may be required. A promising new method, reported by Ubelaker et al. (2006), uses elevated **carbon-14** (^{14}C) levels resulting from atmospheric testing of thermonuclear devises between 1950 and 1963 to define the "bomb curve." Radiocarbon analysis of dental enamel and bone samples drawn from selected elements of the same individual can determine whether the death occurred before or during the bomb curve (pre-1963). The method recognizes that different tissues form at varying rates during the human life span and that, by testing both tooth enamel and bone, this technique provides a more accurate assessment than traditional radiocarbon dating (Ubelaker et al., 2006). The method is of considerable forensic value, since deaths that occurred before 1950 may not be considered of medicolegal significance. The bomb-curve method expands on prior studies involving the use of radiocarbon dating to differentiate recent from archaeological bone (see, for e.g. R.E. Taylor et al., 1989).

Other destructive methods have been proposed to identify human remains of forensic interest. Swift (1998) tested the viability of measuring the equilibrium between two naturally occurring **radioisotopes** (^{210}Po and ^{210}Pb) as a means of estimating time-since-death. The pilot study was sufficiently encouraging to warrant a larger study but, to date, no further results have been reported. **Amino acid racemization** dating has been tested as a means of dating teeth and bone that are outside the range of radiocarbon dating (Bada, 1987).

Finally, exploratory studies suggest that DNA testing may provide evidence of the relative age of remains. Identification of the DNA sequence in ancient bone contaminated with modern DNA has been reported (Bouwman et al., 2006), based largely on the presence of extensive degradation products in the endogenous DNA sequence. Alonso et al. (2004) tested a real-time detection system based on the **polymerase chain reaction (PCR)** to evaluate the mitochondrial DNA preservation (copy number and degradation state) of samples harvested from human bone aged 500 to 1,500 years and from 4- to 5-year-old bone recovered from forensic cases. The ancient bone showed low copy number and highly degraded mtDNA compared to the recent bone, indicating that the method could be used to differentiate the two samples. To better understand the DNA degradation process in bone, von Wurmb-Schwark et al. (2003) extracted and amplified nuclear and mtDNA from ancient and artificially aged bone (incubated for up to 30 days in water heated to 90°C). After 12 hours of incubation, DNA extracted from the artificially aged bone was totally degraded but could still be typed. After 36 hours, no reproducible amplification of DNA was possible. The results mirrored those seen in the sample of 24 ancient bones, indicating two defined phases of DNA degradation that may contribute to differentiating recent from archaeological bone. Because of the influence of numerous environmental factors on DNA preservation, the accuracy of this method, which relies on relative DNA degradation, will probably be context specific.

DIFFERENTIATING HUMAN FROM ANIMAL BONE AND NONBIOLOGICAL MATERIALS

In addition to distinguishing modern and prehistoric remains, forensic anthropologists frequently differentiate human and nonhuman bones as well as nonbiological materials that may be mistaken for bone. Anthropologists are often asked to offer such assessments at scenes to prevent the needless investigation of nonhuman material.

The first question is always whether the material is even of biological origin. Many commonly encountered materials, including roots, rocks, and gypsum, as well as altered materials such as burned insulation or foam (Figure 4.6) can be mistaken for bone by untrained observers. Even plastic or plaster replicas of bone (including Halloween decorations and dog chew toys) can initiate police or medicolegal investigations (Figure 4.7). Normally, careful examination of the item quickly differentiates these mimics from actual biological material. If doubt remains, additional testing, including radiography, histology, or dissection, may be warranted.

The best course of study in differentiating animal and human bones is an exhaustive knowledge of human osteology, including familiarity with a number of different human populations or regional groups. Given the large variety of animal species, it is more efficient to learn the human skeleton than to become familiar with the osteological elements of all species. If a particular case requires the identification of animal remains, one can consult comparative faunal osteology collections, which are housed at many major universities, often in departments of zoology or in museums. Comparisons against such collections allow for the identification of skeletal elements to at least the genus level, if not to species level. Should field identification of animal bones be required by job description, training is available in the form of zooarchaeological or faunal osteology courses in some departments of anthropology.

Most morphological differences between animal and human bone relate to differences in locomotion. The bipedal posture of humans is reflected in virtually every element of the body. In contrast, the quadrupedal stance

(a)

(b)

Figure 4.6 Examples of nonbiological materials initially mistaken for bone. **(a)** burned electrical wire casing and **(b)** heat-altered insulation. The materials were recovered by law enforcement from a house fire and a car fire, respectively. (*Photos by Debra Komar*)

of most mammals results in features such as posterior placement of the foramen magnum and elongated pelves. Several osteological elements are unique to animals, including a baculum (penile bone), extended caudal vertebrae, and clawed distal phalanges.

Aside from these elements, human and mammalian bones are functionally analogous. For example, Figure 4.8 illustrates tibiae from four species,

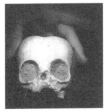

Figure 4.7 Example of a type of nonbiological material mistaken for bone. This plaster skull, a teaching specimen stolen from a local college, was found buried in the yard of a trailer park and reported to police. Both police and a deputy medical investigator on the scene believed the well-cast skull was real, despite its considerable weight. (*Photo courtesy of the New Mexico Office of the Medical Investigator*)

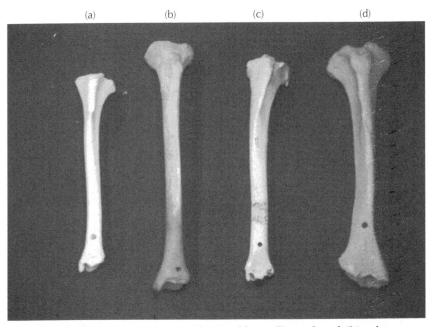

(a)   (b)   (c)   (d)

Figure 4.8 Differentiating human and animal bone. Examples of tibiae from several mammalian species: **(a)** coyote, **(b)** human, **(c)** deer, and **(d)** bear. (*Photo by Debra Komar*)

including human. As the bones are functionally analogous, their gross morphology is unsurprisingly similar. Differentiating human and nonhuman bone relies on differences in size, joint morphology, dental morphology and formulas, and internal structure. The osteology of the genus *Ursus*, commonly known as bear, most closely resembles human bone in both size and morphology. Bear paws without fur and claws are commonly encountered in areas where bears are indigenous and hunted. Gross and radiographic examination of fleshed or articulated specimens readily identifies each species (Figure 4.9).

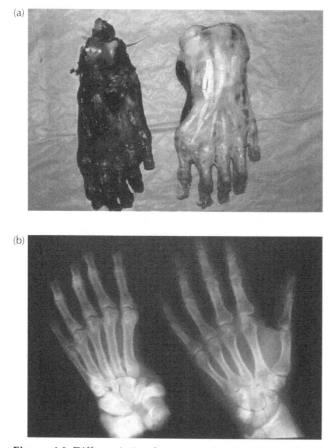

(a)

(b)

Figure 4.9 Differentiating human and animal bone. **(a)** Bear front and rear paw. Without claws and fur, both paws resemble human hands and feet. **(b)** Radiographic comparison of a bear front paw, absent distal phalanges and claws, and a human hand. (*Photos courtesy of the New Mexico Office of the Medical Investigator*)

Although a key to identifying skeletal elements, even for common species, is beyond the scope of this book, some general guidelines can assist anthropologists in separating human from animal bone. These are presented in Box 4.2.

With highly fragmentary or burned bone, it may be necessary to use histological or destructive analyses to determine species. Cortical bone of humans and of certain other mammalian species contains **osteons** and is irregular and disorganized in appearance. In contrast, most nonhuman mammalian cortical bone is organized in horizontal, regular layers, similar

BOX 4.2

General structural distinctions between animal and human bone

- *Articular surfaces* Articular surfaces, particularly of the limbs and extremities, are rounded and smooth in humans but often trochleated in mammals (Figure 4.10). Bear paws, often mistaken for human appendages, have a distal central ridge that bisects the articular surface of the metacarpals and metatarsals (Figure 4.11).
- *Joint morphology* In humans, major joint surfaces are typically smooth, allowing for greater range of motion. Mammalian joints, in contrast, have greater topography or interlocking morphology that decreases the range of motion but increases stability (Figure 4.12).
- *Nutrient foramen* The nutrient foramen of most human long bones is small and singular, while mammalian bones often feature pronounced grooves, leading to large and sometimes multiple foramina.
- *Rib profile* Typical ribs in humans are curved, with an inferior change in angle toward the sternal ends. Typical ribs in mammals are straighter, with little or no change in angle (Figure 4.13).
- *Costal grooves* In typical human ribs, a costal groove is present on the inferior–medial aspect of the rib. In animals, the costal groove is absent, or two grooves may be present, one on the cranial aspect, the other on the caudal aspect.
- *Thoracic vertebral spinous processes* In humans, the spinous process angles inferiorly, while in animals it projects dorsally (Figure 4.14).
- *Lumbar transverse processes* In humans, the transverse processes of a lumbar vertebra are short compared to the width of the centrum. In

Figure 4.10 Animal joint morphology: note the trochleated and bifurcated morphology of the joint surface of this deer metacarpal. (*Photo by Debra Komar*)

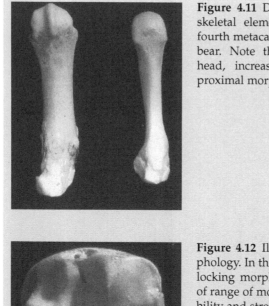

Figure 4.11 Differentiating human and bear skeletal elements. Both specimens are left fourth metacarpals. The element on the left is bear. Note the trochleated distal articular head, increased robusticity, and differing proximal morphology. (*Photo by Debra Komar*)

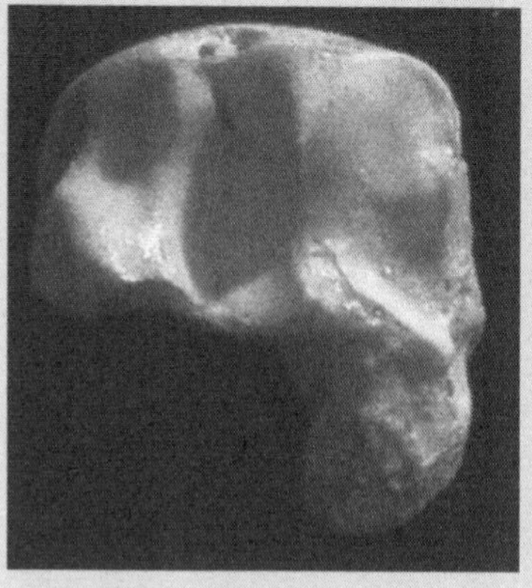

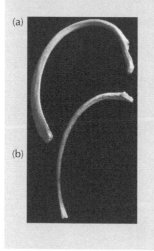

Figure 4.12 Illustration of animal joint morphology. In this deer tarsal, the rugged, interlocking morphology represents the sacrifice of range of motion for the sake of greater stability and strength. (*Photo by Debra Komar*)

(a)

(b)

Figure 4.13 Differentiating human from animal ribs in isolation. Rib (**a**) is a typical human rib; (**b**) is a typical canine rib. Note the greater degree of curvature of the human rib, as well as the change in shaft angle toward the ventral ends. (*Photo by Debra Komar*)

(a)

(b)

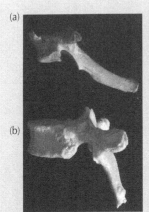

Figure 4.14 Differentiating human and quadripedal vertebrae. Note the differences in centrum size and shape, as well as the change in spinous process angle in these mid-thoracic vertebrae of a coyote (**a**) and a human (**b**). (*Photo by Debra Komar*)

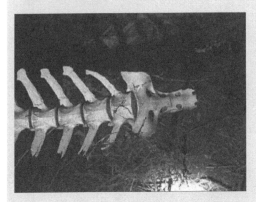

Figure 4.15 Lumbar vertebrae and sacrum of a deer. Notice the length of the transverse processes relative to the centrum widths, as well as the caudal constriction of the sacrum. (*Photo by Debra Komar*)

mammals, the transverse processes are significantly longer than the width of the centrum (Figure 4.15).

- *Femoral head angle* The angle formed by the femoral neck and the diaphysis is approximately 90° in animals but greater than 90° in humans.

- *Dental morphology and formulas* Although highly fragmentary dental remains may prove problematic, intact dentition can readily be differentiated. Mandibular and maxillary fragments containing portions of the alveolar process can be identified through dental formulas and morphology.

- *Internal bone morphology* In cut or fragmentary long bones, the transition between the exposed cortical and cancellous bone provides a clue to the bone's origin. In animals, uniform cancellous bone and a sharp transition between cortical and cancellous bone is typical (Figure 4.16). In humans, the boundary between cortical and cancellous bone is poorly defined.

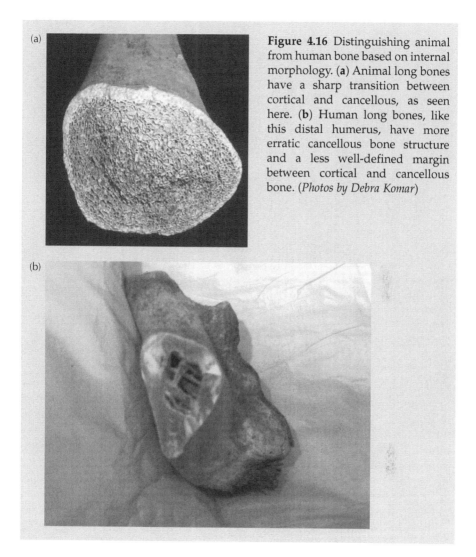

(a)

(b)

Figure 4.16 Distinguishing animal from human bone based on internal morphology. (**a**) Animal long bones have a sharp transition between cortical and cancellous, as seen here. (**b**) Human long bones, like this distal humerus, have more erratic cancellous bone structure and a less well-defined margin between cortical and cancellous bone. (*Photos by Debra Komar*)

in appearance to tree rings. This cellular arrangement is called **plexiform**. Histological differentiation of human and animal bone is examined in detail, with excellent illustrations, in Mulhern and Ubelaker (2001). **Solid-phase, double-antibody radioimmunoassay** has also been proposed to differentiate human and animal bone fragments (Ubelaker et al., 2004).

Two variables can potentially confound efforts to identify skeletal elements: the age of the organism and the presence of pathological conditions. It is crucial that forensic anthropologists acquire thorough training and familiarity with both immature and adult human bone to successfully differentiate human from nonhuman material. Size is among the factors

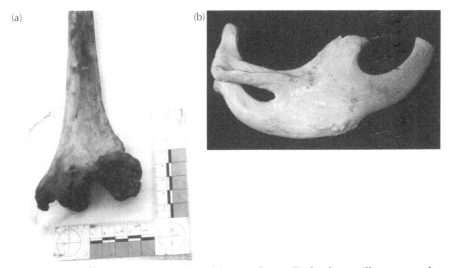

Figure 4.17 Differentiating animal and human bone. Both photos illustrate pathological human bone initially mistaken for animal remains. Traumatic remodeling of the (**a**) distal humerus and (**b**) first and second ribs resulted in bones that didn't "look human" to investigators. (*Photo by (a) Debra Komar; (b) New Mexico Office of the Medical Investigator*)

considered in the differentiation of human and faunal osteology. Both human and animal subadult remains can be misidentified. The presence of unfused or actively fusing epiphyses is a reliable indicator of juvenile remains in all species. Pathological conditions resulting in extensive bone remodeling can also present challenges to investigators. Abnormal bone growth can alter or obliterate morphology (Figure 4.17), making element and species identification difficult. Any indications of pathology or bone remodeling warrant a careful anthropological review of the material prior to excluding it from further medicolegal analysis.

Differentiating human bone from that of avian and fish species is relatively simple (Figure 4.18), since overall bone size, weight, morphology, and structure provide ample evidence for identification. With fragmentary or burned remains, histological examination may be necessary. Students wishing more information on faunal identification should review Gilbert (1990), Hesse and Wapnish (1985), Lyman (1979), O'Connor (2003), and G.S. Smith (1979).

Traditionally, differentiating human remains of medicolegal significance from both animal remains and nonmedicolegal materials has relied on observation of morphology, condition, and context. While the introduction of newer, destructive forms of analysis (such as DNA or radioimmunoassay testing) provides opportunities to evaluate equivocal cases, such methods inherently come with greater costs and introduce potentially significant time

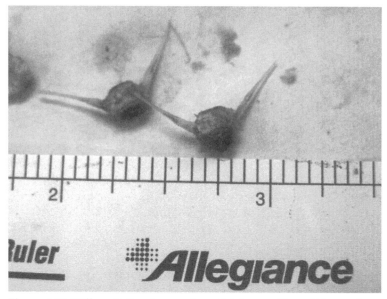

Figure 4.18 Differentiating human from fish bones. These fish vertebrae, along with other isolated fish skeletal elements, were discovered in a bucket at a crime scene by police, who thought the remains might be those of a baby. In addition to their size, the presence of the fused, elongated transverse processes clearly indicate the remains to be nonhuman. *(Photo by Debra Komar)*

delays in investigation. Whether a particular case warrants such levels of analysis depends on the nature of the remains, as well as the resources available (both monetary and equipment). Of paramount importance, however, is whether any criminal charges may be pending based on the evaluation of the material. While "testing for testing's sake" may answer questions of academic interest, medicolegal investigations focus on assessing the forensic significance of recent deaths, as represented in human remains. If the question of medicolegal significance can be adequately addressed through gross observation, subsequent destructive forms of analysis will typically prove unnecessary. Only those cases in which legal questions remain warrant further analysis within a medicolegal context.

After the anthropologist has established that the remains are human and of medicolegal significance, investigators at scene must address issues relating to jurisdictional authority.

JURISDICTION AND THE CRIME SCENE

Jurisdiction was introduced and defined in Chapter 2. In this section, we will see concepts of jurisdiction put into practice. Scene investigators should be mindful that, in terms of jurisdictional expectations and obligations, all regulations pertaining to the handling of fleshed bodies also apply to skeletal remains. Medicolegal cases occur on a spectrum representing the full decompositional process, ranging from fleshed at one extreme to skeletal remains at the other. Cases requiring anthropological consultations merely represent one portion of the continuum. Lesser standards of investigation are not acceptable.

Establishing jurisdiction begins with geography. Anthropologists working with a medicolegal authority or a law enforcement agency have a specific geographic boundary that defines the limit of their jurisdiction. The first step in establishing jurisdiction is to ensure that the crime scene or location of death falls within these geographic boundaries. Maps and **global positioning systems (GPS)** determine exact locations. Agencies or their representatives found working outside their geographic jurisdictions can face penalties. In cases involving criminal acts, failure to respect jurisdictional boundaries could result in the exclusion of evidence or dismissal of an indictment due to an avoidable technicality.

The second step in establishing jurisdiction involves the concept of **superseding authority**. Depending on circumstances, jurisdiction over a local event may transfer to a higher agency. Hierarchies exist within the judicial and law enforcement communities that dictate the order of jurisdictional transfer. For example, the authority of state police agencies supersedes that of local law enforcement such as city police or sheriff's offices. Federal agencies typically supersede both state and local agencies. Although many investigations involve the collaboration of multiple agencies, determining which agency has the ultimate authority over a case improves communication among all parties and provides a defined chain of command.

The jurisdiction of a medicolegal authority depends on governing statutes. For example, medicolegal authorities in centralized states (listed earlier: Box 2.1) participate in all death investigations involving local and state police. However, should the case involve federal authorities, such as the Federal Bureau of Investigation (FBI) or National Transportation Safety Board (NTSB), continued participation by the local ME or coroner or the specific role in the investigation of this official is determined by the protocols of the national agencies. Such agencies typically have the right to call upon consultants of their choice, rather than being statutorily obligated to use local investigators. The same applies to events occurring on federal lands, such as military bases or parks, or tribal lands that may be contained within the medicolegal authority's geographic jurisdiction. Deaths occurring on Native American lands normally do not fall under the jurisdiction of the local medicolegal authority. A tribal group may request the involvement of the ME or coroner, often for a fee, much as families may request autopsy consultations.

Local statutes grant jurisdictional authority to the district attorney to prosecute crimes and to law enforcement to investigate and gather evidence on behalf of the district attorney. As such, the law enforcement agency with jurisdiction is responsible for securing and maintaining the integrity of the crime scene, as well as documenting the scene and collecting all relevant evidence. Whether police need to secure a warrant prior to entering a building, seizing evidence, or making an arrest varies greatly depending on circumstance. The situation can become even more confounding in jurisdictions where statutes do not require warrants for investigations by the medicolegal authority. This important yet complex aspect of criminal justice is considered in detail by Bloom (2003); J.R. Davis (1984), and the American Bar Association (1990). It is, however, important for anthropologists attending crime scenes to recognize the need for warrants and to ensure that law enforcement has obtained a warrant, when necessary, prior to actively participating at a scene (Box 4.3).

Although statutes grant jurisdiction over a crime scene to law enforcement, statutes also give jurisdiction over the body or human remains to the medicolegal authority. Accordingly, an intriguing and sometimes troublesome dynamic is created at scenes. The scene belongs to law enforcement, while the body is the responsibility of the ME or coroner. Medicolegal investigators, including anthropologists, must respect the time frame and guidelines established by law enforcement while at the scene. In return, law enforcement personnel should minimize handling the body and allow the medicolegal authority to conduct its own investigation as needed. While the responsibilities of law enforcement and the medicolegal authority may differ at a scene, the fundamental goal of the investigation is shared. Respect, communication, and cooperation result in the effective and proper handling of a death scene and the mutual exchange of information between agencies.

JURISDICTION

At the Scene	Of the Body
Law Enforcement	Medicolegal Authority

This statutory division of jurisdiction has important implications for anthropologists who participate in crime scenes and respond to requests for consultations. Local statutes define which agency has the authority to engage an anthropologist and for what activities. For instance, statutes state that decisions regarding which deaths fall under medicolegal jurisdiction rest entirely with that authority, not law enforcement. Law enforcement agencies that employ anthropologists or contact local professors to evaluate

BOX 4.3
Understanding the need for warrants

The following case study illustrates why anthropologists must remain mindful of the need for warrants while working on scene.

A police informant reported that, while attending a party at the rural home of a drug dealer, the dealer bragged of killing and disposing of a man who had failed to pay the dealer for drugs. Based on the witness's statement, police obtained a warrant to search the dealer's home. The warrant expressly permitted police to search the house and property (roughly a quarter-acre on a country road). While serving the warrant, police noticed a number of unrestrained dogs on the property. The dogs moved freely between the dealer's property and that of his neighbors. In the dealer's yard, police found a large number of bones, extensively chewed by the dogs. Police requested a consultation from an anthropologist to determine whether any of the remains were human. After examining several bones, the anthropologist identified two large pieces of a human femur. The anthropologist also commented that the dogs were the likely agents of transport and suggested a search of the neighboring properties. However, because the adjoining properties were not named in the warrant, neither police nor the anthropologist had the authority to search those properties. An additional warrant was needed before any further searches could be conducted. The following day, after the second warrant was issued, the anthropologist found a set of partial human remains on the neighbor's property. Had the remains been found prior to securing the second warrant, the search would have been illegal and the remains (and all information derived from them) would have been inadmissible in court.

bones found at scenes may be exceeding their authority. Anthropologists engaged by law enforcement agencies to exhume remains at scenes could find, for example, that their involvement must stop once the body is exposed, as the remains are under the control of the medicolegal authority. Again, respect and communication are essential to avoiding conflicts.

Even so, anthropologists frequently have significant professional interactions with law enforcement agencies. Many anthropologists maintain formal and long-standing relationships with law enforcement, such as the ongoing collaboration between the physical anthropologists of the Smithsonian Institution and the Federal Bureau of Investigation (Grisbaum and Ubelaker, 2001). Forensic anthropologists may be approached by both medical examiners and law enforcement agencies to assist in identification, trauma assessment, or search strategies. An anthropologist who is mindful of jurisdictional issues and well informed of local statutes can successfully develop working relationships with both police agencies and medicolegal authorities.

Responsibility versus Authority

At a crime scene, the forensic anthropologist must appreciate the difference between responsibility and authority. **Responsibilities** are the duties and tasks assigned to an individual as outlined in a job description or terms of employment. An anthropologist's responsibilities may be to exhume, document, and recover remains at a scene, in cooperation with police and medicolegal investigators. **Authority** is derived from laws and statutes granting specific powers to specific individuals. Authority, like jurisdiction, is geographically and circumstantially bound. Expressions of authority may include the right to enter private property without a warrant, to pronounce death, to seize and retain evidence, or to question a witness or family member.

Unless expressly received through formal appointment, anthropologists attending crime scenes rarely have authority. An anthropologist answering a request to attend a scene by any agency must understand the conditions and limitations of his or her participation. For example, a forensic anthropologist may be both a professor at the university and an employee of the medical examiner. In addition to the title of forensic anthropologist, she holds a formal appointment as a deputy medical investigator. As such, she would have the authority to seize evidence, establish or terminate jurisdiction, and pronounce death, among other actions. At a scene, she also has a number of responsibilities reflecting her title as anthropologist. In her absence, students or other forensic anthropologists may assume her responsibilities but not her authority. They may recover the body and document the scene, but all acts of authority, such as evidence seizure or pronouncement, must be carried out by duly appointed medical investigators.

The consequences of overstepping jurisdictional boundaries can be significant. Critical evidence may be excluded at trial because of improper collection procedures. Anthropologists themselves may be subject to criminal penalties or civil lawsuits based on their actions. If violations should occur, an informal or consultant status relationship with a medicolegal office or law enforcement agency will not necessarily afford an anthropologist protection. The anthropologist must fully understand the scope of his or her responsibilities and authority. If either the job description or the responsibilities of an anthropologist entail duties that represent acts of authority, it is in his or her best interest to ensure that there is a formal and express understanding, preferably in writing, with the agency that extends such authority. Thus authority is derived through formal appointment; responsibilities, on the other hand, are defined by protocols.

Protocols

Protocols are standard operating procedures or guidelines established by an authority or agency. These guidelines dictate the steps investigators must take and the methods they must use to successfully complete tasks. For example, a skeletal remains protocol will list the steps police or an investigator must follow at scenes involving skeletal remains or burials. Failure to follow protocols may result in penalties, including job termination and other disciplinary actions.

It is not the goal of this chapter to provide sample protocols. Most jurisdictions already have protocols in place that address specific issues relating to statutes, staffing, budgets, and other needs. The focus of this section is to inform anthropologists of the existence and purpose of protocols and to advise practitioners to seek out and follow the protocols of the agencies for which they consult. Anthropologists working in jurisdictions without such guidelines should consider drafting, in cooperation with the relevant local agency, protocols that outline or define the role of the anthropologist and his or her interaction with the agency.

Having established the importance of jurisdiction, authority, and protocols, we will now focus on individual scene processing. Processing begins with an evaluation of the scene.

EVALUATING SCENES

Investigators frequently encounter a variety of scenes that would benefit from the involvement of an anthropologist. Death scenarios warranting the participation of a forensic anthropologist include the following:

- scenes involving decomposed or skeletal remains
- clandestine burials
- fire scenes
- disarticulated remains recovered from water
- intentional dismemberments or attempts to alter or destroy the body post-mortem
- scenes with extensive animal scavenging and dispersal of remains
- natural or mass disasters

As noted earlier in the discussion regarding authority and responsibility, it falls to the anthropologist to understand the limits of his or her authority. Never confuse consultation or evaluation with authority. Whether the anthropologist has the authority to establish or terminate medicolegal jurisdiction is dramatically different from fulfilling a duty to recognize the nature of the remains. All consulting anthropologists must be cognizant of the legal difference between offering an opinion and having the authority to act, or cause others to act, on that opinion (Box 4.4).

BOX 4.4
Responsibility and authority

Forensic anthropologists are frequently consulted to assist in differentiating modern from prehistoric human remains. The following case study details such a scenario to demonstrate the difference between authority and responsibility.

Hikers in a state park find skeletal remains exposed by erosion. The hikers contact police, who in turn call out an anthropology professor from a local university to evaluate the remains. Because the police requested the consultation, the anthropologist's responsibilities include examining the remains and offering an opinion on their relative age or **cultural affiliation**. However, only the ME or the coroner has the authority to determine whether the remains are of medicolegal significance. The anthropologist has no authority to terminate the investigation, invoke jurisdiction over the remains, or to instruct law enforcement as to how to proceed. By clearly recognizing the boundaries between responsibility and authority, anthropologists can contribute to scene evaluation without overstepping their legal authority.

Anthropologists who wish to participate at crime scenes should consider offering two forms of consultations: on-scene and distance. An **on-scene consultation** entails the presence and involvement of the anthropologist at the death scene. The duties or responsibility of the anthropologist will vary depending on jurisdiction as well as the nature of the scene. **Distance consultations** require the cooperation of investigators at scene. Police or field deputies transmit digital photos (Figure 4.19) and scene information via e-mail or cell phone to the anthropologist. The anthropologist then evaluates the remains and decides whether a site visit is necessary. With the widespread availability of digital photography, the possibility of visually assessing remains prior to their recovery should greatly reduce needless collection of animal remains or the disturbance of prehistoric remains. Both types of consultations have clear benefits as well as disadvantages, detailed in Box 4.5.

If the anthropologist must rely on photographs to evaluate the scene, their quality must be sufficient to allow assessment. Crucial to effective photography is scale. **Scale** relies on introduced objects within the frame against which the size and dimensions of the evidence can be judged. While a commercially prepared scale or ruler is ideal, any available object that is of a uniform, recognizable size may be used. Examples include coins, pens, and trowels (Figure 4.20). Choice of background also impacts the ability to visualize detail. Although partially buried remains must be photographed in place, isolated skeletal elements may be placed on a contrasting background

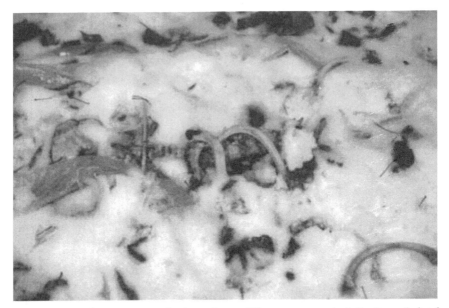

Figure 4.19 A distance consultation photo. Witnesses reported finding skeletal remains in plastic sheeting, wrapped with rope, partially obscured by snow in a farmer's field. Seeing the remains prior to attending the scene allows the forensic anthropologist to determine whether an on-scene consultation is necessary. As the remains were *in situ*, manipulation of the background was not possible. Lighting and focus were adequate, but the photo would benefit from the presence of a scale. As evident in the photo, the remains had been scavenged and dispersed. Because snow obscured the ground and the area of dispersal was unknown, an on-scene consultation was required in this case. (*Photo by Debra Komar*)

to provide clarity (Figure 4.20). Suitable background materials readily available at the scene include car hoods, paper, binders, or garbage bags. Finally, the importance of adequate lighting and focus cannot be underestimated. Anthropologists who effectively communicate these specific needs to on-scene investigators will reap the benefits by avoiding time-consuming, unnecessary travel.

In addition to digital photography, on-scene investigators should be prepared to provide the anthropologist with the following information.

- *Environment* What is the depositional environment? Are the remains on the surface, buried, or under water?
- *Preservation* Is soft tissue present? Do the remains have an odor? Is the tissue desiccated or still fresh? What color are the bones?
- *Scavenging* Is insect activity evident? Have canids or other animals interacted with the remains, chewing or scattering the bones? Are paw prints or scat (feces) evident?

BOX 4.5

Advantages and disadvantages of on-scene and distance consultations

On-Scene Consultations

Advantages	Disadvantages
Improved scene processing	Time delays due to travel
Expert assessment of medicolegal significance	May prolong scene investigations
Assistance in search and recovery	Not necessary or appropriate in all cases

Distance Consultations

Advantages	Disadvantages
Speed, in that assessments can be made quickly	Requires more involvement from on-scene investigators
No need for law enforcement to guard scene or process nonhuman remains	Requires digital camera and means of communication
Saves needless travel	Reliance on incomplete or secondary information
May prevent non medicolegal material from entering the system	

- *Associated materials* Is clothing present? Are the bones directly associated with the clothes (still contained within clothing) or is clothing merely present at the scene? Are modern materials such as watches, jewelry or dental repairs visible?

Such questions allow anthropologists to better assess the need for an on-scene consultation by providing information not readily obtained from photographs.

When determining what type of consultation is needed, two variables should be considered: the context of the remains and the scene type. **Context** is the physical environment and location of the remains, as well as any associated goods or artifacts. Remains that have been disturbed are out of context, while undisturbed remains are in context. A subtle yet important distinction must be made when determining context. The designation *in situ* describes remains that are exactly as deposited, or where all alterations are the result of natural processes. This contrasts with the designation *as found*. *As found* indicates that accidental or intentional human interaction with the remains has occurred. Such interaction could involve witnesses, informants, or investigators. Law enforcement agents, in the course of initially evaluating the scene, occasionally uncover, move, or modify the remains. Witnesses often collect or move remains, believing they are helping police. Identifying

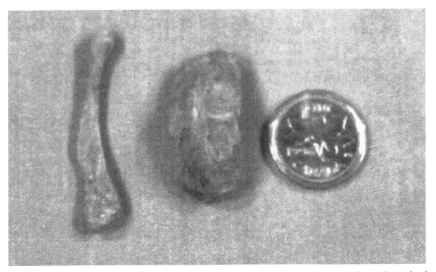

Figure 4.20 Illustration of the importance of scale and background in digital photography. In lieu of commercial scales or rulers, one can use standard-sized objects such as coins or pens. The photo shows the wing and body of a human hyoid, with a penny to provide scale. The choice of a background similar in color to the objects reduces the clarity and detail provided by the photo. (*Photo by Debra Komar*)

when this has occurred is crucial to accurate scene assessment. If the remains are completely out of context (e.g., a skull discovered at a flea market), there is normally no need for an on-scene consultation by the anthropologist. Photos may be sent or the material can be collected and sent directly to the medical examiner or coroner.

Determining whether an on-scene anthropological consultation is necessary also depends largely on the type of scene. Scenes are divided into two types: open and contained. With **open scenes**, the location, distribution, or dispersal of remains is unknown. Examples of open scenes include an animal-scavenged body in a rural area, body parts found in water, and a burned body in a collapsed building. In **contained scenes**, the remains are isolated in a single location or defined area. Examples of contained scenes include graves, a body in a car trunk, and body parts in plastic bags.

With open scenes, an anthropologist on scene greatly contributes to efficient scene processing. Anthropologists assist in differentiating human from animal remains, developing search strategies and, in cases involving scattered remains, determining the portion of remains still unaccounted for. In scenes involving fire, the anthropologist aids in discerning tissue from other materials and again can determine the percentage of remains recovered. We recommend that protocols for open scenes include a request for anthropological support.

The necessity of an on-site consultation in contained scenes must be evaluated on a case-by-case basis. Burials are best excavated with the assistance of an anthropologist, even if the elapsed time since burial is short and the body is still fleshed. With other contained scenes, the need for an on-scene consultation will depend on the complexity of the recovery and whether the remains are *in situ* or disturbed. Often the remains have already been disturbed, such as skeletal elements collected by witnesses. For remains that are still *in situ* or in context, the value of an anthropologist at scene will be case specific. Protocols involving contained scenes should be written with sufficient flexibility to accommodate all potential scenarios. Typically, burials should automatically initiate a consultation, while other scenes should be evaluated remotely by the forensic anthropologist prior to processing by law enforcement to ensure proper collection methods.

When a location has been classified as an open scene, a search is needed to locate remains, evidence, and areas of interest. The development and execution of a plan to identify evidence is called a **search strategy**.

SEARCH STRATEGIES

With open scenes, the location of the remains or the extent of their dispersal is unknown. As such, the search area must be defined. Law enforcement has jurisdiction over the scene, and thus the responsibility for developing search strategies falls to police. However, in scenes involving human remains, the responsibility for recovering the remains lies with the medicolegal authority. Typically in such cases, search strategies are jointly developed by both agencies. Experienced forensic anthropologists are often called on to develop and execute search plans, particularly in cases involving scattered remains.

Developing search strategies relies on information provided at the scene as well as an initial evaluation of the terrain, personnel and other resources available, and daylight remaining. Searches of open scenes are best conducted in daylight, since reliance on artificial light sources such as flashlights limits visibility and increases the potential for missed evidence.

The identification of a death scene occurs when partial remains are discovered or when informants lead law enforcement to the body's location. This information provides a starting point to begin the search. Among the variety of methods or technologies available to assist searchers are the following.

- *Visual or pedestrian searches* This is the most basic form of search. Investigators and volunteers perform visual assessments of prescribed areas. These searches are often referred to as "shoulder to shoulder," "hands and knees," or "sweeps," depending on the level of detail required. The success of visual searches relies on the observational skills of the participants and the search pattern used to cover the defined area.

- *Scent detection dogs* Commonly known as **cadaver dogs**, these civilian or police canines have been specially trained to detect and alert to the presence of human remains. Dogs may be trained using actual remains, soiled clothing from decomposed bodies, or **pseudoscent**, a commercially available chemical signature that simulates the odor of decay. The level of training and accuracy of both the dog and the handler fluctuates dramatically, and excessive claims of ability or success should be viewed with caution (Komar, 1999b; Lasseter et al., 2003). For example, the ability of dogs to differentiate human from animal bone has likely been untested, and all remains must be examined by the consulting anthropologist.

- *Probes* Probes are either solid or coring. Solid probes may be used to detect changes in the consistency of the soil or substrate. Hollow cores are tubes inserted into the soil that permit extraction of a sample or **core** when removed. Examination of the sample reveals changes in soil composition or the presence of bone fragments or evidence.

- *Metal detectors* These commercially available sensors alert to the presence of various metal compounds. Although they cannot detect bone, metal detectors can locate associated materials such as zippers, keys, weapons, or projectiles.

- *Ground-penetrating radar (GPR)* GPR uses electromagnetic wave propagation and scattering to image, locate, and quantify changes in electrical and magnetic properties in the ground. GPR can be used from the surface (Figure 4.21) or from a borehole or core. Detection of a buried feature depends on the contrast of electric and magnetic properties and on antenna placement (Mellet, 1992; M.L. Miller, 2003; P.S. Miller, 1996). Other proposed geophysical methods include **resistivity**, which can identify disturbances in soil by examining soil variables such as water content and particle size (Wessling, 2001). The utility of resistivity in mass graves has also been considered (Tuller and Sterenberg, 2005).

- *Cesium magnetometer* Magnetometers measure the earth's magnetic field in a given location; subsurface features have a magnetic signature that differs from that of the ground surrounding them. Buck (2003), who compared ground-penetrating radar, magnetometry, and electrical resistivity technology for their utility in locating recently buried human remains, found that while the equipment and results may exceed the skills sets of typical forensic investigators, each of these methods showed some promise in detecting buried remains.

- *Infrared photography (IP)* IP produces images not possible with conventional film. In practice, there is no significant difference between infrared and normal photography. Both employ the same cameras, light sources, and processing but use different film. The infrared range of the electromagnetic spectrum is invisible to the human eye. Infrared detects differential heat emission, revealing disturbances in soil or heat-emitting objects obscured by surface debris.

Figure 4.21 A commercially available surface ground-penetrating radar unit. A monitor mounted on the hand grip allows the operator to detect changes in soil, indicating the presence of buried features (including graves). *(Photo by Sean Worrell)*

- *Thermal scanning* Similar to IP, thermal scanning converts infrared radiation emitted from bodies or other heat sources into electrical impulses that can be visualized on a monitor. Handheld units are especially useful.

- *Aerial photography* Cameras mounted on fixed-wing aircraft, helicopters, satellites, or other controlled air vehicles can provide a unique perspective on a scene, allowing investigators to visualize objects too large or too subtle to be detected from the ground. Satellite imagery has been instrumental in detecting mass graves in genocide investigations. It is interesting to note that aerial photography is considered outside the realm of privacy. Anything witnessed or documented from the air is considered public domain. The exception is the use of airborne thermal scanners. In 2001 the Supreme Court found the use of such scanners unconstitutional (*Kyllo v. United States* [99–8508] 533 U.S. 27, 2001). Law enforcement must secure a warrant prior to the use of aerial thermal scanners.

- *Divers, underwater cameras, and sonar* Water scenes present unique difficulties to investigators. Divers can perform the equivalent of a visual search, while the use of sonar accomplishes many of the same objectives as ground-penetrating radar. Waterproof cameras (either handheld or mounted on remote-controlled submarines) may also be used.

Search strategies are developed based on the depositional environment of the remains. Categories include burials, surface scatters, water deposition, fire scenes, and mass or natural disasters. Each category presents a different set of challenges to anthropologists. For example, surface deposited remains may be obscured by snow (Figure 4.19) or leaf litter (Figure 4.22). A summary of potential search techniques relevant to each category follows.

- *Burials* Visual search, scent detection dogs, probes, ground-penetrating radar, magnetometry, metal detectors, infrared and aerial photography, and thermal scanning (depending on the length of interment).
- *Surface scatters* Visual search, scent detection dogs, solid probes, aerial photography, metal detectors, infrared and thermal scanning.
- *Water* Solid probes, divers, underwater photography, and sonar. The use of dogs has limited potential, depending on the depth of the water.
- *Fire scenes* Visual search, scent detection dogs, probes, and metal detectors.
- **Mass disasters** Visual search, scent detection dogs, probes, metal detectors, and aerial photography. Thermal scanning and infrared photography may be limited by the nature of the event, such as scenes involving plane crashes or fire (see also Chapter 9).

Figure 4.22 Illustration of the difficulties inherent in strategies used to search for evidence obscured by leaf cover. Note the right front bear paw (initially mistaken for a human hand) in the center of the photograph. The remains, located at the base of a tree, were also poorly visualized because of shade. (*Photo by Debra Komar*)

Regardless of the strategy employed, the objective is to identify evidence. The next sections address evidence collection and documentation.

EVIDENCE

As noted in Chapter 3, the primary goal of scene investigation is to collect all available evidence. As time passes, evidence can be dispersed or modified by natural and human agents. Evidence may undergo alteration, destruction, or transport. Regardless of the condition of the evidence, its presentation in court requires a continuous chain of custody. **Chain of custody** is synonymous with *evidence continuity* and *chain of evidence*. Chain of custody requires investigators to maintain evidence integrity and security. The investigator responsible for collecting and cataloging evidence either retains custody or formally relinquishes custody to another investigator or agency. If such a transfer of evidence occurs, it must be documented. Successful admission of evidence in court requires an unbroken and well-documented chain of evidence. An example of the ramifications of a poorly maintained chain of evidence is given in Box 4.6. A discussion of chain of custody specific to anthropological investigations is found in Melbye and Jimenez (1997). Maintaining chain of evidence requires

- the investigator to be aware of the exact location of the evidence at all times,
- the evidence to be maintained in a secure location,
- the evidence to be sealed to prevent tampering,
- that access to the evidence be restricted to the responsible investigator or authorized designates,
- that any handling, transport, analysis, or examination of the evidence to be acknowledged in a written log.

The chain of custody begins with the discovery of a piece of evidence at the scene. The discovery must be witnessed by a second individual and documented prior to the disturbance or collection of the evidence. As multiple pieces of similar evidence, such as footprints or bullets, are recovered at a scene, police give each piece of evidence a catalog or evidence number to allow tracking. All documentation should reference this number, and all investigators, regardless of agency, should use the same number. Photographs and videotape documentation should visually include the number, when possible. The exact location, time of discovery, names of the discoverer and witness, time of collection, collecting investigator, and evidence number are entered in the evidence log upon collection. This information must also be included on the evidence bag or container. Each bag is sealed with tamperproof tape and the collecting investigator signs the edge of the tape to provide clear evidence of any subsequent tampering. **Tampering** is all unauthorized or undocumented handling of evidence. A master log of evidence collected and photographs taken is maintained at each scene.

BOX 4.6

Maintaining the chain of evidence—a cautionary tale

The trial of football legend O.J. Simpson, accused of murdering Nicole Brown and Ron Goldman, provided multiple illustrations of how problems quickly arise at trial when evidence collection has not been carefully documented. One of the most contested pieces of evidence—the bloody glove found at Simpson's home by Detective Mark Fuhrman—was challenged on the grounds that the discovery of the evidence was not witnessed and that the item had not even been noticed by 16 other officers present at the scene. Had Fuhrman called for another officer to witness the discovery, and had the glove been photographed *in situ*, evidence admissibility would have been uncontested.

Another problem resulted from an inaccuracy in note taking. A nurse tasked with drawing a blood sample from Simpson initially noted that 8 cc of blood was drawn, but the prosecution could only account for 6.5 cc of blood in evidence. Even this small a discrepancy allowed the evidence to be challenged, as the inconsistency raised the possibility of tampering (Aaseng, 1995; Schuetz and Lilley, 1999).

Although most trials do not receive the public attention given to the O.J. Simpson trial, the media spectacle reflected negatively on the evidence collection protocols of the California law enforcement agencies involved. This resulted in a loss of public trust in the validity of forensic investigations.

Jurisdictional issues must be considered during evidence collection. The body or remains, as well as all physical evidence in direct association with the remains, are the responsibility of the medicolegal authority. The medicolegal investigator is responsible for seizing all such evidence. This includes clothing, wallets and identification, jewelry and personal effects, as well as weapons or projectiles directly associated with the body, such as the knife embedded in the victim. The medicolegal authority also seizes all physical evidence recovered from a grave, including body wrapping or associated materials. Other physical evidence at the scene, including finger- and footprints or blood spatter, will be collected by law enforcement.

Evidence collected at a scene will ultimately be separated. For instance, the body and associated materials are transported to the morgue, while other physical evidence is taken to the crime laboratory. When scene processing is complete, each agency's designee responsible for maintaining chain of evidence signs for all evidence seized by that agency. This designee is then responsible for transporting the material to the facility, where it will be logged in and placed in a secure evidence-holding area. Within all

BOX 4.7

Maintaining chain of evidence at the morgue

The level of security varies from area to area of a medicolegal morgue facility. Not all areas are considered secure. For instance, refrigerated body storage may include large communal holding areas for noncriminal cases, while the remains of homicide victims are locked in individual coolers. Valuables and personal effects are stored in secured areas, separate from body and clothing storage. The autopsy suite is considered a secure area where evidence can be handled and examined. Once the examination is complete, articles identified as physical evidence, such as projectiles or weapons, should be sealed and stored separately. This physical evidence, along with biological evidence such as DNA or semen samples recovered at autopsy, will be released to law enforcement for further analysis in the crime lab.

The ME or the coroner retains custody of the body and nonevidentiary personal effects until the entire autopsy process is complete and the remains have been released to the next-of-kin. When biological materials, including skeletal elements, have evidentiary value, they can be retained indefinitely by the medicolegal authority. These materials continue to be classified as evidence, and the chain of custody must be maintained. Care must be exercised to ensure that biological materials are preserved or stabilized and are not allowed to degrade or decompose.

medicolegal and police facilities, evidence storage must be secure, and access must be restricted to authorized personnel (Box 4.7).

Although the chain of evidence begins in the field, it must be maintained until the case is resolved in court. At trial, a senior law enforcement officer is assigned to provide evidence continuity for the court. Anthropologists or medicolegal investigators who intend to present evidence at trial should assume custody of the evidence, transport it to the court, and then release the evidence to the continuity officer. At this point, the evidence becomes the property of the court and the responsibility of the continuity officer. The evidence will be maintained in a secure location and be made available to the witness at trial. Once presented to the court, the evidence will be entered and assigned a court evidence number. With few exceptions, evidence admitted to the court is not returned to the witness or investigators (Box 4.8).

Failure to maintain and document chain of custody at any point in the investigation carries serious consequences. Should the continuity of a piece of evidence be successfully challenged in court, the item will be dismissed and all subsequent conclusions drawn from it will also become inadmissible. Chain of custody begins with scene documentation.

BOX 4.8

Admission of supporting evidence in court

Anthropologists preparing to testify at trial should be mindful of the rules governing the admission of supporting evidence in court. **Supporting evidence** is material not collected at the scene. It may include reports and maps, photographs taken at autopsy, or diagrams and illustrations intended for use in court. Simply put, never bring anything to court that you are not prepared to relinquish to the court permanently. Anything introduced at trial becomes the property of the court, and all photographs, reports, books, or other items brought to the court's attention will be seized and retained. If you intend to use specimens for illustrative purposes, cite published materials, or introduce photographs, retain a copy for your files and/or ensure you can relinquish the material. If you take it to the witness stand, be prepared to leave it there.

DOCUMENTATION

The purposes of crime scene documentation include (1) creating a permanent record of the scene and the activities of the investigators, (2) allowing for reconstruction of the scene in court, (3) illustrating the spatial relationships among evidence, (4) providing a source of information for defense experts and other analysts who did not attend the scene, and (5) supporting the chain of evidence. Scenes are documented through the use of still and video photography, mapping, note taking, and report writing.

Scenes are processed and documented according to protocols. First, the perimeter of the scene is secured by law enforcement with crime scene tape, delineating the internal boundary of the area of interest. Once established, all investigators crossing the tape will have their names, agency affiliations, and times of entry and exit recorded in a master log by a scene continuity officer. This establishes a permanent record of all individuals who entered the scene and the nature of their participation. A second, wider perimeter will be established by law enforcement to maintain overall scene security. The purpose of the second perimeter is to limit access to the scene to authorized personnel. Bystanders and members of the media will be required to remain outside the widest perimeter.

Documenting the scene begins with photography to record the scene as it was discovered by law enforcement. Criminalists equipped with still and video cameras fully document all aspects of the scene prior to any evidence collection. Still photography provides a visual record of the scene, while the videotape documents the activities of the investigators and allows for audio commentary. The introduction of digital photography in forensic investigations raised concerns over the possibility of tampering, but such concerns have been addressed (Box 4.9).

BOX 4.9

Digital photography in forensic investigations

In 1996 Horner et al. described the value of digital dental radiographic images to forensic investigations. The digital format permits the easy and rapid transfer of images to remote sites, allowing forensic odontologists to compare the antemontem digital images to postmortem images for identification purposes. However, the report also identified the potential legal problems of image manipulation and called for the manufacturers of the computer imaging program to alter the system to prevent tampering. Boscolo et al. (2002) later tested this idea, purposely altering digital dental radiographic images and asking odontologists to identify the altered images. They concluded that professionals had difficulty in identifying altered images and again called for controls on the abuse of digital technology. Concerns have also been expressed over the legal requirements of electronic data storage and record keeping (Figgener and Runte, 2003; Goldthorpe and McConnell, 2000).

The problem extends beyond odontology records. Computer-based medical imaging records, such as those obtained from CT and MRI, are stored and viewed electronically. The use of digital photography at crime scenes and autopsy has also become prevalent. In response, the rapidly evolving discipline of forensic computing has introduced safeguards for electronic data. **Forensic computing** focuses on issues relating to electronic data and technology as it applies to the forensic sciences. For example, forensic computing addresses Internet crime and exploitation. Although the technology may be relatively new, the theory behind it is not. Locard's principle is the basis for forensic computing and its application to criminal investigations—every contact leaves a trace (Bouhaidar, 2005). Now all manipulation, alteration, or destruction of digital images or information can be identified and documented. At the very least, if modification cannot be stopped, at least it can be recognized. Encryption programs have been introduced to prevent tampering, and the encoding of all electronic information provides a chain of evidence. At present, the introduction of any form of electronic or digital evidence requires that the judge, as gatekeeper, be satisfied that an adequate chain of evidence has been maintained.

Once the criminalists have completed initial photography, additional investigators including the medicolegal team and anthropologists enter the scene and begin their own photographic documentation. Although information is shared among agencies, law enforcement and the medicolegal authority are responsible for independently photographing and documenting the scene. This duplication of effort provides an alternative source of information should problems occur, such as film damaged in developing, as well as providing each agency a permanent record of the scene. As the focus of police at scene differs from that of medicolegal investigators, each agency will record information not required by the other. When all photography is complete, both agencies begin mapping the scene.

Although photography creates a visual record of the scene, mapping provides a better understanding of the spatial relationship between objects. Both mapping and photography are necessary at every scene. While these forms of documentation augment each other, neither can replace the other. The number of maps needed to document the scene will vary depending on the scene's complexity, but all scenes require a **plan view**. This map orients the scene relative to major structures, such as buildings or roads, allowing investigators to place the scene within a larger geographic context. Details recorded on the plan view include any named roads within the area (including street address, if applicable), all buildings or permanent structures, an arrow indicating north, as well as the date, case number, and the name of the investigator generating the map. For simple scenes involving a single set of intact surface remains, this plan view may prove sufficient, provided the locations of all significant evidence are noted on the map (Figure 4.23). More complex scenes require additional maps.

To document the spatial relationship between objects, a single defined point known as the **datum** must be established. A **datum** is a fixed point to which all other evidence or objects must be referenced. When selecting the datum at a scene, investigators should consider the following issues:

- The datum should be a permanent point. Preference is given to structures such as houses, utility poles, or other objects that are stable, stationary, and permanent. It may be necessary to return to the scene during court proceedings, and significant delays often occur between scene processing and trial. In rural locations where no permanent object is available, investigators can establish a datum by pounding a survey spike into the ground. This spike must remain after the scene has been released.

- The datum must be a single, well-defined point. Poorly established points, such as the edge of a road or building, create inaccuracy during data collection. Existing fixed points, such as the upper right-hand corner of a door frame or the northeast corner of a building at ground level, serve as reliable and easily identifiable points of reference. Alternatively, the datum can be established by driving a nail into a telephone pole or painting a datum point on a permanent object.

- The datum should be at a higher elevation than the evidence. The height of the datum is important, because it provides not only a measure of distance between objects but also the horizontal distribution of objects. Selecting a datum point that is lower than the evidence creates difficulties in mapping.

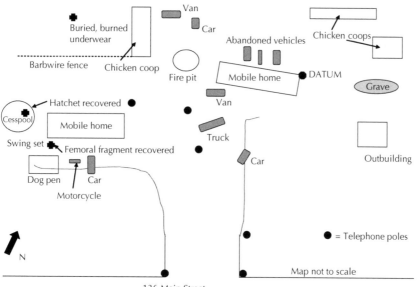

Van
Car
Buried, burned
underwear
Abandoned vehicles   Chicken coops
Barbwire fence   Chicken coop
Fire pit   Mobile home   DATUM
Grave
Hatchet recovered
Van
Cesspool
Mobile home
Swing set   Femoral fragment recovered   Truck
Dog pen   Car   Car   Outbuilding
Motorcycle
N
= Telephone poles
Map not to scale
136 Main Street

Figure 4.23 A scene plan view map. Note the inclusion of all major roads and structures, the datum point, and a north arrow, as well as a statement that the map is not to scale.

- The geographic location of the datum must be known. The datum must be identifiable on a map. Once the datum has been established, a GPS unit provides the latitudinal and longitudinal coordinates of the datum, as well as its elevation.

- The datum should be centralized within widely distributed evidence or located adjacent to the majority of evidence. Since all evidence must be measured back to the datum, establishing the datum a considerable distance from the evidence is impractical. At scenes involving very widely dispersed evidence, the creation of a subdatum facilitates mapping. In these cases, it is necessary to establish the distance and relationship of the datum and subdatum only once. All evidence from a specific area can then be measured to the nearest subdatum.

- With some burials or large-scale scenes such as plane crashes, it is often necessary to establish grids. A **grid** is a series of equal, measured squares superimposed over a scene through the use of stakes and string (Figure 4.24). Grids allow investigators to record the specific location of evidence within each grid square, providing greater resolution to the mapping. When a grid is established and documented on the plan view map, it is only necessary to measure the relationship of the closest corners of the grid to the datum. Noting the location of all evidence within each grid square then provides its location relative to the datum.

Figure 4.24 A 1-meter square grid laid out over a grave prior to excavation. The grave outline is visible in squares 3, 4, 6, and 7. The grid was extended beyond the grave outline to capture a number of features, such as suspect footprints and cigarette butts. (*Photo courtesy of the New Mexico Office of the Medical Investigator*)

- Select one system of measurement, either English or metric, and use it consistently throughout a scene. All maps must show the datum and an arrow identifying true or magnetic north.

Documenting the scene also requires note taking. Collecting information regarding the names and agency affiliations of participants is standard. A written summary of witness statements and information provided at the police briefing facilitates accurate recall of information after the fact. Noting the time and nature of each task or activity provides a time line for the overall processing of the scene that may be required in court.

Taking notes at the scene provides investigators with a source of information that cannot be discerned from photographs and maps. Ambient temperature and weather conditions, verbal statements by other investigators or informants, and descriptions of matter out of place or circumstantial evidence noted at scene are appropriate.

Caution is also warranted regarding note taking at the scene. Notes are not private. All notes and documentation generated at scene are subject to discovery and subpoena at trial (see Chapter 3). Recording information and observations is necessary and appropriate. However, anthropologists and investigators should resist making notes that reflect initial or unsubstantiated

BOX 4.10

Maintaining a professional voice in report writing

In 1995 a forensic anthropologist attended a death scene at the request of the local medical examiner. The anthropologist, accompanied by some graduate students, was required to wait several hours while law enforcement secured a warrant and did their initial photography. In the report he submitted to the medical examiner, the anthropologist expressed his frustration and anger at the police for being kept waiting. In 2002 the body was identified and two men were charged with murder. The original forensic anthropologist who authored the report had retired and another anthropologist was called to testify. The defense lawyer used the harsh and critical language of the anthropology report to challenge the credibility of the police at the scene, suggesting that law enforcement personnel had "bungled" the scene and were incompetent. As the testifying anthropologist had not been present at the scene and could not contradict the report based on actual observation, nothing could be offered to refute the claims.

This example illustrates the need to remove any personal or emotional observations from a report, no matter how innocent or irrelevant they may initially appear. A statement such as "the team waited three hours while police secured a warrant and performed their photographic documentation of the scene" accurately reflects the events without assigning blame or expressing emotion. Any other personal observations are unnecessary and could prejudice the case.

opinions, personal comments, or criticisms of other agencies, as they may be called on in court to explain or justify these comments. For example, describing the nature and placement of matter-out-of-place evidence is a necessary part of the documentation process. Adding an unequivocal note regarding its interpreted meaning or relevance is not. Further investigation and testing may prove such opinions incorrect. In the hands of defense lawyers, such statements could prove troublesome for the investigator who made them, no matter how apt they seemed to be at the time. Anthropologists should limit note taking to observations, not opinions, ideas, hunches, or emotions (Box 4.10).

BURIALS

Documenting burials is a common task for anthropologists at scene. However, capturing the three-dimensional information necessary to document a burial requires more than simple measures of distance from object to datum. For each burial, a minimum of two maps is required. The first is a plan view, showing the relationship of the grave to other significant features of the scene such as buildings or roads (Figure 4.23). The second is a **grave profile**, a vertical representation of the body in the grave (Figure 4.25).

Clandestine burials are unmarked graves that are not in a cemetery. Although the individual in the grave cannot be assumed to represent a homicide or even an unnatural death, the improper burial alone is considered a criminal offence (excluding historic and prehistoric unmarked graves). Such graves may also reveal the intent of an assailant to conceal a crime. Digging a grave, transporting the body, and backfilling the grave takes considerable time and effort. In addition, prolonged interaction with the body increases the probability that the offender will be discovered during the commission of the crime. Anthropologists who excavate clandestine burials are often asked in court to offer an opinion about how much effort the grave's formation represents, as a statement of the intent of the accused. Evidence in support of this opinion is provided through maps and documentation recorded at scene.

Minimally, investigators need to know the overall dimension of the grave: its depth, width, and length. It is also important to know the depth of the body relative to the ground surface. This requires three separate sets of measurements. The first occurs prior to the excavation of the grave and documents the ground level. The second set of measurements is taken when the body is exposed within the grave. These measurements provide the overall depth of the body relative to the surface. The final set of measurements is taken after the remains have been removed and the grave fully excavated to

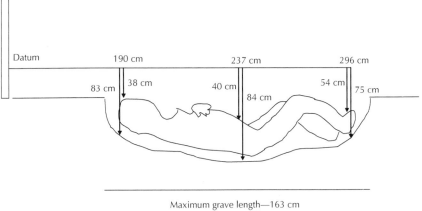

Maximum grave length—163 cm
Maximum grave width—74 cm

Figure 4.25 Grave profile showing depths to the body and to sterile soil as well as overall grave dimensions. Note the position of the datum point: both body and sterile soil measurements were taken at identical distances from the datum.

sterile soil. **Sterile soil** is the undisturbed ground representing the lower limits of the grave as it was originally formed. These final measurements reveal the overall dimensions of the grave as it was dug, including grave length and width. Depth measurements should be taken at consistent distances from the datum. After measuring to sterile soil, we recommend evaluating the floor of the grave with a metal detector or through further excavation to ensure that no projectiles or other pieces of evidence were introduced into the soil while the body was in the grave.

All soil removed from the grave is screened to ensure that no evidence is overlooked. Screen mesh size varies; the appropriate size is determined by soil type and consistency as well as the size or type of evidence being sought. The anthropologist responsible for excavating the grave should also recognize the potential for ephemeral evidence. **Ephemeral evidence** is a physical feature that cannot be collected in a traditional manner. A **feature** is any non-portable remnant of human activity, such as a grave or tire track. The nature of the feature precludes simply picking it up and sealing it in an evidence bag. Ephemeral evidence associated with burials includes shovel marks indicating the tool used to dig the grave and footprints in the soil of the bottom of the grave created when the perpetrator stood in the grave. Careful excavation techniques, combined with the determination to identify and preserve such features, ensure their survival during exhumation. While the anthropologist discovers and reveals such evidentiary features, it is law enforcement's responsibility to document them. Capturing ephemeral data relies on photography and casting techniques using a variety of media (detailed in Bodziak, 2000). Once excavation is complete, investigators are required to backfill the grave as a matter of public safety.

Archaeological versus Forensic Approaches to Excavation

While many of the questions addressed and methods used in archaeological and forensic contexts are the same, it is important to remember that the goals of each type of inquiry are distinct. As such, methods and approaches will vary. Many forensic anthropologists receive their initial training in excavation techniques from archaeological field schools, which emphasize excavation by standardized units and the recognition of cultural, natural, or arbitrary levels of deposition. Although suitable for archaeological applications, rigid adherence to textbook archaeological methods in forensic contexts is not appropriate. Forensic recoveries should adopt modified, flexible excavation strategies, developed on a case-by-case basis (Hoshower, 1998). In a mass grave, for example, the mandatory use of an excavation strategy based on a 1-meter-square grid and a 10-centimeter depth interval potentially introduces artificial and arbitrary division of evidence, space, and time and does not contribute to the interpretation of the grave (although an alternative, elaborate approach is considered in Duday and Guillon, 2006).

Both archaeological and forensic investigations requiring excavation are fundamentally concerned with reconstructing and understanding human behavior. Both types of inquiry also require accurate documentation of the location and relationship of features, artifacts, and evidence, as well as the ability to reproduce the scene for future analysis. It is how these goals are met that varies between forensic and archaeological contexts. Flexibility in excavation strategy in forensic investigations should not be seen as license for poor methodology or less rigor. Rather, it acknowledges that forensic scenes differ from prehistoric sites both temporally and spatially. While methods and strategies may vary between archaeological and forensic sites, theories relating to evidence interpretation, such as those relating to depositional sequence, are shared.

Theories of Deposition

Interpreting scenes involving burials requires an understanding of the laws of stratigraphic succession and superposition. Stratigraphic succession encompasses both geological and archaeological stratification. *Geological stratification* reflects the natural deposition of strata within a region; *archaeological stratification* represents the effects of human actions within that region. The **law of stratigraphic succession** states that the deposition of strata represents the accumulation of materials over time and that the sequence of strata reflects that temporal relationship.

The **law of superposition** states that strata or materials in their original depositional sequence will reflect the age of the strata relative to other strata. Layers located at the bottom of a sequence were laid prior to those above it and are therefore older. This law does not indicate the exact ages of the individual strata, only their ages relative to one another. This law applies to both geologic and archaeologic stratification. To illustrate the relevance of these laws to forensic anthropology, a case study is presented in which the investigation relied on superposition to establish the age of the remains.

CASE STUDY 4.1

In June 2003, a New Mexico public works crew excavating a trench to expose a city water line encountered human remains inches below the gravel shoulder of a road. The crew contacted law enforcement, who requested the medical examiner and state forensic anthropologist evaluate the scene. Scene assessment was complicated because a water line had been installed through a historic cemetery (circa 1800s) and many of

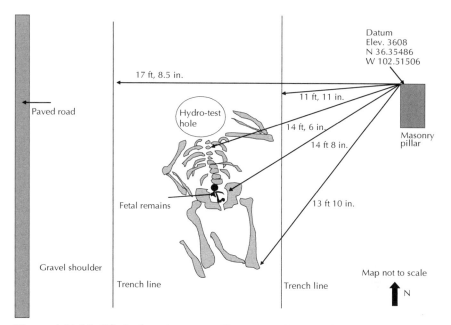

Figure 4.26 Modified plan view map illustrating Case Study 4.1, with a detailed view of the remains *in situ*. The map illustrates the presence and location of the fetal remains in greater detail than is possible in photographs.

the inhumations had been left *in situ*. The discovered remains were those of a pregnant adult female, with fetal remains present in the pelvic cavity (Figure 4.26). The remains were fully articulated, and no indications of coffin materials or body wrapping were recovered. Complete excavation of the scene revealed the outline of the original trench dug to install the water and gas lines in 1984. The body of the pregnant woman was recovered within this excavated trench (Figure 4.27), directly above the service lines. Additional excavation uncovered historic cemetery materials outside the original trench, including coffin fragments and remains.

Discussion: The articulated state of the remains, including the fetal remains in the pelvis, indicates that the remains were fleshed when deposited in the ground. This, along with the absence of coffin materials or containers, argues against the possibility that they represent historic cemetery remains that had been unearthed by the previous excavation and subsequently reburied. The law of superposition indicates that the

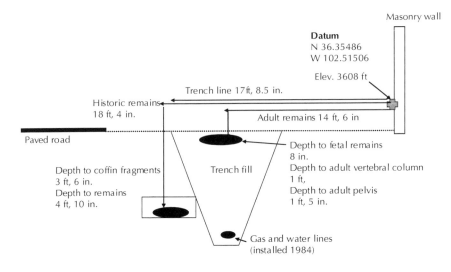

Map not to scale

Figure 4.27 The grave profile map for Case Study 4.1. According to the law of super-position, the remains must be contemporaneous with or more recent than the gas line at the bottom of the trench, installed in 1984.

burial of the pregnant woman must have occurred during the 1984 instal-lation of the pipelines or sometime thereafter, indicating the remains to be of medicolegal signficance.

REPORT WRITING

Report writing is the formal documentation and presentation of the events surrounding processing a scene. The report should include a summary of the information provided at scene, a list of participants from other agencies, a chronological accounting of the events and activities in which the author participated, information on methods used, and a list of evidence collected. Copies of maps and selected photographs should also be included. If it was necessary to deviate from established protocols during scene processing, an explanation should be provided in the report. An anthropologist who par-ticipates in both the scene and postmortem examination of a case may be asked to produce a separate report for each or to combine the two into a sin-gle report, depending on the protocols of the agency for which he or she is consulting.

A report should be concise but comprehensive. Maps and illustrations must be informative and referenced within the text. If literature is cited, a bibliography must be provided. Language should be professional but simple: readers may include pathologists, law enforcement personnel, lawyers, and possibly a judge and jury. Avoid jargon, unexplained abbreviations, acronyms, and undefined technical terms. Humor and personal comments are never appropriate and should be avoided. Like an autopsy report, a scene report becomes a legal document subject to dissemination and review.

Schedules permitting, investigators should begin report writing immediately upon leaving the scene, when details are still fresh. It is permissible to create a preliminary report, describing the activities and analysis of the scene to date. Such reports must be clearly identified as preliminary. Additional information and investigation can be documented in an addendum to the report. Once the investigation is complete, all previous reports should be incorporated into a final report. Although all reports are subject to discovery at trial, it is understood that information and opinions may change between the preliminary and final reports, as long as the reason or basis for such changes is clearly detailed in the final report.

Reports are considered to be confidential documents until deemed otherwise. Most medicolegal authorities are governed by both transparency and confidentiality guidelines. **Transparency** requires that the information generated become a matter of public record. However, such information is not public until it has been finalized and rulings on the cause and manner of death have been made (see Chapter 2). Further, the medicolegal authority receives reports and information from outside agencies, including law enforcement and the district attorney, as well as the decedent's medical records. Such information is confidential and must be protected. Reports generated by consultants, including anthropologists, are also confidential and are not included in the public record section of the file. Anthropologists who author reports should not release copies of their reports to any outside person or agency without understanding the rules regarding confidentiality. Unless otherwise specified by a court order to seal, cases that have been resolved at trial become matters of public record. As this point, dissemination of reports is permitted.

Finally, we emphasize a note of caution. The inclusion of a case, investigation, or file number on an anthropologist's curriculum vitae (CV) or résumé—student or professional—implies authorship of a report. If a person participated only as observer or trainee at scene or autopsy but did not write the report, such information must be clear on the CV to prevent confusion or the appearance of deception. Always retain a copy of any report you have written. Certification, state licensing, or conditions of employment often require copies of some or all reports. In such circumstances, the rules regarding confidentiality apply. The forensic anthropologist must not release sealed or confidential reports to any outside agency, regardless of the nature of the request.

CONCLUSION

Death investigation must be viewed in its entirety. Scene investigators who remain mindful of the downstream needs of the autopsy and judicial processes can facilitate other aspects of the investigation by maintaining a proper chain of evidence and thoroughly documenting the scene.

Scene investigation expands the role of the forensic anthropologist beyond analysis of bones. Anthropologists bring a wealth of knowledge vital to the effective processing of certain types of death scenes. Following the identification of material as having medicolegal significance, the process begins with the appropriate agencies establishing jurisdiction over the scene. Successfully integrating the anthropologist into scene investigation relies on protocols that initiate consultations and search strategies that maximize recovery, as well as experience in proper evidence handling and thorough documentation. Because of the need for more complex mapping and recovery methods, anthropologists play a particularly vital role in scenes involving buried or dispersed remains. The process culminates in a written report, allowing investigators to summarize evidence and present their findings and opinions.

CHAPTER 5

Beginning the Identification Process
Developing a Biological Profile

In creating a biological profile, sometimes termed an **osteobiography** (Box 5.1), the forensic anthropologist generates a set of attributes that facilitate the process of individual identification to be discussed in Chapter 8. They represent initial descriptors that permit investigators to narrow the pool of missing persons by excluding individuals who do not share the same physical attributes. As Chapter 8 emphasizes, missing persons reports vary in the level of detail entered into databases. A most basic format is that of Florida, which requires only race, sex, and age. California adds stature and weight, while Texas includes these five items plus "unique characters," which could include skeletal features. An emphasis on sex-age-race-stature leads us to focus on these attributes here as we begin to construct a biological profile from skeletal remains. Other, more specialized aspects are considered at the close of the section. These include estimates of parity, occupation, handedness, and weight.

While few forensic anthropologists may find themselves testifying in court about the methods raised in this chapter, it is important to appreciate the accuracy of any given procedure and not to pretend to more than the method at hand warrants. This is crucial during the early stages of profile development because a too narrow range for age-at-death, for example, can lead investigators to exclude from consideration the person whose remains are the subject of inquiry. If the biological profile becomes a point of contention during a trial, the forensic anthropologist should also be prepared to discuss methods in relationship to the *Daubert* criteria, as discussed in Chapter 3.

The need to speak in terms of error rates, falsifiability, and hypothesis testing reflects an inherent tension within the discipline, in as much as forensic anthropologists may not be used to speaking in probabilistic terms. It is much better to indicate within a biological profile that an unknown set of remains has a 95% probability of being male than to simply provide a conclusion.

The courts are wise in advising rigor and scientific testing, especially in the development of new methods and in resolving contested approaches. That this is crucial can be illustrated with one well-known but unfortunate forensic anthropological example, illustrated in Box 5.2. This is an important cautionary tale for forensic anthropologists and for expert witnesses generally. Certainly, Dr. Robbins did not intend to provide inaccurate information or to limit the freedom of innocent parties. Yet, the system in place at the time permitted this to happen, and perhaps even encouraged it.

BOX 5.1

Osteobiography

While the term "biological profile" is commonly used to reference core osteologiocal attributes (e.g., the PAC CIL, Reichs, 1998b; Steadman, 2003), *osteobiography* may also be used. Frank Saul (1972) coined this term in his study of the ancient Maya, referring to the reconstruction of individual and group characteristics from the study of human remains. The individual features are those most important to the forensic anthropologist, though he or she must be able to evaluate population data to place cases in context.

BOX 5.2

Louise Robbins as expert witness

Professor Louise Robbins held a PhD in anthropology (1968) and was a provisional member (1976) and subsequently fellow (1978) of the Physical Anthropology Section of the AAFS until her death in 1987. She studied not only human remains but also variation in gait. This led her to focus on footprints and shoe prints as individualizing features, and thus she developed a theory that linked not only bare footprints, but also shoe prints, insole indentations, and shoes to individuals. Robbins also inferred that stature could be estimated from foot and footgear dimensions. Alluded to as "Cinderella analysis," since it fitted shoes to the alleged owner, her procedures were presented to professional audiences several times, beginning in 1978 with an article in the peer-reviewed *Journal of Forensic Sciences* (*JFS*). In 1984 she held a workshop at the First Inter-American Congress of Forensic Sciences, and a year later an article appeared in *Law Enforcement Communications*. A book-length treatment published by a press long associated with forensic science followed in 1985, and yet another article appeared in *JFS* during 1986. Professor Robbins also presented papers on her method at professional meetings of the AAFS. Under the *Frye* criteria, which pertained at the time of her testimony, she certainly had professional credentials as a forensic anthropologist, and her work had been published in peer-reviewed journals. The more liberal rulings under Federal Rule of Evidence 702 would have emphasized her status as an expert. Thus, it could be argued that she herself was qualified to testify and that her research was acceptable to the forensic anthropological community.

While no experts validated Robbins' conclusions, and several physical anthropologists argued against them under oath, it was not until the year of her death (1987) that a panel of forensic scientists, including anthropologists and criminologists, concluded that her methods were flawed. None of her inferences were formally tested until later, after her death (Giles and Vallandigham, 1991; Gordon and Buikstra, 1992). Some convictions strongly influenced by her testimony were later overturned, but in the meantime, innocent people had been imprisoned. Had the more stringent *Daubert* criteria been applied, it should have become more immediately apparent that some of Robbins' results could not be replicated through scientific procedures and that those amenable to testing presented unreasonably high error rates.

(Bodziak, 2000; Frisbie and Garrett, 2005; Klepinger, 2006; Moenssens et al., 1995; Powell et al., 2006).

While it is true that *Daubert* applies only in federal courts, its standards have influence well beyond the bounds of the federal court system (Keierleber and Bohan, 2005). As emphasized in Chapter 3, judges are asked to serve as gatekeepers for expert testimony based on the degree to which the methods employed fulfill the *Daubert* criteria: (1) are falsifiable and testable, and have been tested and proved reliable; (2) have a known, low error rate; (3) are generally accepted within the appropriate scientific community; and (4) have been subjected to peer review. Interpretations of these criteria vary, with some adding the presence of standards by which the method is applied to the list (Keierleber and Bohan, 2005). As Gatowski et al. (2001) discovered in their survey of trial judges concerning their understanding and application of the *Daubert* criteria, there is no consensus among judges as to the relative weights to be given the different criteria. Further, while general acceptability (criterion 3) and peer review (criterion 4) appear to be fairly well understood, it was clear that the meaning of the term "falsifiable" was known only to a minority of those surveyed, and there was little agreement concerning the standard to be used in evaluating error rates.

Another, more recent decision has extended the *Daubert* criteria beyond scientific expert testimony to technical expertise. Importantly, as noted in Chapter 3, *Kumho Tire Co. v. Carmichael* also suggests that not *all* the *Daubert* criteria apply to every form of technical testimony (see also Wellborn, 2005). This gives the judge, as gatekeeper, latitude in applying the four cornerstones of the *Daubert* decision. *Kumho* also permits experts to develop theories through observation and apply them to the case at hand, as in technical expert testimony (Chapter 3; see also Grivas and Komar, 2007).

To engage in research and to evaluate that of others, the forensic anthropologist must understand certain basic features of research design and statistics. We therefore preface our consideration of attributes useful in generating biological profiles with some basic information about research design and statistical methods. Our goal is to present information sufficient for the evaluation of methods for developing biological profiles, with emphasis on issues relevant to the *Daubert* criteria. The conduct of original research requires, of course, considerable additional statistical expertise, which is beyond the scope of this volume. We begin by discussing the manner in which theories are built and then turn to issues related to reliability, error estimation, and statistical testing.

THEORIES AND METHODS IN FORENSIC ANTHROPOLOGY

The theories and methods of forensic anthropology have been derived through the process of scientific reasoning known as the **hypothetico-deductive method** (see Bowling, 2002; Hailman and Strier, 1997; Hulley et al., 2001 for detailed discussions of research design and methods). By observing phenomena, the researcher engages in **induction** and thus creates a hypothesis or more generally, a model about the way the world works. This hypothesis is then used to make predictions, which are subsequently tested. If the hypothesis fails the test, another set of predictions should be generated and further tests conducted. **Deduction** is the term applied to the process of creating expectations and testing them though observations. A hypothesis can never be proved true, though positive results from numerous tests reinforce one's confidence. Theories develop through the testing of multiple hypotheses with many lines of evidence (Box 5.3).

An important concept in hypothesis testing is **falsifiability**, which is frequently linked to the issue of testability (or actually testing) when the *Daubert* criteria are discussed. To be falsifiable, an empirical statement must be amenable to testing through the presentation of counterexamples. The proposition that dorsal pitting of the pubis is linked to parity is falsifiable because the presence dorsal pits in females who have not given birth would constitute a counterexample. As developed in the *Daubert* ruling, the issue of falsifiability relates to decisions about whether a method is or is not scientifically derived. To be scientific, it must have developed in a context in which

---

BOX 5.3

What is a theory?

Formally, a scientific theory can be defined as "a set of statements or principles devised to explain a group of facts or phenomena, especially one that has been repeatedly tested or is widely accepted and can be used to make predictions about natural phenomena" (*American Heritage Dictionary of the English Language*, 4th edition, 2000).

---

the theory or technique is testable or falsifiable. A judge may, of course, extend this to the issue of actual tests of the method and the issue of reliability, which are more explicit in the criteria concerning peer review, error, and general acceptance.

Many of the techniques discussed in this section are based on theories of growth and development. Sex diagnosis is developed through knowledge of the relative sizes of males and females in human groups and distinctive physical differences that are expressed differently in males and females. Theories of developmental and degenerative change form the basis of our predictions of age-at-death. Laws of inheritance and population genetics underlie studies of ancestry, although the genetics of complex structures is imperfectly known. The breadth of knowledge required to fully understand the implications of such theories requires broad training in the biological and related sciences.

Whether actively engaged in creating new forensic anthropological knowledge or applying methods developed elsewhere, the practitioner must be sensitive to the robustness of his or her methods. This includes knowledge of the design that guided the original research, including assumptions made, sources of bias, statistical testing, and the degree to which the results of the basic research are appropriate for the case at hand. A model may have **internal validity**,—that is, it may be true for one human group or set of remains—but lack **external validity** and thus not be generalizable to other human groups or sets of remains. It is crucial that all methods be subject to rigorous testing for both internal and external validity prior to their application in forensic contexts. In the following sections, we consider methods that have strong external validity along with examples of new techniques and some that, to date, have not achieved consensus.

The methods discussed here are based on various scales of measurement, primarily **interval** and **ordinal** (Box 5.4). Before the results of any set of observations can be accepted, they must be subject to statistical testing appropriate to the scale of measurements and the mathematical nature of the distribution of observations. Statistical tests for interval and ratio data are generally more powerful than those for ordinal or nominal data.

Statistical tests appropriate for normally distributed interval data are termed parametric and include a variety of univariate and multivariate

---

BOX 5.4
Scales of measurement

- *Nominal* A name or number is used to define a category, such as male or female. Nominal categories have no inherent order: for example, the categories red, blue, and yellow.
- *Ordinal* Observations are placed in rank order, such as small, medium, large, or 1, 2, 3.
- *Interval* These units of measurement are fixed or standard, such as millimeters or degrees Fahrenheit. Note that 0 is arbitrary—0° is not the absence of heat.
- *Ratio* Ratio scales have the attributes of an interval scale and in addition, an absolute zero, such as degrees Kelvin, in which a temperature of 0 K is the absence of molecular motion (heat).

---

methods, some of which will be discussed in the following sections. Non-normal interval data may be normalized through mathematical transformation; failing that, nonparametric tests should be used. Nonparametric tests are also appropriate for nominal and ordinal data (Blalock, 1972; Siegel, 1956).

As emphasized earlier, forensic anthropologists must be concerned with the **accuracy** of their methods, the degree to which they are correct. This is crucial both in the development of original research and in the evaluation of the work of others. Two related terms, which are sometimes incorrectly used as synonyms, are **precision** and **reliability.** Whereas accuracy assesses the degree to which research results are a true representation, precision and reliability have to do with repeatability. A measurement that is highly repeatable, that is, has high precision (reliability) may be a poor representation of, for example, sexual dimorphism of a skeleton, and thus be inaccurate. Another important distinction is between instrument precision and the precision of a given measurement. Researchers may use the term "reliability" to refer to measurement precision and "precision" in relation to the instruments used to make measurements. In this usage, reliability is considered to be the stability or reproducibility of a given measurement during tests and retests.

In assessing the reliability of a measurement or other observation, researchers should conduct and report tests of inter- and intraobserver error. One commonly used statistic for measuring observer error is the kappa statistic (k.) An example detailing the calculation of such error is given in Box 5.5.

Questions of precision, reliability, and accuracy have been faced directly by various researchers cited in the following sections. Williams and Rogers (2006), for example, find that certain mandibular traits such as symphysis height and chin form show high precision but low accuracy and are therefore less valuable than certain features of the cranium such as mastoid size that show high accuracy and moderately high precision.

BOX 5.5

Reliability testing, percent agreement, and calculating kappa

A commonly used measure of reliability is percentage agreement—the percent of concordant results in repeated tests. An inherent problem with percentage agreement is that it does not account for agreement that occurs by chance alone. To correct this shortcoming, many researchers utilize the kappa statistic (κ). Kappa is used to measure the agreement between tests by the same observer (intraobserver error) or between two observers (interobserver error). It is important to clearly understand what is being measured, namely, observer variation. Kappa is a measurement of reliability, not validity. It is a test of diagnostic criteria and its inter- or intraobserver reproducibility, undertaken to evaluate whether the criteria are sufficiently well defined (Svanholm et al., 1989).

The kappa statistic corrects for agreement by chance by subtracting the proportion of agreement that would be expected to occur by chance alone (Fleiss, 1971). The formula is as follows:

$$\kappa = \frac{P_o - P_e}{1.0 - P_e}$$

where P_o = the proportion of observed agreement and P_e = the proportion of agreement expected by chance (Cohen, 1960). To demonstrate how these values are generated, the following example is offered.

The presence or absence of the ventral arc was recorded on 493 pubic bones by two separate researchers (interobserver error test), for a total number of 986 observations. To calculate kappa, an agreement table must be generated:

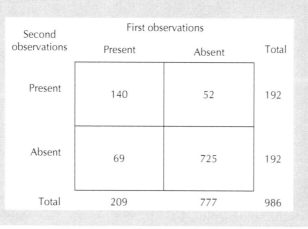

Second observations	First observations		Total
	Present	Absent	
Present	140	52	192
Absent	69	725	192
Total	209	777	986

First, the percentage agreement must be calculated, using the following formula:

$$\frac{\text{total of concordant observations}}{\text{total number of observations}}$$

In this example, the percentage agreement would be:

$$\frac{140 + 725}{140 + 52 + 69 + 725} = 0.877$$

The chance agreement is the agreement expected if both observers rated the responses at random. The total chance agreement is the sum of the chance agreement for each cell on the diagonal:

- Chance agreement for the "present/present" cell is $(209 \times 192)/986 = 40.7$.
- Chance agreement for the "absent/absent" cell is $(777 \times 794)/986 = 625.7$.
- Total chance agreement is then $(40.7 + 625.7)/986 = 0.676$.

Kappa is therefore:

$$\kappa = \frac{0.877 - 0.676}{1.0 - 0.676} = 0.62$$

A kappa value ranges between 1.0 and –1.0, with 1.0 representing excellent agreement and 0 through –1.0 indicating poor to no agreement. Scales for interpreting the kappa statistic can be found in Landis and Koch (1977) and Fleiss (1971), among others. The kappa value of 0.62 calculated in the example would be interpreted as "substantial" (Landis and Koch, 1977) or "fair to good" (Fleiss, 1971).

Error is inherent in any measurement. Forensic anthropologists seek accurate methods that minimize error, thus ensuring that casework is of high quality and that the *Daubert/Kumho* criteria are satisfied. A second, equally important concern is to understand how to characterize error correctly and not to presume more accuracy than is justified by a given technique. This is especially important in developing quantified attributes of the biological profile such as age-at-death and stature.

Later in this chapter we illustrate that there are many possible methods for estimating sex of an individual. No method is perfectly accurate, even when applied to closely related groups or individuals. Thus, error is present and should be estimated. A major source of error in sex diagnosis is variability in expression of shape and size in men and women. As Klepinger and Giles (1998) and Klepinger (2006) emphasize, such variability can be quantified in several standard ways, including the **range, coefficient of variation, variance**, and **standard deviation**. Of these, range and standard deviation are most important to the forensic anthropologist. An example that illustrates both the manner in which measures of variability are commonly interpreted and may be misinterpreted appears in Box 5.6.

A further significant point made by Klepinger (2006) and Klepinger and Giles (1998) involves the distinction between percent correct classifications

---

BOX 5.6

Measuring and mismeasuring variability

Diane France (1998: 70) reports the following summary statistics for the vertical diameter (in mm) of the humeral head for Terry Collection Euro-Americans:

Sex	N	Mean	Standard Deviation	Minimum	Maximum
M	84	48.4	2.9	40.8	56.9

Thus, the range for males is 40.8–56.9, which encompasses all the male values. The standard deviation (SD) describes deviation around the mean. One standard deviation, assuming the data are normally distributed, includes 68% of the observed cases. A common standard for reporting variation is ±2 standard deviations, which characterized 95% of the cases in the population or sample. In this case, 95% of the cases fall between 42.6 and 54.2.

As Klepinger (2006) and Klepinger and Giles (1998) emphasize, it is important for measures of variation to be reported accurately and interpreted correctly. All too frequently the standard deviation is confused with **standard error (SE)**, which is an ambiguous term without further definition, though it frequently references the standard error of the mean. As such, it is a measure of the variation in means when a single population is sampled repeatedly. The SE is computed as the SD divided by the square root of sample size. In the above sample, from the Terry Collection, the SE is approximately 1, a number considerably smaller than the SD. A researcher who mistakes the SE for the SD will presume greater methodological accuracy than the method warrants. The standard error of the mean (SEM) is a useful measure for assessing the level of confidence researchers may have in their statistical inferences, since as it varies with sample size. A large standard error of the mean suggests that researchers should interpret results conservatively.

in a population sample and the application of criteria to a single unknown. While percent correct classification is useful in **paleodemography**, for forensic casework one needs to know how frequently one may misclassify individuals, according to a given technique. In Box 5.7, we provide an example, drawn from research by Mittler and Sheridan (1992), who explored the possibility of sexing juvenile remains through observations of auricular surface elevation. While these investigators used an archaeological sample, the remains were mummified and the genitalia were preserved. The study sample may therefore be considered to be of documented sex.

Statistical testing is a crucial aspect of forensic research design, with special attention to accuracy, reliability, and error rates. For example, statistical testing frequently addresses the **null hypothesis:** that there is no difference between two groups. The researcher must decide prior to the analysis what probabilistic criteria will be used to reject the null hypothesis and thus accept that a significant difference exists between two groups or between an individual and a reference sample. For example, a probability value of .05 or less is frequently designated as the level whereby one can reject the null hypothesis, meaning that there is one chance in 20 of rejecting the null hypothesis when it is actually true (**type I or alpha error**). In other words, there is one

---

BOX 5.7
Estimating sex for an unknown case

In Mittler and Sheridan's (1992) study, an elevated auricular surface is hypothesized to be a feature of young women, while its absence (nonraised) is projected as a male feature. The researchers report that observations of the auricular surface correctly classified 85% male and 58% female infants and children. We now apply the procedures of Klepinger (2006) and Klepinger and Giles (1998), which are drawn from a medical analogy.

If we assume an equivalent number of males and females in the sample, we can calculate the probability of correct classification for an unknown case in the following manner. Looking at females first, we find that 29% (58% of 50%) will test female because they are female. In addition, 7.5% (15%, which is 100% minus 85%, of 50%) are males who incorrectly classify as females. Thus, 36.5% of the sample will test as female, but only 79.5% of these are true females. The same procedure calculated for males finds that 42.5% (85% of 50%) will be classified as males because they are male. An additional 21% (42% of 50%) will be misclassified females. It follows that 63.5% of the sample will test as male, but only 66.9% will be correctly classified. This relatively simple procedure thus establishes the level of certainty for an individual unknown. If an auricular surface is elevated, there is a 79.5% chance the remains are female. In the absence of elevation, there is only a 66.9% chance that the remains are male.

chance in 20 that the researcher will think that two groups are different when in reality they are not. **Type II or beta error** reflects the power of the test, that is, the probability of accepting the null hypothesis when it should be rejected. Other statistical approaches include **Bayesian methods**, which incorporate additional probabilities. Examples of Bayesian approaches are illustrated later when we discuss methods of estimating age-at-death through the "transition analysis" (Boldsen et al., 2002) and of generating a probabilistic overall biological profile in individual identification (Steadman et al., 2006a). Schmitt and colleagues (2002) also utilize a Bayesian approach in evaluating the potential of new standards for estimating adult age-at-death.

In forensic anthropological research, study samples of large size may not be available. It is therefore important that both the statistical test and its power be reported. A **power calculation** measures the likelihood that the statistical test will detect true departures from the null, given sample size, test chosen, and differences between groups. It also allows one to estimate how large a sample size should be to permit one to reach conclusions in which one may validly place confidence. Ideally, power calculations should be made early in the research design, prior to selecting sample size, but they can also be applied retrospectively. In general, a power of 80% or better is desired when using a $p \leq .5$. Suppose that p is less than or equal to .5 and the power of a test of a new method for estimating sex in skeletal remains is calculated as .86, this means that if there is a true difference between the two groups, the researcher will be able to detect it 86% of the time. Numerous software packages are available for engaging in power analysis.

Several biological profile attributes, including sex diagnosis and ancestry estimation, require the use of **discriminant function analysis** (see also Klepinger, 2006; Pietrusewsky, 2000). As described by these two authors, the method derives mathematical functions that maximize the differences among two or more groups (e.g., male/female or different ancestral groups), by creating a linear function that maximizes the between-group variance in comparison to total variance. The number of functions is one less than the number of groups or number of measurements, whichever is smaller. Measurements from an unknown, multiplied by the coefficients generated by the procedure, permits classification and also the generation of probabilities for group membership. **Posterior probabilities** assume that the unknown belongs to one of the reference groups used to construct the discriminant functions and explores the strength of that relationship. Based on average variability of all groups in the analysis, **typicality probabilities** evaluate how likely it is that the unknown belongs to any or none of the groups.

The generalized form taken by a discriminant function is

$$Y = a_1 X_1 + a_2 X_2 + a_3 X_3 + \cdots + a_n X_n$$

where a = the coefficient or weight and X = measurement. The computed value of Y is compared to a predetermined sectioning point that maximizes the accuracy of assignment in the study sample. A number of equations with distinctive combinations of measurements may be generated from a given database, each with stated confidence limits. When assessing sex, especially in elements other than the pelvis, the prudent researcher will choose discriminant functions generated from a sample closely related to the unknown, if at all possible. It is also advisable to select the equation with the narrowest confidence limits that can be taken on the unknown individual.

Having now reviewed concepts relating to research design, reliability, error, variability, and statistical testing, we turn to elements of the biological profile. We begin by considering attributes commonly recorded in missing persons reports: the estimation of sex, followed by age-at-death, ancestry, and stature. Subsequently, methods proposed for inferring parturition, handedness, occupation, and body weight are also evaluated. We emphasize that each of these exercises involves estimation, *not* determination. In each example, we are dealing with probabilities of success and error, which each practitioner should have in mind during both casework and research initiatives.

THE ESTIMATION OF SEX FROM SKELETAL REMAINS (SEX DIAGNOSIS)

In the following two sub sections we consider the estimation of sex through the observation and measurement of male and female features of the skeleton. As we emphasize, the most accurate attributes for sex diagnosis begin to develop in the bony pelvis during adolescence. Later in the aging process, other sex-specific dimensions and features develop in other parts of the body. We also consider accuracy in evaluating the utility of both standard and newly developed techniques.

An ability to accurately attribute sex in juvenile materials has long been sought by both forensic anthropologists and bioarchaeologists, and a number of methods have been proposed for evaluating juvenile remains. Here we critically review those that have been proposed for the immature pelvis, the dentition, and DNA.

Sex: Juvenile Remains

Ossification of the human skeleton begins at approximately the sixth fetal week; sexual differentiation at the eighth. It is therefore theoretically possible that sexually distinctive attributes will be identified in juveniles. However, studies of those elements that are dimorphic in adults, principally the bony pelvis, the skull, and dentition, have failed to find convincing evidence of sexual dimorphism in preadolescent skeletons. While Weaver (1998) remains optimistic that a skeletal or dental method will be found, others are more skeptical (Scheuer and Black, 2000, 2004). A statistically sophisticated, critical study of this issue is summarized in Box 5.8.

BOX 5.8

Case study: sexual dimorphism in human fetal remains?

One of the definitive and methodologically sophisticated studies of sex in juvenile remains is that of Holcomb and Konigsberg (1995), who focused on the human fetal sciatic notch. Using 133 fetal ilia of known maturation status and sex from the Trotter Collection, the researchers photographed an ilium from each individual and digitized the photographs to establish a replicable method for characterizing the depth of the sciatic notch. Tests for bivariate normalcy led to the deletion of six outlying values. Since the variance–covariance matrices were not equal for males and females, an iterative scheme and randomization approach were used to test significance. Holcomb and Konigsberg identified significant dimorphism in notch shape, specifically the location of the maximum depth of the notch when evaluated in the anterior–posterior position. The authors speculate that this pattern will be obscured by postnatal growth and will not reappear until adolescence. They also appropriately evaluate possible sources of measurement error and underscore that their misclassification rate of ~40% does not inspire confidence for forensic applications. At this time, there appear to be no convincing methods for estimating sex from skeletal attributes in immature remains.

Models developed for dental dimorphism in adults can be applied to juvenile secondary dentitions, and there is some evidence that deciduous teeth are dimorphic as well. As Scheuer and Black (2004: 20) emphasize, however, the differences are very slight, especially for the deciduous dentition. In one of the most extensive studies including mixed dentitions, De Vito and Saunders (1990) produced a series of equations, based on a living Canadian reference sample, for predicting sex (see also Saunders, 2000). The authors, using combinations of three to five measurements of deciduous teeth and mixed dentitions, produced discriminant functions that correctly classified between 76 and 90% of the validation sample. The use of deciduous teeth in this study approaches the accuracy of functions developed for permanent teeth. A test on identified historic pioneer children's remains did not yield similar levels of accuracy, a result Saunders attributes to mortality bias against size and/or maturity. Thus, while promising, discriminant functions based on dentition require testing on other samples before they achieve the level of certainty consistent with the *Daubert* criteria. Whether working in forensic or archaeological contexts, practitioners must recognize that deceased children may have suffered from conditions that would predispose them to small tooth size, thus leading to an artificial and erroneous elevation in the number of remains attributed to the female sex.

DNA analysis also holds potential for distinguishing between remains of girls and boys, provided nuclear DNA is preserved. To avoid false negatives, a method that amplifies fragments of a gene that is found on both the X and Y chromosomes and differs slightly between the two is required. Stone and colleagues (Stone, 2000; Stone et al., 1996), for example, have had good results in millennium-old remains by focusing on the **amelogenin gene**. Slightly fewer than half (46%) these individuals permitted sex diagnosis with good concordance between morphological sex and DNA in adults. Amelogenin has been accepted in court as an indicator of sex (Taylor et al., 1997), although it may not be universally applicable, especially in individuals of South and East Asian ancestry (Chang et al., 2003). Other methods, such as DNA capillary electrophoresis, have also been applied to infant human bone recovered 16 years after death (Yamamoto et al., 1998). Archaeological applications of the amelogenin method to neonates have been promising as well (Faerman et al., 1997; Lassen et al., 1997).

Care must be exercised before accepting any new method for estimating sex in immature remains. For example, in 2001 Loth and Henneberg published an analysis of dimorphic shape features in young children's remains in a peer-reviewed journal, reporting an accuracy of 81%. A subsequent blind test (Scheuer, 2002) reported an accuracy of only 68%, which is unacceptably low. Franklin and coworkers (2006) also found little evidence of sexual dimorphism in young juvenile mandibles. Obviously, while this novel approach appeared to meet some of the *Daubert* criteria, including publication in a peer-reviewed journal, it has failed to gain general acceptance by the scientific community owing to its low accuracy and failure to achieve external validity.

Sex: Adult Remains

The most predictably dimorphic bones in the human body are those of the bony pelvis, which become distinctive during the adolescent growth spurt (Coleman, 1969). Researchers have developed methods for estimating sex through both measurements and observations of **morphological variation**, such as the shape of the pubic bone and the presence of a ventral arc (France, 1998; Klepinger, 2006; Phenice, 1969; White and Folkens, 2005). Many morphological observations and measurements used in estimating sex are largely unmodified from Hrdlicka's pioneering handbooks (1920, 1939) and Krogman's classic work (1939, 1962). On the other hand, as we report here, new attributes are continually being proposed. As we emphasize, it is crucial that these receive critical review for both internal and external validity. We will first consider sexually dimorphic morphological features and then turn to the measurement of sexual dimorphism.

1. Morphological Features in Sex Diagnosis

In Box 5.9 we present a listing of morphological structures of the bony pelvis drawn from the research of Rogers and Saunders (1994), anchored by study of a documented nineteenth-century skeletal sample from St. Thomas

BOX 5.9

Evaluation of morphological features of the bony pelvis for sex diagnosis

Numbers refer to order defined by combined accuracy and precision scores (after Rogers and Saunders, 1994).

Pubis	Male Expression	Female Expression
5. Subpubic concavity angle	V-shaped	U-shaped
16. Ischiopubic ramus ridge	Ridge absent	Ridge present
1. Ventral arc presence	Arc absent	Arc present
6. Shape of pubic bone	Narrow	Broad/rectangular
8. Dorsal pubic pitting	Absent	Present
Ilium		
12. Sciatic notch shape/size	Small/close/deep	Wide/shallow
17. Auricular surface height	Not raised	Raised
10. Preauricular sulcus presence/shape	Absent or thin grooves	Large/circular depressions
12. Ilium shape	High/vertical	Laterally divergent
Overall pelvis		
15. Pelvic inlet shape	Heart shaped	Elliptical
3. True pelvis size and shape	Small	Shallow/spacious
2. Obturator foramen shape	Large/ovoid	Small/triangular
8. Acetabulum size/ orientation	Large/directed laterally	Small/directed anterolaterally
7. Development of muscle markings	Marked	Gracile/smooth
Sacrum		
4. Shape	Long/narrow	Short/broad
14. Number of segments	6+	5
10. Posterior view: visibility of sacroiliac joints	Visible	Not visible

Anglican Church. While we recognize that testing or developing methods in archaeological samples of unknown individuals is to be discouraged, the availability of information from coffin plates and parish records that identify individuals by age-at-death and sex effectively make this study comparable to those conducted on nineteenth-century autopsy collections. While a modern sample would be preferable, for forensic purposes, this example is cited here because it represents an attempt to assess accuracy and reliability in the use of morphological features in sex diagnosis.

The numbers to the left of all observations reflect the authors' combined ranking of accuracy and reliability/precision. Accuracy is determined in relation to documentation, and reliability scores reflect intraobserver consistency in repeated trials. The ventral arc, for example, ranked fifth in accuracy and first in precision, placing it first in the list, followed closely by the shapes of the obturator foramen and the true pelvis. When the ventral arc is combined with obturator foramen shape, accuracy increased to 98%. While the entire trait list has an accuracy rating of 95.9%, levels of precision were low for several of the features. The authors recognize that their results differ from those of some other researchers, undoubtedly reflecting population and methodological variability. The first six features had both high accuracy and precision: ventral arc presence, obturator foramen shape, true pelvis size and shape, sacrum shape, subpubic concavity angle, and shape of the pubic bone. Moreover, Rogers and Saunders found that the Phenice feature accuracy (combined ventral arc, subpubic concavity, and ischiopubic ridge) was only 88%, while Phenice had reported 96%. Lower figures of 83 and 70% have been reported by Lovell (1989) and by McLaughlin and Bruce (1990), respectively. For the ventral arc alone, the Rogers and Saunders figure of 86.9% (accuracy) is lower than that of Sutherland and Suchey (1991), who reported 96%. In contrast with assertions concerning, for example, age-related changes in expression of the ventral arc (Lovell, 1989), Rogers and Saunders found no age effect. Thus their study underscores the need for rigor, including elements of both accuracy and precision in the selection of morphological features, the necessity of being sensitive to population variability, and the importance of using multiple traits of high precision and accuracy when evaluating remains.

In companion studies of morphological features of the skull (Box 5.10), Williams and Rogers (2006) and Rogers (2005a) have examined attributes commonly recorded by forensic anthropologists, such as occipital and mastoid process rugosity, along with several features of the facial skeleton that appear to perform well and should be considered for further study. Rogers (2005a) studied skulls from the same historic series investigated by Rogers and Saunders (1994). Williams and Rogers (2006) turned to a sample of 100 crania (50 males and 50 females) of Euro-American origin drawn from the Bass Collection. The research design for each of these studies paralleled that of Rogers and Saunders (1994), explicitly seeking features that provide figures of ≥ 80% accuracy and ≥ 90% precision (intraobserver error rate ≤ 10%). The numbers to the left of each feature listed in Box 5.10 represent combined accuracy and precision scores for Williams and Rogers (2006) and Rogers (2005a).

Unsurprisingly, there is not perfect concordance between the rank orders from the two studies of Euro-Americans from distinctly separate centuries and—as is likely—gene pools. The overall accuracy of 14 preferred features

BOX 5.10

Evaluation of morphological features of the skull for sex diagnosis
Numbers refer to combined accuracy–precision ranking from
Williams and Rogers (2006), Rogers (2005a)

Feature		Male Expression	Female Expression
2, 5	Size and architecture	Big/rugged	Small/smooth
7, 5	Frontal shape	Low, slopes posteriorly	Rounded, full, vertical
7, 7	Frontal eminence	Small	Large
1, 1	Supraorbital ridges	Medium to large	Small to medium
7, 7	Orbits	Squared, low	Rounded, high
5, —	Orbits	Rounded margins	Sharp margins
3, 1	Nasal aperture	High, thin sharp margins	Low, wider rounded margins
6, 4	Nasal size	Large	Small
6, 1	Malars	Posterior–lateral, rugged	Anterior–lateral, smooth
2, 1	Zygomatic arches	Extends	Does not extend
9, 7	Parietal eminences	Small	Large
1, 3	Mastoid	Medium–large	Small–medium
5, 2	Occipital/Nuchal crest	Rugged	Not rugged
8, 7	Occipital condyles	Large	Small
7, 6	Palate	Large, U-shaped	Small, parabolic
4, 4	Mandible	Large, high symphysis	Small, lower symphysis
5, 4	Mandible	Broad ascending ramus	Narrow ascending ramus
5, 4	Mandible	Gonial angle < 125 degrees	Gonial angle >125 degrees
8, 4	Mandible	Gonial angle everted	Gonial angle not everted
4, 2	Chin	Square, two eminences	Rounded, one eminence

studied by Williams and Rogers (2006) was 92%. Focusing on 17 attributes, Rogers (2005a) achieved a slightly lower accuracy of 89%. These bracket Hrdlicka and Krogman's expectations of 90% for an experienced student of osteology; Stewart's figure is 90 to 95% (reported in Krogman and Iscan, 1986: 198–90). In fact, when Krogman and Stewart engaged in blind tests of their sex diagnosis accuracy in cranial materials, their number were considerably lower (82–87 and 77%, respectively [Giles, 1964]). Five of the six most highly ranked features for the Bass Collection skulls (mastoid process size, supraorbital ridge, cranium size and architecture, zygomatic process extension, and nasal aperture shape) performed similarly well in the study of historic period Canadians. As the researchers emphasize, the first three of these are features traditionally observed by physical anthropologists in sex diagnosis. The other two are less commonly considered to be sexually dimorphic and merit testing beyond groups of Euro-American ancestry. Two mandibular features, symphysis height and chin form, achieved somewhat high composite rankings, but in general accuracy was low, and the researchers conclude that attributes of the mandible are less useful than those of the cranium. In neither study did the researchers find age to be strongly associated with the accuracy of cranial features, though this may be an artifact of small sample size. Other workers have suggested age associations, including increased accuracy in sexing male skulls with age and decreased accuracy in females (Meindl et al., 1985; P. L. Walker, 1995).

As noted shortly, many other bones of the body have been measured to provide estimates of sex, while morphological observations have received much less emphasis. Recently, however, Rogers (1999) has focused on the carrying angle of the arm and selected four morphological features of the distal humerus, which when combined yielded an accuracy of 92%. The attributes include olecranon fossa shape and size, orientation of the medial epicondyle, and shape of the trochlea. Having tested her results in three skeletal samples (Grant Collection, University of New Mexico Documented Collection, and Bass Collection), Rogers considers her methodology to be robust. When tested in a documented U.K. skeletal collection from the seventeenth to nineteenth centuries, the St. Bride's Collection, a lower level of accuracy (79.1%) was achieved (Falys et al., 2005), although these authors remain optimistic about the method in forensic contexts.

Konigsberg and Hens (1998) have also attempted to add methodological rigor to sex diagnosis through the observation of cranial morphology,

---

BOX 5.11

Global versus local reference standards

A key issue facing any practitioner in developing biological profiles is the choice of an appropriate reference algorithm. Should it be local or global? As we discuss methods for sex diagnosis, we emphasize that only a few attributes of the bony pelvis are accurate across our species, while others vary considerably across the world's populations. Thus, researchers have sought to develop local or regional standards for dimorphism, measured in many bones of the body. In parallel, standards for observing age-related changes at the sternal end of the fourth rib vary by ancestry and sex. Yet, what is the forensic anthropologist to do when faced with hundreds or even thousands of victims of mass disasters, such as airplane crashes or the thousands of deaths on September 11, 2001, at the World Trade Center? Practitioners therefore need to have general standards that characterize variability across human groups, as well as regional models. Chaillet and colleagues (2005), in their applications of Demirjian's dental method for estimating juvenile age-at-death, faced a related issue in age assessment: the requirement to establish the degree to which accuracy is compromised when the more general model is applied.

including dimorphic size and shape differences. Albeit in an archaeological context, they combine visual assessment with statistical models to generate classifications. Their correct classification rate fell within the low 80s, at the low end of correct classification figures for discriminant functions based on the skull. This method is appropriate for fragmentary material, as is that of Holland (1986), who focused on measurement of the cranial base.

In closing this section, we wish to emphasize that very few skeletal attributes, notably those of the bony pelvis, appear to present predictable patterns of dimorphism with high accuracy across human groups. Others vary by geographic region, due to shared heritage of physical attributes. As emphasized in Box 5.11, it is important for the forensic practitioner to appreciate the need for defining locally sensitive patterns while also recognizing that on occasion the decedent's ancestry will be unknown. In such examples, the best choice will be a model that produces the greatest accuracy across many human groups. The identification of such models should be a goal of forensic anthropological research.

2. The Measurement of Sexual Dimorphism

Metric assessments of sexual dimorphism have generated models for many parts of the skeleton, as suggested by Box 5.12. Generally, these studies fall into two types: those that generate discriminant functions and others that establish a threshold for a single measurement or a ratio. Some few single measurements, such as femoral head diameters, can be quite accurate. Data presented in France (1998) suggest that correct classification percentages for African Americans and European American from the Forensic Data Base (FDB) maintained at the University of Tennessee (see Chapter 1) range between 83 and 93%. When combined with pelvic measurements, this observation can be an important part of discriminant functions that achieve even higher group assignment, as reported shortly.

While sexual dimorphism is a notable feature of adult bony pelves, measuring this dimorphism has proved problematic due to imprecision in establishing landmarks within the acetabulum, necessary for measuring the superior ramus of the pubis and the height of the ischium. Albanese (2003), in a rigorous study based on two collections, the Portuguese Coimbra and the Terry identifies a measurement of pubic length and ischial height with high precision and accuracy. The pubic landmarks are the superior margin of the pubic symphysis and the superior–anterior apex of the lunate surface of the acetabulum. The lunate point and the most inferior point on the ischium, not necessarily perpendicular to the pubic measurement, are used for the ischium. The pubic length measurement, when combined with four others of the pelvis and femur, correctly classified 98.5% of the sample. This study is particularly important as a rigorous treatment of pelvic indicators in sex diagnosis.

The head of humerus is also quite dimorphic, as is the epicondylar breadth. France (1998) reports formulas based on measurements of the

BOX 5.12
The estimation of sex through measurement

Sample	Element(s)	% Correct	Reference
Terry and Coimbra	Os coxae and femur	91–98.5	Albanese, 2003
Chinese	Femur	82.3–87.2	Wu, 1989
Croatian	Femur	91.3–94.4	Slaus et al., 2003
Spanish	Femur	85.6–99.1	Trancho et al., 1997
Thai	Femur	85.6–94.2	King et al., 1998
Dart (South African black)	Calcaneus	79–86	Bidmos and Asala, 2004
Dart (South African black)	Talus	80–89	Bidmos and Dayal, 2004
Terry	Talus/calcaneus	79–89	Steele, 1976
South Italy	Calcaneus	76.3–85	Introna et al., 1998
Dart (South African black)	Patella	78.3–85	Dayal and Bidmos, 2005
Japanese	Tibia	80.3–87.2	Iscan et al., 1994
Terry	Tibia	77.2–87.3	Iscan and Miller-Shaivitz, 1984 a, b
Japanese	Tibia	95.7–96	Hanihara, 1958
Terry	Humerus	85.5–93.5	France, 1983
FDB (Tennessee)	Humerus	86–93	France, 1998
FDB (Tennessee)	Scapula	94	France, 1998
FDB (Tennessee)	Clavicle	88–96	France, 1998
FDB (Tennessee)	Radius	86–99	France, 1998
Japanese	Radius	97	Hanihara, 1958
Japanese	Ulna	88.9–90.5	Hanihara, 1958
FDB (Tennessee)	Ulna	91	France, 1998
British	Metacarpals and hand phalanges	74–94	Scheuer and Elkington, 1993
UNM documented	Metacarpals	75–90	Stojanowski, 1999
Pennsylvania	Metacarpals	63–95.7	Burrows et al., 2003
Terry, UNM	Metacarpals	84.4–92	Falsetti, 1995
documented and Royal Free Medical Society of West Africa	4th rib	74–80	Wiredu et al., 1999
Florida	4th rib	82–89	Iscan, 1985
Turkey	4th rib	86–90	Cologlu et al., 1998
Taiwan	Skull	100	Hsiao et al., 1996
Terry	Cranial base	70–85	Holland, 1986
Terry	Petrous temporal	66–74	Kalmey and Rathbun, 1996
Terry	Skull	82.9–87.9	Giles and Elliot, 1962, 1963
Japanese	Skull	83.1–86.4	Hanihara, 1959

vertical and transverse diameters of humerus head, epicondylar width, and articular width (distal) for the African Americans and European Americans in the Terry Collection that yielded percent classification accuracies between 85.5% (epicondylar width) and 93.5% (articular width). Maximum humeral head diameters from the FDB range between 86 and 93% accuracy. Other large and small bones of the appendicular skeleton have been measured, providing a wide range of accuracy determinations, as seen earlier in Box 5.12. These underscore the population-specific nature of sexual dimorphism beyond the pelvis and a few articular surfaces.

Cranial sexual dimorphism varies by population and is not as accurate in sex diagnosis as estimates based on the pelvic bones. Most metric algorithms are based on discriminant functions, beginning with Giles and Elliot's seminal contributions (1962, 1963; also Giles, 1964). Correct classifications for the skull typically range between 80 and 90% when the unknown is known to be morphologically similar to the reference sample. Measurements of other parts of the skeleton show varying levels of accuracy, frequently population dependent. Tests in the FDB report accuracy rates between 80 and 90%, with some exceeding 90% (France, 1998). Certain commonalities in dimorphism patterns across African and European Americans have been observed in the FDB, suggesting that ancestry may be less important for some dimensions (e.g., height and breath of the scapula) than for others (France, 1998). Dimorphism standards developed in North American forensic contexts for the clavicle, however, perform poorly in contemporary Guatemala (Ríos Frutos, 2002).

A secular trend in cranial dimorphism has been reported by Bennett (1993, see also Ousley and Jantz, 1998). Bennett observed that the Giles and Elliott (1963) discriminant function assigns female sex for all of Jeffrey Dahmer's late-twentieth-century African American male victims. A function derived from the FDB provides a correct classification. This pattern may also reflect ancestry in different parts of Africa or other genetic difference, perhaps due to admixture.

Box 5.13 presents a case study example that develops a **likelihood ratio** for estimating the probability that a decedent was indeed a male, using the Phenice features. This type of statistical approach has the advantage of providing probability estimates that can be combined with likelihood ratios from other aspects of the biological profile to provide a unified probability that identification is correct.

We close this section by once again emphasizing the population-specific nature of sexual dimorphism in bones other than the pelvis and a few long bone articular regions, and we therefore urge caution in the application of discriminant functions generated on a population sample far removed in ancestry. This applies to the Giles and Elliot equations, models derived from the Forensic Data Base (FDB) and available in the computer package FORDISC, as well as other statistical algorithms. The more universally tested the models are on global databases (Murail et al., 2005; Ousley and Jantz, 2005), the more confident we can be that they are generalizable.

BOX 5.13

Likelihood ratios in estimating sex

A recent article by Steadman et al. (2006a) demonstrates a method for developing probability estimates in sex assessment, which can be linked statistically to other aspects of the biological profile in developing a single figure to represent the probability that the deceased fits the biological profile of a known individual. The authors underscore that one of the most important aspects of the analysis is the choice of an appropriate reference sample, in this case using national population statistics on sex ratios for missing persons. The equation for the likelihood ratio is simply the probability of the correct identification divided by the probability of an incorrect identification.

$$\frac{P(\text{sex}|\text{correct ID})}{P(\text{sex}|\text{wrong ID})} = \frac{P(\text{"M"}|M)}{P(\text{"M"}|M)*P(M') + P(\text{"M"}|F)*P(F)}$$

Thus, the numerator is the probability of assessing sex as male ("M") if the decedent is truly male (M). The denominator is the probability of identifying a male, $P(M)$ times the probability of obtaining "male" ("M") features in a true male plus the probability of getting a female, $P(F)$, times the probability of assessing "male" characteristics incorrectly.

All three Phenice features of the decedent's pelvis were male. Coauthor Konigsberg had previously assessed Phenice features in the Terry Collection, using 432 males and 361 females (Konigsberg et al., 2002). He classified 427 of the actual 432, or 98.84%, of the males as male. Similarly, of actual females, Konigsberg misclassified 7 of 361 (1.94%). Thus, the likelihood ratio becomes $0.9884/[(0.5 \times 0.9884) + (0.5 \times 0.0194)]$ or 1.9615, very near the upper limit of 2.

As the authors emphasize, the real power of the likelihood ratio is the ability to multiply probability across other features of the biological profile. After considering age, stature, and skeletal and dental pathology, the authors conclude that it is more than 3 million times more likely to obtain these features if the decedent is the identified missing person than if the identification is incorrect. This provides convincing evidence that the identification is correct.

THE ESTIMATION OF AGE-AT-DEATH FROM SKELETAL REMAINS

We now turn to the estimation of chronological age-at-death through assessment of developmental and degenerative skeletal changes. As methods appropriate for skeletally immature juvenile remains differ from those appropriate for adult materials, we again divide our discussion into these two age-related categories. Initially, we focus on skeletal and dental attributes that facilitate juvenile assessments. We then turn to those appropriate for adult remains, considering both invasive and noninvasive strategies. As before, we are concerned with the accuracy of each technique, both locally and globally (Box 5.11). We also acknowledge that in general accuracy will decrease as the decedent's age advances, since aging rates vary in a cumulative fashion through a cohort's life span.

Age: Juvenile Remains

Excellent summaries of methods appropriate for estimation of age-at-death in immature remains and the current status of the field appear in Saunders (2000), Scheuer and Black (2000, 2004), B. H. Smith (1991), and Ubelaker (1989). All underscore unresolved methodological issues and biases that make choosing methodologies a difficult matter. In general, however, all agree that age estimates based on the developing dentition are frequently more accurate and generalizable than those made on skeletal maturation. Sources of error include choice and size of reference samples, the use of standards developed from clinical x-rays for examining dry bones or isolated teeth, and the increased variability in aging rate that occurs as a cohort ages. Even among juveniles, those dying young may be aged more accurately than those dying in later childhood.

Two of the most widely used dental methods for estimating age-at-death in juvenile materials are those of Demirjian (1978) and Moorees and colleagues (1963a, 1963b, see also Smith, 1991). Each involves recognizing developmental stages from direct observation or x-rays and then evaluating this maturity score against the reference sample. While Saunders (2000) finds the Moorees method more readily adapted to North American or European samples, clinicians appear to prefer modifications of Demirjian's eight-stage model. Moorees' data apply to mandibular permanent teeth only, though they have the advantage of providing both information about **age of attainment** and mean age within a given stage. Smith (1991) recommends developing a maturity score, based on independent observations of each tooth and then averaging mean values representing the midpoint between the beginning and end of a stage. While this has the advantage of providing a point estimate, without a basis for estimating confidence intervals, it may pretend more accuracy than is warranted. Another way of approaching the issue is to identify the most recent stage attained as a lower limit and define the upper limit as the next stage not yet achieved. From such data both a narrow and a broad encompassing estimate can be generated.

Saunders (2000) accepts Smith's inference that in young children (4–10 years of age); Moorees' (1963a, 1963b) method permits an age estimate to within 2 months of age. Recent tests, however, such as that of the Branch Davidian compound victims (see Box 5.14), suggest that ±1 year is a more reasonable estimate of accuracy (Houck et al., 1996). Recent research using the Demirjian method (Chaillet et al., 2005) based on a multiethnic sample

BOX 5.14

Age estimates of remains from the Branch Davidian compound

As a result of the 1993 siege of the Branch Davidian compound by the FBI and the U.S. Bureau of Alcohol, Tobacco, Firearms, and Explosives, 82 of the church's members lost their lives, including leader David Koresh. Forensic anthropologists were assigned the task of providing biological profiles for 44 sets of remains (27 adults, 17 children), and identity was later established through PCR-based DNA typing. Preservation varied, and some of the remains were fragmented and burned. An important aspect of biological profile development was the estimation of age, which was accomplished by a team of forensic anthropologists who used a suite of standard age estimation techniques, including observation of long bone length, epiphyseal appearance and union, dental development, pubic symphyseal morphology, auricular surface morphology, sternal rib morphology, and degenerative changes. The observers developed a consensus method for establishing age ranges based on indicators appropriate for children or adults, with observations sometimes limited by incomplete remains. An average age range was calculated across all possible indicators. As the remains were subsequently identified, this provides an opportunity for the researchers to evaluate accuracy.

The overall age patterning showed close statistical similarity to true ages, leading to the conclusion that a consensus method utilizing multiple indicators can provide very accurate results. Females, on average, showed higher levels of inaccuracy than males. This was true for both adults and children. The Moorrees, Fanning, and Hunt standards appear, at worst, to provide accuracy of plus or minus one year. Even so, estimates for specific individuals departed notably from their true ages. Three males were underaged by a factor of 12.5 to 13 years, while another was overaged by 13 years. Two females were overaged by 12.5 and 13.5 years. In each of these cases, the remains were charred and incomplete, thus limiting the range of possible applications. Another contributing factor is probably the construction of confidence interval ranges that were much narrower than the methods warranted. None of the methods appeared more misleading than others.

The lessons learned from this exercise include the need to use multiple indicators, to consult with other observers, not to construct overly narrow age ranges—especially in incomplete materials—and to constantly reevaluate one's skill level with knowledge based on success rates in age estimation in documented materials, including forensic cases such as these.

(Houck et al., 1996)

reports a mean accuracy of ± 2.15 years for children ages 2 to 18. The difference between the accuracy estimates may be explained by a number of differences in research design, including the larger age range in the Chaillet samples and its broader diversity. Chaillet and colleagues (2005) reported that, in all groups, females were more developmentally advanced than males. Geographic variation is also reported, with Australians showing the most rapid dental maturity and Koreans the slowest. Importantly for forensic applications, this study develops general models, useful when dealing with an unknown decedent, while also recognizing the increased accuracy possible in restricted geographic settings.

Turning to the use of skeletal dimensions to estimate age-at-death, Scheuer and Black (2004) report methods for estimating age in the developing embryo (first 8 weeks of uterine life) and fetuses, as well as for the postnatal period. Model standards link measurements of "greatest length" via ultrasound to development during the full prenatal period (adapted from O'Rahilly and Müller, 2001). They note that other workers prefer age estimates based on skeletal measurements from radiographs, and others emphasize crown–rump length as the measurement of choice. Warren (1999) develops a radiographic approach, with explicit comparisons of his North American sample with the European fetal remains studied by Fazekas and Kósa (1978).

Weaver (1998) provides a review of fetal skeletal development. He includes the cranium and the postcranial skeleton, ossification centers, and length standards. Ossification centers, their initial appearance, their size and morphology, and ultimate fusion to the adjacent center also figure in Scheuer and Black's (2000, 2004) treatments, which provide extended bone-by-bone discussions and should be a vital reference for anyone estimating age-at-death in immature remains. Scheuer and Black emphasize that standards for initial ossification centers observed in x-rays may be useful in intact but decomposed bodies. These standards are unlikely, however, to prove helpful in skeletonized remains. The morphology and size of centers should be valuable in forensic contexts, as should ultimate times of fusion, especially if sex can be estimated independently.

While chronological age can be estimated by observing ossification centers across the skeleton, the most commonly used standards are those for the hand and wrist. Forensic anthropologists frequently apply either the standards developed by Gruelich and Pyle (1959) or one of several variants of the method proposed by Tanner and colleagues, now most commonly referenced as either TW-2 or TW-3 (1975, 2001). Gruelich and Pyle published an atlas that presents hand–wrist x-rays of children at intervals of 6 months. Though a more elaborate procedure can be followed, normally observers simply utilize a pattern fit approach and assign an age based on the best match. First published in 1950, the Gruelich-Pyle standards are anchored by studies of middle class Euro-American children. In the Tanner-Whitehouse method (TW-2 or TW-3), each bone of the hand and wrist is scored numerically according to stage of development, and a composite skeletal maturity score (SMS) is generated which is then compared to a set of standards. An advantage of the Tanner-Whitehouse methods is that, since skeletal maturity is first assessed independent of skeletal age, scales can be shifted to accommodate global variation in maturation rates (Vignolo et al., 1999). A Web-based version of a digital hand atlas based on large, diverse samples is also promising (Cao et al., 1999)

Scheuer and Black (2000, 2004) extensively document and provide equations for estimating age from long bone length and other skeletal dimensions. These references provide an extensive, valuable guide to human growth and development. While forensic anthropologists generally agree that dental eruption is a better predictor of age-at-death than dental maturation, in the case of childhood deaths of well-nourished children, bone dimensions may provide comparable accuracy (Hoffman, 1979).

Scheuer and Black (2000, 2004) have also summarized an extensive literature on ages of epiphyseal appearance and fusion. In assigning an age range, identifying the most recent maturational event to have occurred as a lower limit and the next that has not yet happened will assist in establishing an appropriate range of values. While it is wise not to pretend to overprecision, age ranges can be narrowed by focusing on a large suite of epiphyses and eliminating outliers. For example, following Scheuer and Black's values for the humerus (as reported in Klepinger, 2006), an unfused humerus head would provide an upper limit of 20 years for males and 17 for females, while a fused distal epiphysis would argue for lower limits of 12 (male) and 11 (female). Observing an unfused medial epicondyle allows the age range to be narrowed to 12–14 (male) and 11–13 (female). If sex is unknown, the appropriate range would be 11 to 14 years.

Recent radiographic studies of epiphyseal fusion have been designed to be maximally useful for forensic anthropological casework. For example, Crowder and Austin (2005) have demonstrated both considerable variation across groups of different ancestries and an apparent secular trend whereby fusion of epiphyses of the distal fibula and tibia may occur as early as 12 years and is virtually complete in all females by age 16 in contemporary samples. They also identify ancestry differences, with Mexican American and African males appearing to present accelerated fusion in comparison to those of European ancestry. Schaefer and Black (2005) report that epiphyseal union occurs earlier among Bosnian males than in the remains studied by McKern and Stewart study (1958), which were those of deceased military personnel from the Korean War era (mid-twentieth century). Obviously, considerable research is needed both locally and globally to develop a coherent picture of appropriate standards for linking developmental and chronological age in a manner that refines current estimates.

While not currently a widely accepted or applied method, age estimation for juveniles based on microstructural dental development has been reported by Fitzgerald and Rose (2000). Dental criteria for estimating postnatal survival, based on identification of the **neonatal line**, have also been proposed (Gustafson, 1966; Skinner and Dupras, 1993; Smith and Avishai, 2005).

Age: Adult Remains

Now we explore a range of methods commonly used in the estimation of age-at-death in adult skeletons. Emphasis is placed on the accuracy and precision of each technique, while also underscoring the necessity of reaching final estimates through a composite approach. Woven through this discussion are critical reviews of methods for estimating age-at-death in both ancient and modern humans that have developed within the field of paleodemography, primarily during the closing decades of the twentieth century.

As Jackes (2000: 418) emphasizes, the estimation of age in adult remains continues to be "an evolving and interesting field of study, not simply the application of standard methods." Compelling critiques leveled in the 1980s (e.g., Bocquet-Appel and Masset, 1982) have forced both forensic anthropologists and bioarchaeologists (paleodemographers) to rethink approaches to estimating age-at-death. Certainly, "standard methods" abound, based on morphological changes in the face of the pubic symphysis , the auricular surface, cranial sutures, rib ends, bone and tooth histology, cartilage ossifications, bone involution, and other degenerative bony changes (for summaries, see Buikstra and Ubelaker, 1994; Cox, 2000a; İşcan, 1989; White and Folkens, 2005). Other, newly proposed techniques, such as observing the acetabulum for features such as the sharpness of the rim and porosity within the fossa, also hold promise (Rissech et al., 2006; Rouge-Maillart et al., 2004), but many challenges remain.

One important issue has to do with choice of standards. Should we be seeking one or a few highly accurate features, or should we employ a wider variety of methods, combining them in a pattern recognition exercise? Which ones should be chosen? Which statistics? How do we generate probability and error estimates? In general, most practitioners agree that multiple attributes should be evaluated in forensic assessments (Baccino et al., 1999; Houck et al., 1996; Lovejoy et al., 1985b; Suchey and Katz, 1998; Ubelaker, 2000b, but see Schmitt et al., 2002). In Box 5.14, we presented an example of such a composite approach applied to the Branch Davidian compound case (Houck et al., 1996). As in the Branch Davidian case study, most forensic anthropologists have their favored set of methods, based on experience. Typically, a core set of attributes are scored: for example, for the pubic symphysis, the auricular surface, the fourth right sternal rib end, and cranial suture closure, and a consensus pattern inferred, as illustrated in Box 5.14. Secondarily, other overall indicators of bone development or degeneration are added to the mix. More specialized, labor-intensive, and expensive methods may be considered when the forensic case warrants such attention.

We divide the following discussion into two sections. The first deals with noninvasive morphological estimation methods such as the core attributes just mentioned. We then turn to the more specialized methods involving histological and chemical analysis. We have placed the Lamendin technique for observing transparency in dental roots along with related features, a noninvasive method, in the second section because it developed historically from a suite of observations that included dental histology.

1. Using Grossly Observable Morphological Attributes for Age Assessment

We turn now to an evaluation of accuracy and precision (when available) for those age-associated features that are grossly observable and do not require bone or tooth destruction. These include the face of the os pubis, the auricular surface, the sternal aspect of the fourth rib, cranial suture closure, acetabulum morphology, and a newly developed technique for mathematically combining observations termed the "transition analysis."

Observing features of the symphyseal face of the os pubis to estimate age-at-death began in the United States with the development of the Todd 10-stage system (Todd, 1920). By the middle of the twentieth century, the method's accuracy had been questioned (Brooks, 1955), leading to an attempt by McKern and Stewart (1958) to develop an approach that scored six stages (0–5) of metamorphosis across three dimensions of variability: the dorsal plateau, the ventral rampart, and rim formation. Recognizing that such standards, developed on the dead from the Korean conflict, did not capture variability in females, Gilbert and McKern (1973) developed a parallel approach for estimating age-at-death in female remains.

Subsequent tests of the Todd, McKern-Stewart, and Gilbert methods in a large sample ($n = 1,225$) of documented remains from the Los Angeles Medical Examiner's Office led to the development of the symphyseal technique of preference for U.S. forensic anthropologists today: the Suchey-Brooks method (Brooks and Suchey, 1990; Suchey and Katz, 1998). Basically a modification of the Todd system, this method changes some of the focal attributes, decreases the number of stages from 10 to 6, and expands age range estimates for each stage. Among other advantages, the Suchey-Brooks method has been well published, with explicitly defined observational criteria, and both photographs and casts available. The large sample also has the advantage of being broadly representative of diversity. An issue raised in critique of the method for establishing biological profiles is that the 95% range for many of the six sex-specific stages is quite broad. Stage 3 for females includes ages 21 to 53 years, for example. One might ask, as did Suchey and Katz (1998), whether the method has any value in forensic contexts. Suchey examined her caseload in Orange County, California, between 1969 and 1995. Of 46 sets of remains, 26 were subsequently identified. All but one had the pubic bone available for identification. Suchey found that pubic observations performed well and that in the cases where her overall age range estimates did not include the decedent ($n = 4$), the ranges had been narrowed through observation of other indicators. This underscores the need for the forensic anthropologist to provide broad age ranges, as appropriate, so that law enforcement does not unduly narrow the search of missing persons data. A depressing fact, however, as noted by Klepinger (2006), is that subsequent tests have suggested that the 95% range, which is not the same as 2 standard deviations owing to the deviation of the sample from a normal distribution, does not encompass sufficient variability to characterize certain populations (see also Schmitt, 2004).

The auricular surface method was developed by Lovejoy et al. (1985a) on archaeological materials and the Todd Collection, wherein a number of ages-at-death are estimated. Over time, the eight stages and relatively narrow age ranges have been judged too constrained, leading to recommendations to collapse certain categories and expand the range of age estimates for each (Osborne et al., 2004). With the suggested modifications in place, this revision mirrors the Suchey-Brooks method's redefinition of the Todd procedures. Thus, Osborne et al. (2004), working with the Terry Collection, have proposed six rather than eight stages, with prediction intervals much larger than those originally proposed. Osborne's work did, however, confirm that for his samples from the Terry and Bass collection, there was no significant sex or ancestry effect. As Klepinger (2006) advises, even though Osborne and colleagues' method yet awaits testing in other samples, it should be considered a method of choice in forensic analyses.

Osborne's results contrast somewhat with those reported by Schmitt et al. (2002), who have sampled a broader ancestral population base that includes African, North American, European, and Asian groups. Using a distinctly different observational system, these authors agree that there is little evidence of sex-specific changes but argue for significant differences across groups of different ancestry, with the Asian samples being markedly distinct. Both Osborne et al. (2004) and Schmitt et al. (2002) report acceptable levels of observer error. Obviously, both methods require testing in additional samples by other researchers before they can be said to have achieved general acceptance.

The medial surface of the (right) fourth rib presents age-associated stages (Loth and Iscan, 1989). At present, the method involves the classification of an individual to one of eight phases, with sex-specific standards for African Americans and Euro-Americans. Loth and Iscan (1989) also report that ribs 3 and 5 conform to the same phase progression and a lack of asymmetry, although a test by Yoder et al. (2001) led these workers to argue for a composite rib score rather than relying on a single rib. As Klepinger (2006) emphasizes, given the small sample sizes in Loth and Iscan's studies, the confidence intervals of these investigators are misleading. We predict that more variability in the relationship between phases and chronological age will be discovered once large samples have been studied.

The observation of cranial sutures in age estimation has received considerable discussion and criticism. Various standards based on endocranial, ectocranial, and facial sutures have been proposed (Mann et al., 1991; Meindl and Lovejoy, 1985; Todd and Lyon, 1924). Each has been critiqued from perspectives of accuracy and precision, with workers frequently attempting to find the most accurate system rather than combining elements from the face and the neurocranium. **Secular trends** are identified as an issue in some studies (Masset, 1989), and error rates are high (Nawrocki, 1998). Nawrocki (1998), following extensive study, opines that once appropriate statistical methods for estimating accuracy have been applied to all skeletal indicators of age, cranial suture closure will not be found to be much less accurate than other methods. He also argues for using multiple areas of the skull, having included ectocranial, endocranial, and palatine sutures in his study. Multivariate statistics are recommended in selecting accurate indicators, in controlling effects of heritage differences, in interpreting test results properly, and in generating appropriate estimates of predictive quality. Klepinger (2006) is less optimistic but endorses Nawrocki's equations as the method of choice, pending further study.

A new method for estimating age-at-death from observations of the acetabulum has been proposed by Rissech et al. (2006). The technique has developed from earlier versions that explored a smaller set of attributes (Rouge-Maillart et al., 2004). Rissech et al. (2006) include seven dimensions of variability located within three regions of the acetabulum: the rim, the outer edge of the acetabular fossa, and the acetabular fossa. The researchers report that their results for males from the Coimbra sample show strong correlation between these acetabular attributes and age. Excellent accuracy is reported, and differences between known and predicted ages fall within 10 years 89% of the time. For this method to achieve general acceptance by the forensic anthropological community, there must be testing by other observers in additional samples, as Hampton (2005) noted for the earlier Rouge-Maillart et al. (2004) standards.

A noninvasive method for combining age indicators mathematically that has emerged from paleodemographers' concern for methodological rigor is the "transition analysis" developed by Boldsen and coworkers (2002). Based on a highly replicable set of developmental stage observations in familiar locations—the pubic symphysis, the auricular region, and cranial sutures— the method develops statistical (likelihood) estimates of age-at-death, along with associated confidence intervals. Assumptions concerning the age-at-death distribution from which an unknown is drawn are central to the method, though a much less thorny problem for forensic anthropologists than for paleodemographers. Limited-scale testing suggests that the transition analysis performs well in older adults (Boldsen et al., 2002; Buikstra et al., 2005). It also avoids the reference sample mimicry issue, whereby aging methods in a population sample tend to replicate the age distribution of the sample on which the method was developed (Bocquet-Appel and Masset, 1982). It also appears to be in line with Jackes' (2000) compelling admonition that researchers who intend to develop standards first free themselves from a preoccupation with the linkage between developmental stage and chronological age. In so doing, they will become able to understand the aging process and, we hope, ultimately to generate better methods for age estimation. Certainly, the ability to develop scientifically sound techniques whereby error estimates can be made is attractive in the face of the *Daubert* challenge.

2. Histological and Chemical Methods for Estimating Age-at-Death

Turning to more specialized, invasive methods, we note that changes observable in the dentition, as Gustafson (1950) recognized, can be used for estimating age-at-death. Gustafson (1950) utilized a suite of six observations, including both macroscopically and microscopically visible features. These include wear, secondary dentin development, level of periodontal attachment (gingival resorption), cementum apposition, root resorption, and root transparency. A four-stage scale was used. While Gustafson emphasized the accuracy of his method, tests by other workers produced uneven results (summarized in Bang, 1989), and his statistical approach has been criticized (Maples and Rice, 1979).

One of Gustafson's macroscopically observable indicators, root transparency, was brought to the attention of the forensic anthropological community by Lamendin and colleagues (1992). The observation of root transparency has the advantage of being nonintrusive and simple to use. Lamendin et al. (1992) combined observations of root transparency and periodontosis, both being measured on the labial/buccal surfaces of single-rooted teeth. The formula proposed by Lamedin worked best for individuals between 40 and 80 years of age, with a mean error (average difference between actual and estimated age) of 8.9 ± 2.2 (1 standard deviation) years, lower than a revised Gustafson method's figure of 14.2 ± 3.4. More recently, Prince and Ubelaker (2002) have revised the technique. The authors have increased accuracy by developing sex- and ancestry-specific equations, with mean errors between 6.24 ± 4.97 and 9.19 ± 7.17 years. If 2 standard deviations is considered, the accuracy of this method appears to be in line with other indicators of adult age. It is, however, limited by large inter- and intraobserver errors (Klepinger, 2006; Prince and Ubelaker, 2002; Soomer et al., 2003). Desirable attributes, however, include the simplicity of the technique and the possibility of estimating age-at-death in the older adult years.

Another dental histological approach involves counting **cemental annuli**, which are circumferential depositions of cementum on the tooth root—a phenomenon common to all mammals. Jackes (2000) believes that this approach holds promise, as do Charles et al. (1989) and Wittwer-Backhofen and Buba (2002), although others are less enthusiastic (Renz and Radlanski, 2006). Further external validation is needed before cemental annuli counts are ready to meet the *Daubert* criteria of general acceptance.

A further destructive method for estimating age-at-death involves bone **histomorphometry** or the assessment of structures visualized in histological sections. In Box 5.15, in a discussion of the principles of histomorphometry in age assessment, we note that various methods of this type exist for age estimation. In general, histomorphometric methods report accuracies similar to those of other techniques for estimating age-at-death in adults, many of which require much less equipment, time, or expertise. Even so, in situations of extreme fragmentation and/or incomplete remains, histomorphometry may produce results where other methods fail.

Researchers who use bone histological methods are sensitive to the need to be minimally invasive, preferably maintaining the longitudinal integrity of the bone (Chan et al., 2007). Accuracy may be compromised, however, due to restricted sampling location(s) (Pfeiffer et al., 1995). One of the least intrusive methods is that proposed by Thompson (1978, 1979), who estimated age-at-death based on observations of histological sections drawn from bone

---

BOX 5.15

Bone modeling and remodeling

During the preadult years, long bones grow in length and breadth, assuming a form that is sculpted by a process called **modeling**. During modeling, bone is added to certain surfaces and subtracted to others, permitting bone to drift and adjust within tissue space. This process commonly affixes primary lamellar bone in parallel sheets to circumferential and endosteal surfaces. Once skeletal maturity has been achieved, a different process that is both subtractive and additive, termed **remodeling**, assumes primacy. Bone remodeling follows an activation-resorption-formation sequence that creates discrete bone structural units that are the basis for most bone histomorphometry. Within bone cortices, these bone structural units (BSUs) are secondary osteons. Ratios of secondary osteons with their **Haversian canals** at center to counts of earlier **non-Haversian structures** developed during the modeling phase form the basis of one histological method for predicting age. Another relies on ratios of secondary osteons to osteons cut by the newly forming BSUs. Other workers have counted secondary osteon lamellae, and yet others have looked at the ratios of smaller osteons (type II) that are embedded in the larger type I osteonal structures. Alternatives to methods that characterize osteons are those that simply characterize a given microscopic field in terms of remodeled bone, that is, secondary osteons. While sex, pathology, ancestry, levels of physical activity, and other factors, influence remodeling rates, various authors report accuracy in age estimation similar to and in some cases better than that for other estimators of adult age. Even so, practitioners should be sensitive to the degree to which standards developed on healthy individuals are appropriate for decedents who have suffered from chronic disease or nutritional stress (Paine and Brenton, 2006).

(after Robling and Stout, 2000)

cores taken from upper and lower limb long bone perimeters. Thompson's regression equations produced standard errors for osteon area from the left humerus (6.21) and the male right femur (6.41). In a test on eight forensic cases, he reported differences between the estimated age and actual age of between 1 and 5 years (Thompson, 1979). Thompson's method also performed favorably in a much later test (Crowder, 2005). Chan et al. (2007) conclude that the anterior diaphysis can be sampled in any location, rather than requiring the researcher to sample at midshaft, as in the Thompson method. Thus, fragmentary remains can be used, provided an anterior location can be verified.

An invasive chemical method for estimating age-at-death involves the rate of amino acid racemization in bones, cartilage, and teeth. The potential of this method has been recognized for over three decades (Helfman and Bada, 1976) and has performed well in limited-scale tests (Ogino et al., 1985; Ohtani et al., 2002, 2005). Issues of expense and expertise suggest that this assay should be applied only when special circumstances warrant.

ANCESTRY

Studying inherited attributes and estimating ancestry has been a contentious issue, leading some detractors to dismiss the whole of forensic anthropology as "racist" and as a force that has acted against the best interests of physical anthropology. Such attributions are unfortunate; they are unfair characterizations of a field whose practitioners alleviate considerable emotional pain through the application of their craft. Within regions, members of groups are frequently classified through folk taxonomies. To the degree that these taxonomies are reflected in measurable differences in skeletal structure, then forensic classifications of socially constructed groups are valid. Recent forensic examples of attempts to create regionally valid ancestry algorithms include Ross' (2002) study of Cuban Americans in South Florida. In another context, Ousley and Jantz (2002: 121) address the issue of "race":

> Depending on the methodology, differences may or may not be found between different populations. Sometimes these populations may correspond to social races, or language, geography, or other subjective criteria. In fact, one would expect to find differences between reproductively and geographically separated populations due to drift and selection. As Howells (1995) said, "There are no races, only populations." "Race" can and should be used in its social context.

While the study of physical differences has considerable time depth in American anthropology, we concentrate here on the methods most commonly used by forensic anthropologists today in assessing ancestry. By far the most commonly used is the FORDISC computer program, now available in version 3.0 (Ousley and Jantz, 2005; see also Ousley and Jantz 1998, 1996). FORDISC has been developed in the context of the FDB discussed in

Chapter 1, where data for nearly 1,600 documented individuals serve as the basis for discriminant functions. While the database is dominated by men and women of European and African descent, persons with Native American, Hispanic, and East Asian heritage are represented. Individuals may also be classified in reference to the Howells database (Howells, 1973, 1989), assembled from globally distributed archaeological samples. A source of ambiguity in this and other approaches to the estimation of ancestry is that references samples may represent a mix of biological, ethnic, and national groups. This is clearly the case in FORDISC 3.0, where an argument can be made that African Americans and Euro-Americans are largely biological categories, Hispanics are an ethnic grouping without necessary genetic coherence, and national samples from Vietnam, Japan, China, and Guatemala represent assemblages containing distinctive ancestries and ethnicities.

FORDISC 3.0 employs an interactive discriminant function approach to classifying unknown crania based on documented samples, employing both measurements and angles. Two to 11 groups may be used in craniometric analysis, while postcranial investigation is limited to 2 to 4; as many as 39 measurements may be used. Both posterior and typicality probabilities provide information in interpreting estimated group membership.

In applying FORDISC to evaluate remains that may be from a group not well represented in the database, it is important to exert caution. Recent applications to a Spanish sample dated between 1500 and 1700 produced results difficult to interpret in terms of ancestry (Ubelaker et al., 2002a). An attempt to classify a Nubian Meroitic sample was similarly disappointing (Williams et al., 2005).

Until recently, the assessment of ancestry through the use of the attributes termed **morphoscopic traits** (morphological or nonmetric features recorded as categorical data, such as nasal guttering or suture form: Rhine, 1990b) lacked the statistical rigor common to metric approaches. In fact, the low frequency of supposedly diagnostic attributes within racial groups has severely compromised the accuracy of such approaches. Hefner and Ousley (2006), Hefner et al. (in press), and Ousley and Hefner (2005) have been working on this issue, introducing a variety of parametric and nonparametric approaches. Most recently they have applied **concordance analysis** and CAP (**canonical analysis of the principal coordinates**), which yields classification rates of between 84 and 87% for two groups (American whites and blacks), similar to rates for logistic regression and linear discriminant function analysis. This approach appears much more rigorous and in line with the *Daubert* criteria than earlier methods.

The study of ancestry and attribution of race remains a contentious aspect of biological profile development. Ongoing criticism of both metric and morphoscopic approaches—for example, Albanese and Saunders (2006)—can be dispelled only through scientific rigor of the type necessary to meet the *Daubert* criteria.

STATURE ESTIMATION

The estimation of living stature has a long history in both forensic anthropology and bioarchaeology. Until recently, the primary methods used by forensic anthropologists in the United States were based on the regression equations published by Trotter (1970, see also Trotter and Gleser, 1958), which related measured stature from a number of sources to long bone length. While measured statures of military personnel are likely more accurate than self-reported statures on documents such as driver's licenses, issues have been raised concerning the methods by which Trotter measured the tibia (Jantz et al., 1994; Ubelaker, 2000b). In addition, Trotter's samples were of relatively small size and the potential for secular changes in the relationship between limb length and stature exist, suggesting that the FDB should be a more valuable source of models for stature estimates than Trotter's standards, even as reworked statistically by Klepinger and Giles (1998), who include standards for estimates in older adults who lose stature with age.

Ousley (1995) explicitly considers the issue of accuracy for the estimation of forensic stature (FSTAT—usually from documents such as driver's licenses) and measured stature (MSTAT) through analysis of Trotter's data and that from the FDB. He argues that the prediction intervals (PIs) are better methods for estimating precision than standard errors, because they include sample size and have an explicit probability; for example, with a 90% PI, 10% of the predictions are outside the stated range. Ousley finds through reanalysis of Trotter's data that estimates are less precise than previously reported. New regression equations based on the FDB are presented, some of which are more accurate for estimating forensic stature than Trotter's data set. Prediction intervals of ± 3 to 4.5 in. are reported. While such intervals may be only marginally useful in narrowing the range of possible candidates for identification (see Chapter 8 for further discussion of the identification process), they have the virtue of minimizing premature, erroneous exclusions, and they have the virtue of scientific rigor.

In an elaborate statistical treatment of the issue of stature estimation in forensic cases, bioarchaeology, and paleoanthropology, Konigsberg et al. (1998) compare five statistical methods for estimating stature: (1) regression of stature on a long bone length, (2) regression of long bone length on stature followed by solving for stature, (3) major axis regression of stature on long bone length, (4) reduced major axis regression of stature on long bone length, and (5) use of a ratio of long bone to stature. Basing their analysis of the WWII and Terry Collection remains studied by Trotter, along with 19 African Pygmies and the fossil A.L 288–1, a Plio-Pleistocene fossil hominid (Lucy), the authors conclude that in forensic contexts, *if* it can be assumed that the case came from the same stature distribution as the reference sample, regression of stature on bone length(s) is preferable, a matter of choosing a Bayesian approach above a maximum likelihood method. If the assumption cannot be made, then regression of long bone length on stature followed by solving for stature, essentially a maximum likelihood process, should be followed. While the latter carries a heavier burden of increased confidence intervals, it also bears fewer risks of erroneous exclusions.

Practitioners using FORDISC can use various long bone combinations to estimate stature. An inverse calibration approach should be available in the near future (personal communication, S. Ousley, 2006).

OTHER PHYSICAL FEATURES

As emphasized in the introduction to this chapter, a key function of the biological profile in forensic anthropology is as a screening device for missing persons investigations. Commonly, the physical features considered earlier in the chapters are those available on missing persons reports. Occasionally weight is also reported. When tentative IDs have been made, other characteristics may become available, such as occupation, handedness, and in the case of women, parity. In this section we critically evaluate the ability of the forensic anthropologist to provide accurate information on these subjects. We argue that none of the associations between the target information (parity, occupation, handedness, weight) and physical features are sufficiently strong to meet the *Daubert* criteria.

Parturition scars, including both preauricular sulci and dorsal pitting near the pubic symphysis, have been proposed as indicators of both parturition and sex. Judy Suchey's rigorous study of women's skeletons with known parity histories concluded that numbers of full-term pregnancies cannot be estimated accurately from severity of pitting, although extreme pitting increases the probability that the remains represent a female who experienced late-term pregnancy (France, 1998; Suchey et al., 1979). A lack of consensus is also evident in two conflicting studies, the first by Cox and Scott (1992; see also Cox, 2000b), who discovered no relationship between either preauricular sulcus or dorsal pitting of the pubis with parity and dismiss the features as reflecting biomechanical influences in their eighteenth-century British sample. Instead, they report that an extended pubic tubercle is significantly associated with both parity and number of births. Snodgrass and Galloway (2003: 1230), using a portion of the Los Angeles autopsy sample, concluded that pubic tubercle height was related to overall pelvic morphology. While there was a weak correlation of dorsal pitting with parity, especially in younger women, this relationship did not reach the level of accuracy necessary to be useful in forensic applications to the individual.

Skeletons and teeth may reflect changes due to occupational stresses, which can reflect both the frequency and the strength of activity. Bones remodel and change shape under stress, especially if the occupational behavior is introduced during youth. Joints show evidence of osteoarthritis; tendinous or ligamentous periosteal attachments may hypertrophy; and teeth become worn as a result of nonmasticatory activities such as chewing to soften hides or passing flexible fibers across the teeth (Bridges, 1989; Capasso et al., 1999; Hawkey and Merbs, 1995; Kennedy, 1989; Merbs, 1983; Milner and Larsen, 1991). Critical but balanced reviews of these approaches by Jurmain (1999) and Wilczak and Kennedy (1998) emphasize the complex and poorly understood links between specific behaviors and forms of skeletal and, to a lesser extent, dental changes. While studies of occupational stresses may be found within the medical and anthropological literatures, there are presently no accurate methods for predicting occupation for an individual skeleton.

A related issue is the estimation of handedness. Stewart (1976, 1979a) recorded an association of handedness with shoulder joint asymmetry and upper limb length. He associated handedness with disproportionate beveling of the dorsal margin of the glenoid fossa, dorsal inclination of the plane of the glenoid fossa, and relatively long limb bones from the favored side. Schulter-Ellis (1980), in a small sample ($n = 10$) of documented individuals, reported that linear dimensions and the deflection angle of the glenoid fossa were associated with handedness. Asymmetries in clavicles and second metacarpals have been hypothesized to be associated with handedness, while clinical studies suggest that upper limb length and bone density are correlated with handedness (Steele, 2000). Steele (2000) also cautions that, while there is potential to estimate handedness based on the convergence of a number of attributes, such conclusions are confounded by a number of factors, including the fact that left-handed individuals tend to have less consistent patterns of lateralization, and cultural preferences for right-handedness. Attempts to link handedness to jugular foramen asymmetry have been similarly unconvincing (Glassman and Bass, 1986). A novel, recently developed approach by Synstelien and Hamilton (2003) focuses on lateral deviation of spinous processes in vertebrae that anchor the major muscles that move the arm. While preliminary results are promising, the method awaits rigorous tests of external validity. At present, therefore, we must conclude that there exists no well-accepted forensic method for assigning handedness in a probabilistic manner, nor would accurate prediction (if possible) contribute significantly to forensic investigations.

Very few researchers have attempted to estimate body weight from either weights or dimensions of the bony skeleton (see Krogman and Iscan, 1986; Porter, 2002, for useful summaries). Baker and Newman (1957) studied the bones from prisoners of war (Korean conflict) and concluded that weight of femur alone predicted body weight as accurately as the total weight of all bones. They concluded that neither parameter appeared particularly useful. More recent paleoanthropological cross-species comparisons of the relationship between body mass and stature or skeletal dimensions (Aiello and Wood, 1994; McHenry, 1992) renewed enthusiasm for studies focusing strictly on *Homo sapiens*. Ruff (2000, 2005), using bi-iliac breadth and stature in relationship to body mass, reported low prediction error in elite athletes and in groups from high latitudes, and thus stimulated interest in the forensic community. Subsequent tests of cranial dimensions as predictors (Stubblefield, 2003), as well an attempt to replicate Ruff's results in a documented collection (Suskewicz, 2004), have failed to discover predictive accuracy sufficient for forensic purposes.

CONCLUSION

Where will the *Daubert* call for scientific rigor lead osteobiography? Doubtless increased statistical sophistication will continue to emerge. One challenge to forensic anthropologists will be finding the most elegant technique that is scientifically valid and then translating both the method and the results into terms suitable for our reports and testimony. A second likely by-product of the trend toward increased concern about reliability is that practitioners will become increasingly sensitized to the strengths and limitations of their methods in scientific terms. As a result, fewer reports will define elements of the biological profile, for example, age, in terms more precise than our techniques warrant.

When research is designed to meet the *Daubert* challenge, care must be taken to focus on issues whose resolution will be maximally useful to the forensic anthropological practitioner. An overview of research on skeletal sexual dimorphism, for example, identifies a strong trend within the anthropological community to develop independent statistical models for estimating sex in every bone in the body for groups from every world area. While such exercises appear well able to keep anthropologists occupied for the remainder of this century, we argue that, for the advancement of forensic anthropology, such efforts seem somewhat misdirected. While the exploration of human physical diversity is a time-honored tradition in anthropology and anatomy, the exploration of diversity alone appears less useful for forensic anthropology than the exploration of the manner in which we can characterize both global and regional diversity in a probabilistic manner (Box 5.11). As a positive example, we cite the Chaillet et al. (2005) study of the Demirjian method for predicting the relationship between tooth development and chronological age in a worldwide sample. How much precision is lost in juvenile age estimates when ancestry or point of geographic origin is unknown, they ask? How much accuracy is gained when we can specify heritage factors? While the physical anthropologist may be focused on describing and interpreting diversity, the forensic anthropologist's attention should be directed toward establishing the accuracy of his or her standards on both global and local scales. Given today's international mobility, the practitioner must be aware of accuracy loss if ancestry is not known, especially crucial in contexts of mass death investigations such as plane crashes.

We have in this chapter addressed the manner in which a forensic anthropologist should apply methods appropriate to the estimation of attributes commonly considered to be part of the biological profile, such as sex, age-at-death, stature, and ancestry. In so doing, we have offered critical review of current techniques and urged both practitioners and students to be sensitive to issues of accuracy and precision. Another emphasis has been the degree to which these methods may or may not meet the *Daubert* criteria and whether they are useful to law enforcement in screening unknown remains against missing persons reports. In this evaluation, certain topics received less emphasis because they are seldom entered in missing persons reports (parity, handedness) and/or because they simply cannot be accurately predicted from analysis of the bony skeleton (body weight, occupation, parity, handedness).

In the following chapters we continue to explore classes of skeletal observations that are essential aspects of forensic anthropological knowledge. Chapter 6 discusses skeletal pathology, focusing on the identification and interpretation of features that will assist the medicolegal authority in establishing the cause and manner of death, especially evidence of trauma. We also provide a discussion of terminology that will facilitate communication with medical practitioners. In Chapter 7, we identify and link, as appropriate, types of postmortem change in skeletal features to estimates of postmortem interval. In Chapter 8, we return to a core issue for the forensic anthropologist, individual identification. Medicolegal procedures are reviewed, and the role of the physical anthropologist in this process is defined.

CHAPTER 6

Pathology and Trauma Assessment

As stated in Chapter 2, the primary goal of the medicolegal system is determining cause and manner of death. While it is the responsibility of the certifying pathologist to make this determination, it is the role of the forensic anthropologist to assist in identifying and documenting all evidence that can contribute to determining cause and manner, as well as possibly establishing personal identification (discussed in greater detail in Chapter 8). To fulfill this role, anthropologists must be able to competently assess pathological conditions and traumatic injury.

It is of equal importance that anthropologists accurately describe their findings, using language that is precise and appropriate in both medical and legal contexts. The correct use of terms increases communication among disciplines and lends credibility to the expert on the witness stand. As specificity in language is key, in this chapter we begin with an introduction to medicolegal terminology, then provide an overview of pathology, and finally focus on the role of the anthropologist in trauma assessment.

THE LANGUAGE OF PATHOLOGY, ANATOMY, AND MEDICINE

In addition to fluency in legal terminology (Chapter 2), all anthropologists participating in casework and producing reports in forensic contexts must be well versed in the terminology of medicine and pathology. While students beginning courses in osteology and anatomy are often overwhelmed by the scope and volume of material, learning a few basic concepts will structure and organize terminology, rendering it accessible. For example, the typical adult has 206 bones with over 2,000 named parts onto which over 600 muscles insert. Students may resign themselves to memorizing all these terms. In reality, there is no need. By understanding how anatomical structures or pathological conditions are named, the need for rote memorization is dramatically reduced and a common language for communication with medical professionals is greatly enhanced.

There are three basic "languages" of medicine: Terminologia Anatomica (formerly Nomina Anatomica), clinical terms, and common names. **Nomina Anatomica** was the system of anatomical terminology adopted in 1955 by the International Congress of Anatomists, who continued to revise the terminology until 1985. Nomina Anatomica was replaced by **Terminologia Anatomica** in 1998. Produced by the Federative Committee on Anatomical Terminology, Terminologia Anatomica (TA) is a system of anatomic nomenclature consisting of approximately 7,500 terms. Choosing the simplest and most exact terms, with preference given to terms that were descriptive of form or function, TA introduced 1,000 new terms to describe structures not already named in other systems of nomenclature. Although English equivalents are known, only the Latin terms have official status. The Federative Committee on Anatomical Terminology is working on similar nomenclature for terms used in histology, embryology, odontology, and anthropology (Pugh, 2000).

Terms are constructed following a basic formula. The first word describes the fundamental component: for example, *os* for bone, *regio* for anatomical region, or *musculus* for muscle. Subsequent terms are descriptive, identifying the action or function. For example, *musculus abductor hallucis* is the muscle responsible for abducting the big toe of the foot. TA often includes descriptors—the use of these terms automatically indicates pairing. Take, for example, the term *musculus abductor digiti minimi pedis*, which is the muscle that abducts the little toe. *Musculus* indicates that the term identifies a muscle, *abductor* identifies the action of the muscle, *digiti minimi* denotes the anatomical structure the muscle is acting upon (in this case a small digit) and *pedis* indicates that the digit is associated with the foot. The pair of this muscle, *musculus abductor digiti minimi manus*, is an anomalous muscle that performs the identical function to the small finger of the hand.

Pairing is also implied by descriptors that relate to anatomical position. For example, *fissura orbitalis inferior* is the TA for the inferior orbital fissure. The use of the descriptor *inferior* indicates that there is also a *superior* orbital fissure. If there were only one, there would be no need for the descriptor, and the lone structure would be referred to as *fissura orbitalis*. Examples of paired descriptors are provided in Box 6.1.

While TA may appear daunting at first, the uniformity of the naming standards ensures that students of anatomy can quickly learn to predict a structure's name by recognizing the names of basic structures (such as bone and muscle), learning the terms describing action or function (such as abductor or foramen), and applying the appropriate descriptors (such as superior or lateral). Basic terms are provided in the glossary.

Clinical terms are typically derived from two sources: (1) they are the English versions of Terminologia Anatomica or (2) they bear the surname of the individual who first described the structure, condition, or clinical test.

BOX 6.1
Examples of paired descriptors

Inferior—Superior
Medialis—Lateralis
Profundus—Superficialis
Anterior—Posterior
Proximal—Distal
Brevis—Longus
Pedis—Manus
Ascending—Descending

For instance, "intermediate cuneiform" is the clinical term given to one of the tarsal bones, which in Terminologia Anatomica is referred to as *os cuneiforme intermedium*. As an example of the second source, fractures of the distal radius with displacement are clinically referred to as Colles fractures, named after the Irish surgeon Abraham Colles, who first described the condition.

Common names are terms or expressions used by the general public as well as clinicians communicating with lay people. The use of common names should also be considered when presenting testimony in court, particularly when addressing a jury, or when communicating with a decedent's family. Common names can often be technically incorrect, however. For example, the general public may use *cancer* to describe all forms of tumors or unintentional growths, while clinical use of the word is restricted to describing malignant **neoplasms** (aggressive or deleterious abnormal tissue growth or tumor). Despite this, in dealings with those outside the medical profession, the use of common terms provides a basis for communication that is easily understood, if not always clinically precise.

Anthropologists working with pathologists, law enforcement officers, lawyers, and next-of-kin must learn to communicate effectively in all three languages. Choosing the appropriate level of discourse enables you to convey the necessary level of specificity to the medical community without alienating or confusing members of the general public. Examples of terms expressed in all three languages are seen in Box 6.2.

Latin roots and the use of descriptive terms also form the basis of pathological terminology. Diseases, disorders, and conditions are named following a similar set of rules. Clinical names typically reference the doctor who first described the condition: for example, Hurler's syndrome, Parsonage-Turner syndrome, Paget's disease. However, all pathological conditions also have a descriptive name. This name is formed by root words, prefixes, and suffixes that convey specific meanings. Consider, for example, the disease *amyotrophic lateral sclerosis*. Known commonly as Lou Gehrig disease, after the famous

BOX 6.2

Translating terms (plural forms in parentheses)

Terminologia Anatomica	Clinical term	Common name
Os femoris (o. femora)	Femur (femora)	Thigh(s)
Norma lateralis	Lateral view	Side
Os coxa (o. coxae)	Innominate	Hip bone(s)
Tendo calcaneus (t. calcanei)	Calcaneal tendon	Achilles tendon or heel
Circulus arteriosus cerebri	Circle of Willis	Blood vessels in the brain

American baseball player disabled by the malady, and clinically as ALS or Charcot disease, this condition also affects famed scientist Stephen Hawking. ALS is a fatal degenerative disease manifested by progressive weakness and wasting of the muscles innervated by affected neurons. This information is contained within its full clinical name. By partitioning the name into its component parts, the details of the disease are easily understood.

- a—not, without
- myo—muscle
- trophic—resulting from interruption of nerve supply or nutrition
- lateral—side, not affecting the midline
- scler—hardening
- osis—a disease or condition

Additional commonly encountered prefixes, suffixes, and root words are included in the glossary. By understanding the components that make up pathological terminology, students and practitioners can quickly grasp the basics of any disease or condition from its name.

PATHOLOGY: AN OVERVIEW

Understanding pathology begins with recognizing the "norm." The **norm**, or **normal state**, describes any tissue, organ, or organism that is functionally intact, free of defect or disorder, and capable of sustaining its functionality. The definition may also be expanded to include that which is statistically common. For example, the arrangement of vessels in the human heart shown in anatomy textbooks occurs in approximately 75% of the population. The various configurations representing the remaining 25% of the population are considered deviations from the norm.

All abnormalities are not inherently pathological. Deviations from the norm fall into two categories: anomaly and pathology. **Anomaly** is deviation from the norm that is nonlethal and does not significantly decrease the functioning of the individual. Examples of anomalies include polydactyly and retained deciduous teeth. **Pathology** is deviation from the norm that compromises or restricts the functioning of the individual. Examples of pathological conditions include cancer and meningitis.

Etiology is the cause or source of pathology, as well as the science and study of the causes of disease and their modes of operation. Pathological conditions are categorized based on their etiologies. **Congenital** conditions exist at birth. They may be hereditary, or they may occur as a result of external influences during gestation, such as the mother's exposure to a specific toxin. **Acquired** conditions are not inherited but develop as the result of exogenous factors. All forms of trauma, such as gunshot wounds or fractures, as well as lifestyle diseases such as obesity, are acquired. To accurately categorize a pathological condition, it is important to focus on the onset of development, rather than the occurrence of symptoms. Diseases that are present at birth but do not manifest symptoms until later in life are still considered congenital.

While the timing of onset is used in categorizing pathological conditions, so too is the duration of the illness or injury. Pathology is categorized as either chronic or acute. **Chronic** conditions are ongoing or persistent pathological states. The affliction cannot be "cured" but may be controlled or managed. Diabetes, asthma, and persistent back pain are examples of chronic conditions. **Acute** conditions represent a single episode of pathology. The individual recovers and returns to a normative state or dies. Examples of acute conditions include gunshot wounds, fractures, and heart attacks. It is possible to have acute episodes of chronic conditions. An asthma sufferer may experience a severe attack that seriously compromises the respiratory system. In this instance, the individual either receives treatment and returns to the normal (albeit asthmatic) state or dies.

Diseases are pathological conditions characterized by the following criteria: (1) recognized etiologic agent(s), (2) an identifiable group of signs and symptoms, and (3) consistent anatomical alterations. Cancer and diabetes are examples of diseases. **Disorders** are a disturbance of function, structure, or both. Disorders can result from genetic or embryologic failure (congenital) or from exogenous factors such as toxins, injury, or disease (acquired). Anorexia nervosa is an example of a disorder. A **syndrome** is a concurrence of symptoms or a collection of signs associated with any morbid process (Pugh, 2000). "Battered wife" syndrome, carpal tunnel syndrome, and Down's syndrome are examples. A **morbid** process or condition is any state of abnormality or deviance, including disease or pathological condition.

All diseases, disorders, and syndromes are diagnosed or identified based on the presence of symptoms or signs. A hallmark symptom or specific diagnostic used to classify a disease is called **pathognomonic**. While general symptoms such as pain, fever, or nausea may be present in a large number of diseases, a pathognomonic feature is specific to a particular disease or disorder. For example, retinal folds are considered pathognomonic of "shaken baby" syndrome. **Retinal folds** result from the transmission of force through the attachments among the lens, vitreous and retina, causing traction on the retina, which splits and creates the surrounding folds or defects (Lantz et al., 2004). **Sequelae** are morbid conditions following or occurring as a consequence of another condition or event. Infection secondary to fracture and an embolism following surgery are sequelae.

All pathology can be studied at the level of the individual or a population. At the population level, **prevalence** is the frequency of a disease or condition within a specific population at a specific point in time. "One out of every 10,000 live births in America" is a statement of prevalence. **Incidence** is the number of new cases of a specific disease occurring during a specified period of time in a specific population. "Over 125,000 new AIDS cases were reported in the United States in 2006" is a statement of incidence. At the level of the individual, **expression** is the physical presentation or manifestation of a disease or disorder. The degree or severity of an affliction is part of the statement of expression. Describing a patient's acne as mild, moderate, or severe is an example of a statement of expression. At the population level, **penetrance** is the proportion of individuals with a specific genotype who express the characteristic or disease in the phenotype. For example, in a study of Belgian newborns, 0.011% were homozygotic carriers of sickle cell disease, while 0.49% were heterozygous for the disease (Boemer et al., 2006).

Classes of Pathology

Traditionally, five classes of pathology are recognized:

1. Genetic disorder
2. Neoplasm
3. Systemic pathology
4. Environmental pathology
5. Infectious pathology

Although natural deaths are common in forensic investigations, identifying systemic, neoplastic, genetic, and infectious classes of pathology will not be the focus of this chapter. Students interested in learning more about the differential diagnosis of pathological conditions in skeletal remains should review the works of Anton (1988), Aufderheide and Rodriguez-Martin (1998), Jaffe (1958), Mann and Murphy (1990), Ortner (2003), Pear (1974), and Steinbock (1976), among others. Although the literature of paleopathology does address forms of trauma common in the archaeological record, such as fractures, the full spectrum of trauma produced by modern weaponry falls outside the training and experience of most physical anthropologists. Such information, however, is crucial in cases encountered by forensic anthropologists. The remainder of this chapter will therefore concentrate on trauma and its assessment, the injuries of specific interest in forensic anthropology. All forms of trauma fall under the class of environmental pathology.

TRAUMA ASSESSMENT

Trauma assessment is similar to blood spatter analysis. Both represent moments or actions frozen in time. Transient physical forces produce permanent evidence of their passing (Symes et al., 1996). It is the responsibility of the anthropologist to recognize this evidence and identify, describe, and interpret these forces.

Proper use of terminology is crucial in describing all traumatic injuries. While these distinctions may be blurred in other contexts, many terms have a specific legal definition, which means that terms are not necessarily synonymous and cannot be used interchangeably. A forensic anthropologist who fails to use these terms precisely risks being discredited on the witness stand or viewed as imprecise by medical colleagues. Pathologists and anthropologists do not always speak the same language. For example, most anthropologists describe the location of a cranial injury relative to an osteological landmark: "the depressed fracture was inferior to lambda." However, most pathologists describe the same injury as a measurement of distance from **vertex** or **sinciput** (the top of the forehead at the midline). Forensic anthropologists may wish to adopt the terms used by pathologists to communicate with medicolegal colleagues.

A **lesion** is any pathological or traumatic discontinuity of tissue. A **defect** is an imperfection, failure, or absence of tissue. A **wound** is any defect or lesion caused by a weapon. To meet the legal definition of a wound, the trauma must break the full thickness of the skin (Garner, 2001). Other forms of trauma result in an **injury**.

During the course of an autopsy, pathologists and anthropologists identify and evaluate any evidence of trauma. The role of the anthropologist in trauma assessment focuses on identifying and describing trauma, particularly injury affecting the skeleton. Determining lethality or the cause of death is exclusively the responsibility of the forensic pathologist. Trauma is categorized according to its source. The categories are sharp force trauma, blunt force trauma (including **asphyxia**), and gunshot wounds.

Sharp Force Trauma

Sharp force trauma includes chopping, stabbing, slashing, or incising injuries inflicted by a sharp object. When the sharp object is wielded as a weapon, the resulting defects are best described as wounds, provided the full thickness of the skin is broken. Classes of sharp force injuries or wounds include abrasions, lacerations, incisions, stab wounds, chattering, and hesitation wounds.

Abrasions result from the rubbing or scraping of the superficial layer of skin or mucous membrane. In teeth, abrasion is defined as the wear of the tooth structure due to mechanical action other than mastication. Abrasions cannot accurately be described as wounds because they do not penetrate the full thickness of the skin. Bite marks (both human and animal) that do not break the skin are categorized as abrasions.

Lacerations tear or split tissue. They occur when the imposed force exceeds the mechanical strength of the tissues. Lacerations can result from blunt force trauma, as well as from sharp force. **Incisions** divide tissue cleanly. Force is delivered along the plane of a sharp edge, and the resulting defect is directly reflective of its sharpness. The groove or incision created by a cutting tool is called the **kerf** (Figure 6.1). **Stab wounds** result from thrusting or impaling force. Unlike incised wounds, stab wounds result from force delivered along the long axis of a narrow or pointed object (Figure 6.2). **Chopping** wounds, such as those produced by a hatchet, are categorized with stab wounds because they result from similar applications of force (Figure 6.3).

Documentation and descriptions of stab wounds should include the following features:

- Number of wounds
- Anatomical position(s)

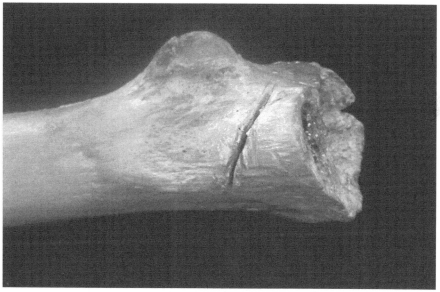

Figure 6.1 Example of an incised sharp force wound on the proximal femur of a dismembered homicide victim. The incised groove is referred to as a kerf. The kerf represents a hesitation wound, indicating that the blade was briefly engaged and then withdrawn (an incomplete action). Note the successful attempt at dismemberment proximal to the hesitation defect. (*Photo courtesy of the New Mexico Office of the Medical Investigator*)

(a)   (b)

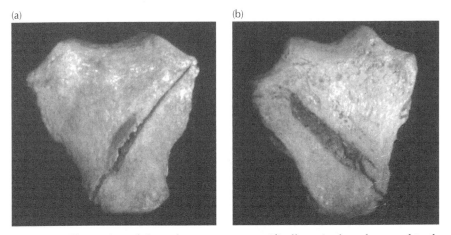

Figure 6.2 Illustration of sharp force trauma, specifically a single stab wound to the manubrium. The reflected bone spur on the anterior aspect **(a)** and the beveling on the posterior aspect **(b)** indicate the blade dimensions and the direction of travel of the weapon, in this case from front to back. The radiating fractures reveal the considerable degree of force with which the wound was inflicted. (*Photo courtesy of the New Mexico Office of the Medical Investigator*)

- Shape of the wound(s)
- Length of the wound(s)
- Depth of the wound(s)
- Direction
- Force

Observations of peripheral evidence, such as defects in clothing or blood spatter at the scene, are also important.

Chattering is defined as a series of discontinuous abrasions or small lacerations occurring when a serrated edge fails to fully engage across the tissue (Figure 6.4). Chattering can result from any serrated tool, including saws, knives, keys, or even bicycle sprockets. Bone is capable of registering chattering defects, although these are most often observed on skin or soft tissue.

Hesitation defects are partial-depth incisions resulting from an incomplete action or attempt to wound. The incision occurs when a weapon engages and then is removed (Figures 6.1 and 6.5). Hesitation marks are often seen in cases of dismemberment as the result of attempts by the assailant to disarticulate the body.

Sharp force injuries are categorized according to the manner by which they occurred: homicidal, nonhomicidal/accidental, defensive, or self-inflicted (Spitz and Fisher, 1980). As with other aspects of death investigation, there is some discrepancy between the approaches used to determine manner

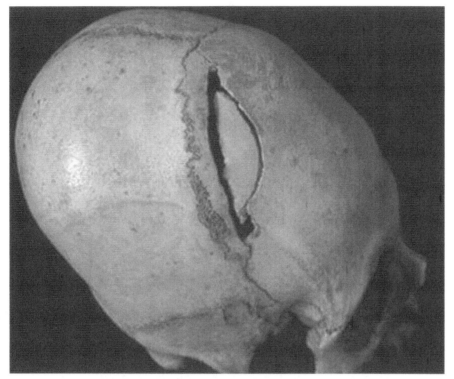

Figure 6.3 Example of a chopping wound to the frontal bone of a prehistoric skull. The wound was caused by a weapon in the ax or hatchet class. The kerf indicates the approximate dimensions of the weapon, while the radiating and concentric fractures suggest the degree of force. (*Photo courtesy of the Maxwell Museum of Anthropology, University of New Mexico*)

by pathologists and by anthropologists. In fleshed bodies, the manner may often be inferred from the location of the defect. For example, defensive wounds tend to occur primarily on the hands and forearms, as victims fend off their assailants. In skeletonized remains, the location of the wounds, coupled with the nature of the defects, may provide clues to the manner. For example, the sharp force defects to the orbit illustrated in Figure 6.6 are consistent with the use of a thin, single-edged blade to remove the eye. The location and bilateral nature of the wounds argue against accidental or defensive injuries. Although such an injury could be self-inflicted in an extreme case, these wounds resulted from postmortem mutilation accompanying a homicide. Absent soft tissue, however, such determinations are often equivocal. Self-inflicted wounds to the arms or wrists can typically be differentiated from defensive injuries by virtue of their location and purposeful arrangement. It is rare for self-inflicted wounds to the arm to impact bone, and thus they are likely underreported in decomposed remains.

(a)

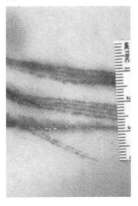

Figure 6.4 (a) Illustration of chattering defects registered on the abdominal skin of a homicide victim. The series of discontinuous abrasions, characteristic of chattering, resulted from the incomplete engagement of a serrated edge across the tissue. **(b)** The key edge used to produce the defect. Any serrated edge is capable of producing chattering. (*Photo courtesy of the New Mexico Office of the Medical Investigator*)

(b)

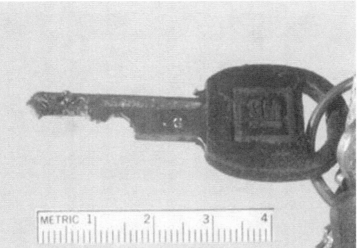

Tool Mark Analysis

Sharp force injuries to bone and other hard tissue offer the potential for tool mark analysis. **Tool marks** are discrete patterns or impressions resulting from contact of a tool or weapon with the impression medium, in this instance bone or tissue. An extensive body of literature details the analysis and interpretation of tool marks on bone. Suggested references include Andahl (1978), Banasr et al. (2003), Bonte (1975), de Gruchey and Rogers (2002), Ernest (1991), Rao and Hart (1983), Rees and Cundy (1969), Silvaram et al. (1977), Springer (1995), Symes (1992), and Symes et al. (1998).

Tool mark analysis examines kerf morphology and residual striae, both subtle characteristics indicating traits of the weapon. Saws, knives, and tools produce measurable traits that aid in narrowing the field of suspect

Figure 6.5 A hesitation sharp force defect on the proximal end of a humerus in a case of homicide with postmortem dismemberment. The hesitation mark indicates that the blade was briefly engaged and then withdrawn. The width of the kerf reflects the width of the saw blade used. (*Photo courtesy of the New Mexico Office of the Medical Investigator*)

weapons. While it is not possible to identify a specific saw or brand name, cut mark analysis can determine the particular class of saw (for more information on individual and class evidence, see Chapter 3). Measurable characteristics include blade dimension, single- or double-edged blade, serrated versus straight edge, manual versus mechanical action (differentiating handheld tools from power tools), and the distribution and arrangement of striae. **Striae** are grooves or defects evident in the surface of the kerf. Striae are produced by teeth as the sharp edge penetrates the bone. All blades have teeth; some may have manufactured serrations, while others are microscopic wear defects (Symes et al., 1998).

Striae are used to establish direction in sharp force trauma. Direction denotes two distinct actions: direction of blade progress and direction of blade stroke. **Blade progress** is the plane of advancement from contact through the termination of the cut. Direction of progress is determined through the overall morphology of the kerf, from hesitation marks indicating the origin of the cut to the break away spur produced by the terminal cut (Figure 6.7). **Blade stroke** is the direction of individual strokes or cutting actions. Direction of stroke produces residual striae on the kerf walls (Symes et al., 1998).

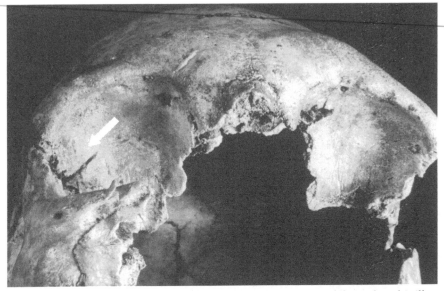

Figure 6.6 This homicidal stab wound to the superior aspect of the right orbit illustrates the interpretation of the manner in which sharp force injuries occurred. The arrow indicates a reflected bone spur resulting from manipulation of the knife as it was extracted. The location of the defect, coupled with the spur indicating extensive manipulation of the weapon, argue against accidental, defensive, or self-inflicted injury. (*Photo courtesy of the New Mexico Office of the Medical Investigator*)

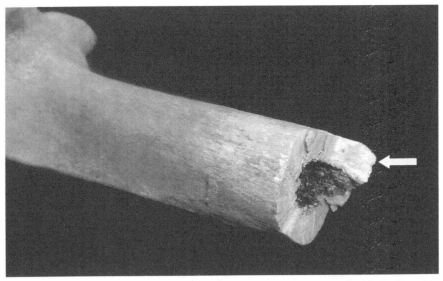

Figure 6.7 In sharp force injury, it is often necessary to determine the blade direction (the path the weapon traveled as it progressed through the element). This humerus was severed postmortem: the handheld blade traveled from bottom left to upper right, as indicated by the direction of the striae evident on the kerf and the breakaway spur (arrow). (*Photo courtesy of the New Mexico Office of the Medical Investigator*)

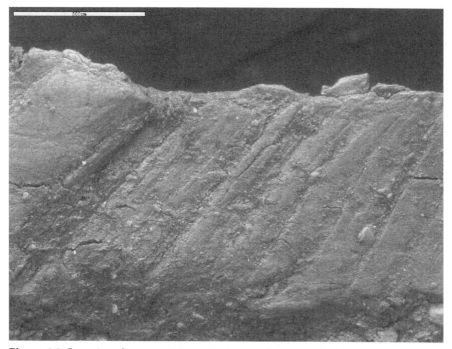

Figure 6.8 Scanning electron microscopic image of a cut mark observed on a cervical vertebra of a homicide victim (magnification 75x). Note the striae, indicating the direction of stroke (the individual cutting action of the blade). The cut was made with a single-edged, nonserrated blade. The irregular spacing of the striae suggests a manual, rather than mechanical, cut. (*Photo courtesy of the New Mexico Office of the Medical Investigator*)

Scanning electron microscopy (SEM) allows a detailed assessment of tool marks registered on bone (Muller-Bolla et al., 2005). The technique is preferable to standard light microscopy for this application in that SEM examines the surface of specimens at much higher levels of magnification and without the need for destructive histological study or preservation. Magnification need not always be greater than light microscopy; with SEM, however, more of the field or object is in focus at one time, allowing for more comprehensive examination. Examples of SEM images showing manual and mechanical forms of sharp force trauma appear in Figures 6.8 and 6.9, respectively.

Blunt Force Trauma

While sharp objects cut and divide tissues, blunt force results in contusions, abrasions, lacerations, and fractures. Although the term *blunt force trauma* conjures images of attacks with lead pipes or other blunt instruments, most blunt force injuries result from vehicular accidents or falls.

Figure 6.9 Scanning electron microscopic image of a cervical vertebra, cut with a mechanical Stryker or autopsy saw (65 x magnification). Note the regular distribution and appearance of the striae, characteristic of mechanical rather than manual cutting action. (*Photo courtesy of the New Mexico Office of the Medical Investigator*)

Description of blunt force trauma involves documenting the resulting fractures and any associated soft tissue injuries (Box 6.3). Classification of fractures includes the following variables:

- Anatomical location of fracture
- Description of the fracture line
- Fracture complexity
- Communication with external environment
- Biomechanical process
- Incomplete fractures

With both single and multiple injuries, description of the trauma must precisely define affected anatomical regions and/or skeletal elements. Terms used to describe the fracture line include transverse, oblique, and spiral. Regardless of the position of the body when examined, or with isolated or fragmentary skeletal remains, the fracture line must be described in reference to standard anatomical position. Fracture complexity is described as either **linear** (a single, simple fracture line) or **comminuted** (crush fractures

BOX 6.3

Fracture terminology

Written reports require proper terminology. Nothing reveals a practitioner's inexperience more readily than a poorly worded description of fractures. These guidelines should assist in describing or documenting fractures.

- The abbreviation for fracture is FX.

- When describing multiple fractures or injuries, begin with the cranium and proceed postcranially in anatomical order.

- Group paired bones such as the radius and ulna or related bones such as the vertebrae and ribs, and describe injuries as a whole.

- Describe a fracture from its point of origin to its termination.

- Recognize that all common fractures have a clinical name—for example, Colles or **Le Fort** fractures—which can be substituted for lengthy descriptions. All practicing pathologists use these names, and all practicing forensic anthropologists are well advised to do so.

- A bone is fractured, not broken. Fractures occur *to* a bone; they should not be described as "on" a bone. Fractures radiate and terminate rather than travel, run, or stop. Fractures intersect; they do not join or meet. Avoid needless adjectives: big, bad, or mangled are common descriptors that have no medical meaning.

- Descriptors such as "hairline," "spider," and "ping-pong" have specific clinical meanings and should be used with caution.

- The best way to document comminuted or multiple fractures is to draw the injuries on a bone diagram. It is not necessary, and may prove needlessly confusing, to label each fracture line and describe it in a written paragraph.

- The object of report writing is to document and describe injuries and events as a whole. A monotonic event should be described in terms of the overall injury. With polytonic events, group injuries and describe them according to sequence, if known. While documenting the full range of trauma is crucial, describing associated fractures individually is unnecessary and may be confusing.

or multiple extensions). Communication with the external environment is described as either **simple**, having no pathway to the outside, or **compound**, with direct communication with the outside, such as a bone projecting through the skin (Figure 6.10). There are some regional differences in terminology, particularly between the United States and England. Alternative terms are *closed* (no communication) and *open* (communication).

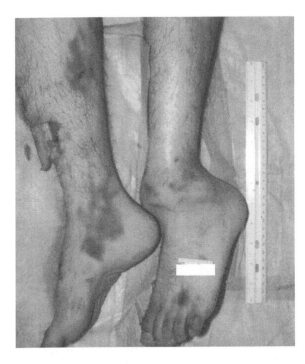

Figure 6.10 Illustration of blunt force trauma involving bilateral, spiral fractures of the distal tibiae and fibulae, with displacement. The individual fell feet first from a significant height. Note the open communication of the fracture of the right leg. (*Photo courtesy of the New Mexico Office of the Medical Investigator*)

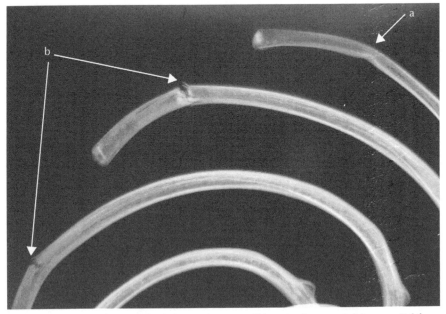

Figure 6.11 X-ray showing both linear fractures **(b)** and a "greenstick" or partial fracture **(a)**. All fractures resulted from the placement of a large boulder on the victim's chest during burial. (*Photo courtesy of the New Mexico Office of the Medical Investigator*)

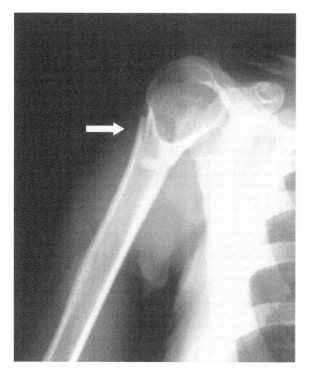

Figure 6.12 Radiographic image showing an impacted fracture of the neck of the right humerus. Note how the lateral aspect of the neck is driven into the diaphysis. *(Photo courtesy of the New Mexico Office of the Medical Investigator)*

Descriptions of biomechanical process detail how the fracture occurred. A **monotonic** injury results from a single traumatic event such as an isolated blow. **Polytonic** injuries are the result of multiple traumatic events, such as a car accident in which the vehicle rolls repeatedly or a pedestrian who is struck by a car, thrown in the air and lands forcibly on the ground. **Repeated loading** fractures include stress fractures produced by repetitive motion. These fractures result from bone fatigue and microtrauma rather than external forces. **Pathological fractures** occur in diseased or weakened bone. Hip fractures in the elderly or those with osteoporosis are examples of pathological fractures.

Not all fractures are considered complete, resulting in linear or comminuted fracture lines. Examples of incomplete fractures include **infractions**, or partial thickness defects, **"greenstick"** injuries (Figure 6.11), and **impacted fractures**, in which the diaphysis is driven into the metaphysis or epiphysis (Figure 6.12). Incomplete fractures are most often seen in juveniles, whose bones are pliable and capable of sustaining applied force without breaking.

Internal and External Factors in Fractures

Internal factors affecting bone fractures are variables inherent in the bone itself. How skeletal elements react to force is determined by the bone's energy absorption capacity, as described by Young's modulus of elasticity, as well as its density and fatigue strength.

Young's modulus of elasticity gives the stiffness of a material, in this case bone, during the elastic portion of its stress–strain curve. The slope of that curve is the modulus of elasticity. The speed at which the stress occurs is defined as the **loading rate**. Slow loading rates are typically seen in blunt force trauma. In **slow loading**, the bone does not fracture until it has passed through the elastic phase of the curve. This can result in simple or linear fracture patterns. Fast or **rapid loading** occurs when high-magnitude stress delivers enormous energy to the bone for a short duration. During fast loading events, bone responds like brittle material. As a result, fracture patterns in fast loading rates are complex and comminuted (Symes et al., 1996). Rapid loading rates are typically seen in gunshot wounds. Fracture mechanics specific to gunshot wounds will be addressed separately in the next section.

External factors are variables that originate outside the bone and describe the force or stress that cause the fracture. Extrinsic factors include the following:

- Magnitude
- Duration
- Load rate
- Direction
- Pulse shape
- Load type

The **magnitude** of the stress is a statement of the amount of force applied to the bone. In blunt force trauma, the spectrum of magnitude ranges from a slap in the face to being hit head-on by a fast-moving train. Magnitude describes the amount of energy being transferred from the blunt object to the body. **Duration** is the length of time the energy is applied. Loading rate encompasses both magnitude and duration. High magnitude for short duration (fast loading) produces complex fractures, while lower magnitudes of force for longer durations (slow loading) result in simple fractures. **Direction** indicates the source of the force relative to the body. **Pulse shape** describes the area over which force is applied. It is essentially the size of the blunt object—recall the example of the slap in the face versus being struck by a train. **Load type** is described as direct, indirect (in which the energy is transferred or diffused), or penetrating. Penetration mechanics will be discussed further in the section on gunshot wounds.

All fractures result from stress following the "weakest link" theory. Fractures originate where local stress first exceeds local strength (Wulpi, 1985). Bone is a composite material of fiber (collagen) and matrix (minerals including calcium). The composite nature of bone produces its viscoelastic properties—the collagen fiber is elastic, while the mineral matrix is viscous.

Viscoelastic materials deform at a rate proportional to the load. The elastic fiber determines the limit of maximum deformity before fracture. The properties of the viscous matrix determine the length of time until fracture (Symes et al., 1996).

Bone is exponentially stronger in compression than tension (Figure 6.13). When a bone bends, it will fail first on the tension side, and the fracture will progress toward the compression side (Symes et al., 1996). In long bones, the secondary fracture radiates toward the tension side; the breakaway spur is on the compression side (Figure 6.14). The spur will occur on the fixed or stable portion of the bone, and the corresponding notch will be found on the mobile portion of bone.

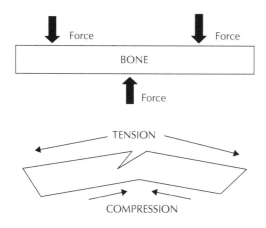

Figure 6.13 Schematic illustration of the response of tubular bone to force and the "weakest link" theory. As the bone bends, the fracture begins on the tension side and radiates toward the compression side. (After Symes et al., 1996)

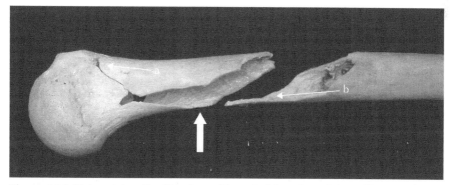

Figure 6.14 Determining the direction of force in blunt force injury, illustrated with a complete spiral fracture of the proximal humerus. The block arrow identifies the direction of force, as indicated by the secondary fracture (a) and spur (b). In long bones, the secondary fractures radiate toward the tension side and the breakaway spur is on the compression side. (*Photo courtesy of the New Mexico Office of the Medical Investigator*)

Asphyxia

Asphyxia is a special form of blunt force trauma, often called blunt neck trauma. Unconsciousness or death is caused by compression of either the airway or major vessels of the neck (Line et al., 1985). Strangulation can result from hanging, and **throttling** or **garrotting** (forms of **ligature** strangulation, involving a rope or other weapon). Manual strangulation or chokeholds (in which the compressive instrument is a body part such as hands, arms, or legs) also can lead to death.

Examinations of neck structures by pathologists reveal hyoid and laryngotracheal fractures, as well as hemorrhages, internal soft tissue injury, and **petechiae** (small hemorrhages of the conjunctive, facial, or periorbital areas). In skeletonized remains, the only definitive clue may be the retained ligature. Much attention has been paid to the frequency of hyoid bone or laryngotracheal cartilage fractures in cases of strangulation (see, e.g., Ubelaker, 1992). Frequencies of such fractures have been reported to range from 14% (Maxeiner, 1998) to 26% (Luke et al., 1985) to 45% (Simonsen, 1988). Hyoid fracture has been shown to be age dependent (Pollanen and Chiasson, 1996; Simonsen, 1988), with fracture frequency increasing with age. Other variables influencing fracture frequency include the nature and magnitude of force applied, the type of constrictive agent (ligature or manual), and intrinsic anatomic features of the hyoid bone such as size and angle (Pollanen and Chiasson, 1996). While postmortem detection of hyoid fracture is relevant to the diagnosis of strangulation, the absence of this finding does not exclude strangulation as the cause of death (Pollanen and Chiasson, 1996).

Although both sharp and blunt force wounds can result from a variety of weapons, the next section deals with wounds resulting from weapons of a specific type—guns. It is impossible to accurately interpret gunshot wounds without first understanding the instruments that cause them (Box 6.4). An exceptional review of guns and gunshot wounds is found in DiMaio (1999). Knowledge of the basic principles of firearms and **ballistics** (Box 6.5) enables anthropologists to better identify and interpret gunshot wounds.

BOX 6.4

Guns: an overview

A **gun** is any device using either pressurized gas or mechanical energy to force a projectile through a tube and expel it out the end. A **firearm** is a gun that uses a spark or flame to ignite a powder charge that releases gas as it burns. The gas creates pressure behind the projectile, forcing it through and out the tube.

Firearms are categorized into two types: handguns and long guns. Handguns are further categorized into revolvers and pistols. **Revolvers** are handguns characterized by a revolving cylinder containing numerous firing chambers; each chamber can be loaded with a bullet, and these can be successively discharged through a single barrel. A **pistol** is a self-loading handgun, the ammunition for which is contained in a removable, spring-loaded magazine housed in the grip. Long gun categories are rifles and shotguns. **Rifles** are firearms with rifled barrels designed to be fired from the shoulder. A **shotgun** is a weapon intended to be fired from the shoulder, but its smooth bore (barrel) is designed to fire multiple pellets. Regardless of type, all firearms have four basic parts: barrel, chamber, breech mechanism, and firing mechanism.

All handguns and most long guns, including rifles and machine guns, have rifled barrels. **Rifling** consists of spiral grooves carved the length of the interior barrel of the gun. The purpose of rifling is to impart rotational spin to the bullet as it leaves the barrel. Much like the gyroscopic action on a well-thrown football, rifling provides greater stability and accuracy. Rifling includes the incised grooves and the metal remaining between the grooves, referred to as **lands**.

Shotguns are normally not rifled but are manufactured with a choke in the barrel. A **choke** is a constriction of the barrel at the muzzle that concentrates the pellets or shot prior to exiting the gun. Shotguns lacking a choke are referred to as cylinder-bore barrel guns. Individuals may choose to remove the choke, producing what are commonly known as "sawed-off shotguns." Removing the choke serves to expand the spread of shot leaving the barrel, resulting in a loss of accuracy but an increase in the likelihood of striking and killing anything directly in front of the shooter.

(DiMaio, 1999)

BOX 6.5

Ballistics

Ballistics is the study of a projectile in flight. It is divided into three categories: internal, external, and terminal ballistics. **Internal ballistics** examines the projectile as it travels through the gun barrel. **External ballistics** focuses on the projectile's flight from the gun muzzle to a target. **Terminal ballistics** deals with the path of the projectile through the target. **Trajectory** is the path of a projectile through the air. Describing the trajectory of a projectile is part of external ballistics. Once the bullet strikes an object, describing its path becomes part of terminal ballistics. The path of a projectile through a body or tissue is no longer described as trajectory, but as the **wound track**.

Several factors determine the behavior of a projectile in terminal ballistics: velocity, bullet type, distance, yaw, and tissue type. When a projectile passes through a body, it transfers its kinetic energy to the surrounding tissue, creating temporary cavitation and a permanent wound track. **Cavitation** is the transient expansion of tissues in response to the transfer of energy from the bullet to the tissue. The degree of cavitation is dependent on the amount of kinetic energy the bullet imparts on impact, the size of the bullet, and the elasticity of the tissue. The kinetic energy possessed by a bullet is determined by its weight and velocity. Although some students might believe it is the bullet that causes the lethal injury in gunshot wounds, it is the cavitation and its devastating effect on the surrounding tissues that often proves fatal. While muscle and skin can sometimes accommodate the tremendous expansion created by cavitation, bone is less capable of such change and responds by fracturing or shattering.

Yaw is the deviation of the long axis of the bullet from its line of flight. With increasing distance, the bullet becomes unstable and begins to wobble, or yaw. The greater the angle or degree of yaw on impact, the greater the loss of kinetic energy. Energy lost during flight means less energy transferred on impact, resulting in less cavitation.

(DiMaio, 1999)

GUNSHOT WOUND INTERPRETATION

Two types of gunshot wound are recognized: penetrating and perforating. **Penetrating wounds** occur when the bullet enters an object but does not exit. With **perforating wounds**, the bullet enters and exits the object. Defining wound type is relative; it is possible for a wound to be both penetrating and perforating. For example, a bullet that passes through the thorax but remains lodged in the spinal column is penetrating to the body but perforating to the lungs. Gunshot wounds are classified as typical and atypical.

Typical Gunshot Wound Morphology

Descriptions of typical gunshot wound morphology include the following characteristics:

- Anatomic location
- Number (single or multiple)
- Range of fire (contact/near contact, intermediate, long)
- Direction
- Wound morphology (penetrating, perforating)
- Wound size
- Caliber estimation, if possible or required
- Bullet type (jacketed, talon, hollow point; Box 6.6)
- Wound track

Range of fire is the distance between the muzzle of the weapon and the object. With **contact** gunshot wounds (sometimes referred to as "hard" contact), the muzzle is in full contact with the skin at the time of discharge.

BOX 6.6

Ammunition

In the United States, the **caliber** of a gun was originally determined by the diameter of the bore, measured between lands. In actuality, caliber can be defined in terms of bullet, land, or groove diameter. European standards are more consistent, using the metric system and a combination of bullet diameter and case length to designate cartridge type. The designation "magnum" describes a cartridge that is larger, producing higher velocity than standard cartridges.

Firearm ammunition comes in two basic forms: cartridges and shot shells. **Cartridges** consist of a casing, a primer, propellant or gunpowder, and the bullet or projectile. Casings, typically made of brass, expand to seal the chamber to prevent the rearward escape of gases when the cartridge is fired. A primer is a shock-sensitive priming mixture used to ignite the gunpowder. Gunpowder may be explosive **black powder**, a mix of potassium nitrate, sulfur, and charcoal, or **smokeless powder**, a highly flammable solid formed through chemical reaction. A powder's **burning rate** is a measure of the speed at which the burning propellant changes to gas.

The most important component of the cartridge is the projectile or bullet. Common handgun variations include the following:

- *Hollow points*, with nose cavities to increase expansion on impact
- *Soft points*, with noses of exposed lead that expand on impact
- *Full* or *total metal jackets*, in which the core of the bullet is encased in a metal covering or jacket
- *Black Talon* or *armor-piercing* projectiles designed to penetrate body armor and maximize lethal injury

Ammunition for long guns or rifles is generally longer than handgun cartridges. Bullets are pointed and more aerodynamic. The combination of pointed projectiles, more propellant, and long, rifled barrels produces greater velocity, longer range, and increased accuracy of rifles over handguns.

Shot shells have a plastic body with a crimped end and metal head that contains a primer (Figure 6.15). Immediately adjacent to the primer is the gunpowder. Separating the gunpowder from the projectiles or shot is a layer of cotton or fiber called a **wad**. Shot consists of uncoated lead balls, and the wad prevents the heat of the gunpowder from melting the shot. Unlike ammunition for handguns, the shot is not given a caliber designation but rather a shot number. The smaller the shot number, the larger the pellet diameter. Shot size ranges from #12 to 000 Buck. For example, #12 shot is 0.05 inch in diameter, while #5 is 0.12 inch in diameter.

Discharging a gun initiates a sequence of events called a firing cycle. Pulling the trigger releases the firing pin, which strikes the primer in the

> base of the chambered cartridge. The primer ignites the gunpowder, producing large quantities of gas and heat. The gas exerts pressure on the sides of the cartridge and on the base of the projectile. The pressure on the base of the bullet propels it down the barrel. Significant amounts of gas, flame, powder, and soot are ejected out the muzzle, along with the projectile. Depending on the type of gun, the spent cartridge is then manually or automatically ejected from the chamber. When a fresh cartridge is chambered, the cycle can be repeated.
>
> (Di Maio, 1999)

Near contact (also called "loose" contact) indicates the muzzle was not in physical contact with the skin but was within 10 mm. Contact and near-contact wounds are identifiable by the presence of the muzzle stamp on the skin. **Intermediate**-range gunshot wounds occur when the muzzle of the weapon is not in contact with the body at the time of discharge but is close enough to result in **"tattooing,"** or the embedding of powder grains into the skin. Powder tattooing is the hallmark or defining characteristic of intermediate-range wounds. Tattooing is also referred to as **stippling**. In **long-range** gunshot wounds, the sole markings or defects on the skin

Figure 6.15 Intact shotgun shell (left) and the same shell type sectioned to reveal its internal structure (right). **A,** the crimped end; **B,** the shot (4 buck, 27 pellet); **C,** the wad, which prevents the shot from melting as the gun powder burns, and **D,** the base, which contains the granulated gunpowder. *(Photo courtesy of the New Mexico Office of the Medical Investigator)*

are those produced by the bullet. No soot, tattooing, or other indications of propellants will be evident (DiMaio, 1999). For anthropologists and pathologists examining skeletal remains, the hallmark signs that aid in determining range of fire will be absent. Estimating range of fire in such cases may not be possible, or may rely solely on scene reconstruction or soot deposits on recovered clothing.

The bullet's path, or direction, is established by differentiating entrance and exit wounds. Penetration mechanics explain how bone reacts to projectiles. Bullets penetrating bone result in **plug** formation and **spall** production. Plug formation occurs when bone fails to shear in response to strong tensile forces. The bullet literally punches out an entrance, forming a plug, and blows out an exit, producing the spall. Spalling produces the beveling or cone-shaped defect that characterizes exit wounds. The presence of **wipe** may also contribute to determining direction. As soft projectiles penetrate tissue, minute fragments of metal are literally wiped off onto the tissue.

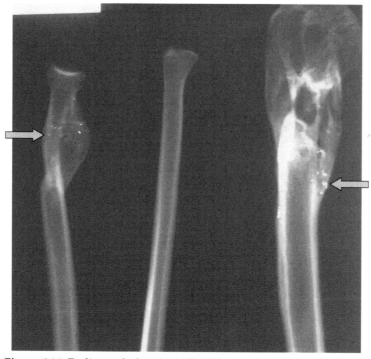

Figure 6.16 Radiograph showing a distal humerus and ulna and the proximal radius of a well-healed gunshot wound to the elbow. The arrows indicate the retained wipe or projectile fragments left behind as the bullet perforated the arm. The pattern of wipe and rearticulation of the elements indicate that the arm was tightly flexed at the time the wound was inflicted. (*Photo courtesy of the New Mexico Office of the Medical Investigator*)

Wipe may be visible with a magnifying lens or through radiography (Figure 6.16). Wipe is concentrated at entrance wounds, becoming progressively less frequent along the wound track.

When describing entrance and exit wounds, it is important to bear in mind the difference between penetrating and perforating wounds. A projectile that penetrates a rib, for example, creates an entrance wound on the lateral aspect of the rib and an exit wound on the medial aspect of the rib (Figure 6.17). Entrance and exit wounds are also described for perforating wounds. In the case illustrated in Figure 6.17, if the bullet had continued through the chest and exited the thorax on the right side, the wound would have been described as perforating the thorax. In this instance, the wound evident on the left rib is described as the entrance, while the defect to the right thorax is the exit. The morphology of the plug and spall, combined with the wound track, allows investigators to determine the direction in which the projectile was traveling and to establish the spatial relationship of

(a)

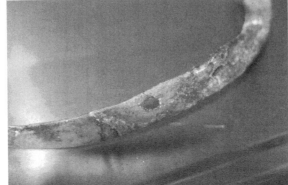

(b)

Figure 6.17 Plug and spall formation and the determination of the direction of a gunshot wound. (**a**) Lateral and (**b**) medial views of a gunshot wound to a left typical rib. The formation of the plug indicates the entrance wound (**a**), and the exit spall or beveling (**b**) indicates the direction of the projectile's travel, in this case from left to right. (*Photos by Debra Komar*)

the shooter to the victim. With entrance defects, the plug and spall are beveled internally; the reverse is true of exit defects.

The ability to estimate caliber from entrance and exit wounds is a matter of considerable debate within forensic pathology (see, e.g., Ross, 1996). Caliber can only be estimated when the bullet is not recovered, as is often the case with skeletal remains. Defects to bone have been argued to be among the better predictors of caliber. Because of the viscoelastic nature of bone, however, projectiles may produce defects in bones that are significantly smaller than their diameters. Bullet diameter is only one of a large number of variables that determine wound morphology. Other important factors include velocity, bullet type, tissue type and thickness, victim age, yaw, and distance.

Atypical Gunshot Wound Morphology

Foreign objects, unusual positions of the body, and shotguns can produce atypical gunshot wounds. Accordingly, special attention must be paid to correctly identify and interpret these wounds. Examples of atypical gunshot wounds include keyhole defects, shored and hidden exits, interposed wounds, and shotgun injuries.

Keyhole Defects

Investigators examining an intact skull notice what appears to be an exit defect but are unable to identify an entrance wound. Although they extract small fragments of projectile from within the skull, they do not recover a bullet. Such circumstances raise the possibility of a "keyhole" or angled defect. **Keyhole defects** are wounds that occur when a bullet strikes the skull at an angle and the projectile fragments. The bulk of the bullet is deflected from the skull, producing a defect that is externally beveled along a portion of its margin (Figure 6.18). The remaining margin of the defect shows internal beveling. The keyhole is both an entrance and exit wound, with only a portion of the projectile entering the skull. Keyhole defects have also been reported in tubular bone (Berryman and Gunther, 2000).

Shored Exits

Sometimes pathologists at autopsy encounter a body with what appears to be two entrance wounds and no exit wounds, and no projectiles are recovered from within the body. This pattern suggests another form of atypical gunshot wound, namely the shored exit. Shored exit wounds strongly mimic the appearance of entrance wounds, in that the defect is rimmed by abrasion. **Shored exits** occur when pressure against the skin, such as from tight clothing or contact with a wall or floor, causes the exiting projectile to abrade the surrounding skin (Aguilar, 1983). Wipe and examination of the wound track, as well as evidence from the scene, aid in identifying shored exits. Anthropologists can also play a role by examining skeletal damage along the

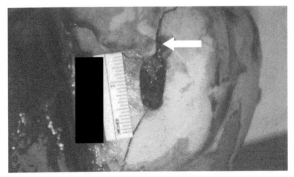

Figure 6.18 An atypical gunshot wound, specifically, a keyhole defect of the skull with radiating fractures. The arrow indicates an area of external beveling on the superior margin of the defect. The bullet struck the skull on the inferior margin of the defect and deflected or "exited" on the superior margin, producing the partial spall or external beveling. Only a small fragment of the projectile was recovered from inside the skull. (*Photo courtesy of the New Mexico Office of the Medical Investigator*)

Figure 6.19 This atypical gunshot wound, a reconstructed "hidden" exit of the lateral cranial vault, was produced by a .50 caliber weapon. Extensive fracturing associated with the entrance provided a pathway for the projectile to exit the skull without plug or spall formation. The pencil tip indicates the actual exit, which corresponded to the defect in the overlying skin. (*Photo courtesy of the New Mexico Office of the Medical Investigator*)

wound track to determine direction. Defects along the wound track should be largely unaffected by the shoring of the exit.

Hidden Exits

Atypical exit wounds also include "hidden" exits. **Hidden exits**, in which no plug or spall is produced, occur when a projectile exits through fractures caused by the entrance wound or other existing fractures (Figure 6.19). Extensive fragmentation, as well as the failure to recover a projectile from within the skull, indicates the potential for a hidden exit. Careful reconstruction of the fragments and examination of the fracture pattern identifies the possible exit.

Interposed Wounds

Intermediary targets also produce atypical wound morphology. The morphology of **interposed** gunshot wounds is difficult to characterize because the projectile's trajectory is altered as the result of passing through one object before striking another. The projectile transfers some kinetic energy to the first object and, depending on bullet type, may deform or separate from its jacket (metal cover). The projectile can also carry fragments of the first object into the second, introducing foreign trace evidence in the wound. The most common intermediary targets are glass, doors, and automobiles.

Shotgun Injuries

Shotgun injuries also differ significantly from typical gunshot wounds due to variability in ammunition types and range-of-fire capabilities. Intraoral shotgun wounds result in massive destruction of the skull from both the projectiles and the associated gas combustion (Figure 6.20). The spread of the shot depends on the presence or absence of a choke and the range of fire. Individual pellets can produce plug and spall formation or may simply embed in bone or other tissues.

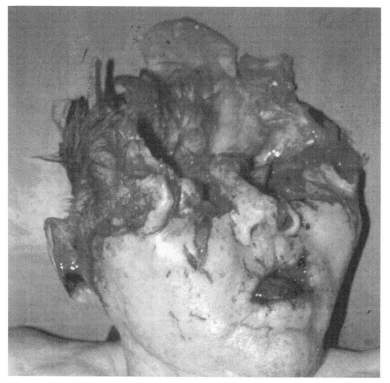

Figure 6.20 The extensive fracturing and tissue loss resulting from an intraoral shotgun wound, considered an atypical gunshot wound. The damage resulted from the expulsion of gas and projectiles from the barrel within the closed confines of the mouth and skull. (*Photo courtesy of the New Mexico Office of the Medical Investigator*)

Anthropologists wishing to gain experience in gunshot wound interpretation would be well advised to observe as many wounds in fleshed bodies as possible. Observing entrance and exit wounds, as well as wound tracks, in soft and hard tissue is invaluable to reconstructing defects in skeletal remains. How bone responds to various projectiles is best understood when the projectile is retained or recovered, as is often the case in fleshed remains. Since small fragments are often not recovered in skeletal remains, studying fragmentation patterns in fleshed remains, as well as through radiographs, provides anthropologists with an opportunity to visualize the effects of penetration trauma.

Determining the Sequence of Injuries in Polytonic Trauma

When trauma is evident, investigators are often called upon to determine the sequence of blows or injuries. Determining the sequence allows investigators to establish the number of blows represented and may allow pathologists to differentiate terminal injuries from nonfatal or postmortem defects.

Establishing the sequence of blows in blunt force trauma requires anthropologists to identify the point of impact and to differentiate concentric

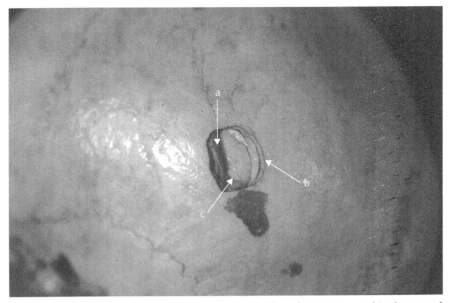

Figure 6.21 Patterns of fractures associated with blunt force trauma; this depressed fracture to the left parietal was caused by a hammer. The arrows indicate the point of impact (**a**), the concentric fractures (**b**), and the radiating fractures (**c**). (*Photo courtesy of the New Mexico Office of the Medical Investigator*)

fractures from radiating fractures. As illustrated in Figure 6.21, the **point of impact** is the point of primary fracture production. **Radiating fractures** disperse from the point of impact. The impact also transmits force in concentric waves, resulting in **concentric fractures** encircling the point of impact.

Sequencing multiple blunt force wounds within an anatomical region relies on the phenomenon of intersecting fractures. Both radiating and concentric fractures terminate when they intersect with an existing fracture. In other words, fractures do not cross other fracture lines. Therefore, wounds with fractures that terminate upon intersection are later in the sequence than those into which they terminate (Figure 6.22).

To sequence multiple gunshot wounds to the same anatomical region, two features must be considered. First, the phenomenon of intersecting fracture lines aids in determining sequence. A radiating fracture terminates when it intersects with an existing fracture. Second, an existing defect and its

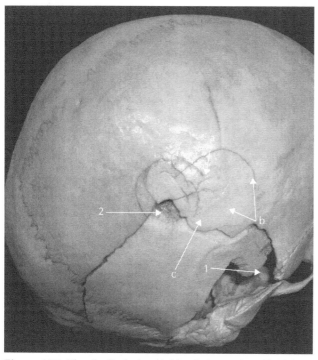

Figure 6.22 The skull of a homicide victim with multiple comminuted fractures illustrates how the sequence of blows in blunt force trauma can be determined. To ascertain the sequence of polytonic injuries, one relies on the phenomenon of intersecting fractures. The arrows indicate concentric fractures (**a**), radiating fractures (**b**), and point of impact (**c**), while the numbers reflect the order of the blows. (*Photo courtesy of the New Mexico Office of the Medical Investigator*)

associated fractures may cause atypical exit morphology in subsequent shots. Once compromised, bone will not respond in the same manner. Figure 6.23 illustrates two adjacent exit wounds to the occipital region of a skull. The first wound produces typical spalling and external beveling. As the bone is now fractured, the second projectile does not produce a typical plug and spall but rather displaces an existing fragment of bone. This results in some external beveling to the ectocranial vault but no plug formation. Such fracture patterns could be misinterpreted as blunt force trauma.

Sequencing multiple sharp force injuries to the same anatomical region proves more difficult. Sharp force wounds often do not produce fractures, depending on the force with which the injury is inflicted. When

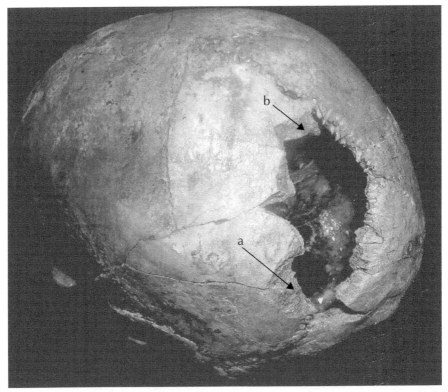

Figure 6.23 Determining the sequence of multiple gunshot wounds to the skull, based on exit wound morphology. Defect **a** occurred prior to defect **b**. Defect **a** is a typical plug-and-spall exit, circular with external beveling, while defect **b** has external beveling but irregular margins. Because the posterior skull was already compromised by fractures associated with defect **a**, the second exiting projectile did not form a plug but rather displaced an existing bone fragment. (*Photo courtesy of the New Mexico Office of the Medical Investigator*)

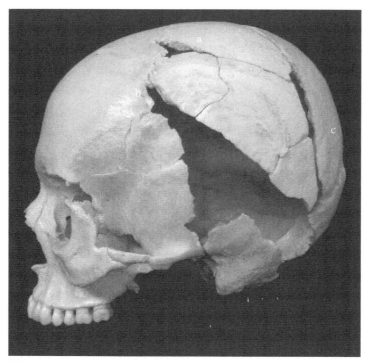

Figure 6.24 Determining the sequence of multiple sharp force wounds inflicted with a small weapon of the hatchet class. Determining the order relies on the phenomenon of intersecting fractures. The sequence of injuries is **c** then **a** then **b**. (*Photo courtesy of the New Mexico Office of the Medical Investigator*)

fracturing occurs, however, the rule regarding intersecting fractures applies (Figure 6.24).

CONCLUSION

As noted in Chapter 1, early practitioners of forensic anthropology often focused on providing biological profiles for skeletonized remains. Today, forensic anthropologists make significant contributions in the assessment of disease and injury at autopsy, regardless of the decompositional state of the remains. While training, research, and experience provide anthropologists with the background to offer such contributions, it is our use of the common language of medicine and pathology that allows us to fully integrate into the medicolegal system. An anthropologist's identification and description of pathological conditions, particularly those affecting the skeleton, should incorporate the classification systems and language used by pathologists and other medical professionals. Written reports of traumatic injuries must employ precise, appropriate terminology, as such reports are often introduced into legal proceedings. When the language is correct, the audience can focus on the ideas presented.

In this chapter, we presented the role of the forensic anthropologist in trauma assessment within the context of a postmortem examination. In Chapter 7, we will continue our discussion of the analysis of remains with issues relating to taphonomic change and time-since-death.

Forensic Taphonomy

Approaches and attitudes in forensic anthropology have changed dramatically in recent years. For example, Stewart (1979a: 31) argued that it was advantageous for anthropologists not to see skeletal remains at the scene or to participate in their recovery; he believed that such exposure to the scene might introduce bias in the determination of biological affinity or other traits. Today, most practitioners would argue that knowledge of the scene and circumstances are crucial in accurately interpreting the evidence. Ultimately, each practicing professional must decide what role he or she wishes to play in forensic investigations. Some responsibilities or skills are considered standard for forensic anthropologists, such as a thorough and competent analysis of remain (Chapters 5 and 6), the recovery of a buried body (Chapter 4), or the identification of decedents (Chapter 8). Other skills, such as time-since-death estimation, are more specialized and require additional training. Regardless, all practitioners should be familiar with the interactions between a body and its decompositional environment.

By focusing on taphonomic analyses, this chapter expands on issues addressed in Chapter 4, which examined the role of the forensic anthropologist at a crime scene. We begin by discussing forensic taphonomy and then examine the five goals of taphonomic analysis in death investigations: time-since-death estimation, differentiating human from animal action, recognizing mechanisms of transport, understanding factors affecting bone preservation, and reconstructing perimortem events.

FORENSIC TAPHONOMY

Taphonomy entails "the laws of burial" (cited in Haglund and Sorg, 1997b: 3). Thus as organisms move from the realm of living organisms (the biosphere) to the earth (the lithosphere), they undergo processes that are called taphonomic. The same processes that ultimately result in the formation of fossils act on human bodies after death. Time, the environment, and human or faunal interaction produce morphological changes in the remains. Interpreting these patterns of change is the basis of taphonomic analysis.

As forensic anthropology has evolved as a discipline, there has been a significant increase in research and literature relating to forensic taphonomy. Two edited volumes by Haglund and Sorg (1997a, 2002) address a wide range of taphonomic issues, as do additional works by Haglund (1991) and Komar (1999a). Much of the literature uses case studies or reviews to document specific processes and their diagnostic characteristics.

Although such discrete studies usefully augment casework, forensic taphonomy remains largely descriptive, rather than theoretical, and relies on actualistic approaches and the theory of uniformitarianism. **Actualistic studies** examine a process through observations of recovered materials or controlled field or laboratory studies. Descriptive data are used to generate diagnostic criteria to identify the process in future casework. Actualism relies on uniformitarian principles. The theory of **uniformitarianism** states that the present is the key to the past. Processes and phenomena observed today are believed to be the same as those seen in the past. As natural taphonomic variables such as weather and environment are immune to the actions of humans, it is argued that the effects of such variables remain constant through time.

Uniformitarianism is largely an accepted theory within paleontology, where it originated. However, its universal application to forensic anthropology remains untested and its validity can be challenged. For example, it is reasonable to argue that the morphological defects of bone produced by canid scavenging are consistent between modern samples and archaeological bone. This reflects the relatively stable dental morphology of canid species over time. However, the onset of scavenging behavior, as well as patterns of disarticulation and dispersal of remains, may differ as a result of changes to canid habitation areas and access to food supplies stemming from human encroachment into previously uninhabited regions. To describe *all* aspects of canid interactions with remains as uniform through time is misleading.

Although it has its supporters, uniformitarianism is not accepted unconditionally as a founding principle of forensic science. For example, there is no expectation that a gun fired today would be identical to the same gun fired two years before. Repeated firing of a gun permanently and progressively alters the weapon. Case-hardening colors, for instance, on a shotgun or rifle receiver are the by-product of a heat-treating process to create a hard wearing surface. The colors are perishable from wear. Subsequent firings literally change the appearance and strength characteristics of a weapon. The same can be said for disease processes in a body. For example, improved detection and treatment options, and increased survivorship, affect the frequency, expression, and distribution of skeletal lesions associated with neoplastic disorders. The current clinical manifestations of the disease do not necessarily mirror such pathological conditions in the archaeological record.

The uncertainties raised by the present lack of generalizing principles should not be seen as grounds for the dismissal of taphonomic analysis but rather as a clear indication that (1) further research is both needed and welcome; (2) practitioners should recognize the assumptions inherent in taphonomic theory and acknowledge such limitations to a jury; and (3) caution in the unrestricted application of uniformitarianism is warranted.

Despite these limitations, taphonomic analysis remains a vital part of death investigation. Identifying taphonomic effects at scenes is crucial to proper interpretation, evidence recognition, and the development of appropriate search strategies. While autopsy focuses on deriving evidence from the body, taphonomic analysis centers on understanding the relationship between the body and its postmortem environment.

THE GOALS OF FORENSIC TAPHONOMY

Forensic taphonomic analysis has five primary goals: (1) to estimate the time-since-death, (2) to distinguish human from nonhuman agents of bone modification, (3) to understand selective transport of remains, (4) to identify variables resulting in differential preservation of bone, and (5) to reconstruct perimortem events and circumstances. Meeting these goals relies on the use of taphonomic models, concepts, and analyses. We will now discuss each goal in greater detail.

Time-Since-Death Estimation

Synonymous with elapsed time, **time-since-death** (TSD) estimation establishes the length of time between the death of an individual and the recovery of the body. Although the term "**postmortem interval**" is also used to describe this period, it is important to recognize that time-since-death estimation identifies the period from death to recovery. As the perimortem period can extend beyond the death of the individual (Chapter 2), the time-since-death estimate may technically include portions of the perimortem period as well.

Investigations require time-since-death estimates for a number of reasons, including the following:

- Establishing a time frame for the decedent, differentiating peri- and postmortem activities, and identifying unaccounted for periods of time
- For identification purposes—establishing TSD can narrow the number of potential missing persons
- As a means of verifying or refuting an accused person's alibi
- Assessing the reliability of toxicology results

BOX 7.1
Decomposition

Decomposition is affected by factors originating inside the body, known as **intrinsic** or **endogenous** factors, or from outside the body, termed **extrinsic** or **exogenous** factors. Gross and chemical decomposition of the body occurs as the result of autolysis and putrefaction. **Autolysis** means "self-digestion." Autolysis occurs through the action of digestive enzymes that break down complex proteins and carbohydrates in the body. The same mechanism that aids in digesting food in life begins to digest the gastrointestinal tract after death. Autolysis is considered a minor component of decomposition. **Putrefaction** results from bacterial activity throughout the body. Endogenous bacteria from the gastrointestinal tract, as well as exogenous bacteria, produce significant chemical changes to the soft tissues. Putrefaction is the major component of decomposition. Putrefaction results in **marbling** (the discoloration of blood vessels and skin), bloating, subcutaneous gas accumulation, and maceration or "gloving." In **gloving** or **maceration**, water is imbibed into the skin, giving the skin the wrinkled appearance commonly seen after prolonged water immersion. Once developed, maceration of the hands and feet can result in skin slippage, such that the skin of the appendages can be removed as a whole.

(Gill-King, 1997; Spitz and Fisher, 1980)

- Safety concerns for investigators, including the viability of infectious diseases or pathogens
- For prosecution purposes, such as establishing negligence or neglect in deaths occurring in prisons or nursing homes

Investigators, including anthropologists, are often called on to estimate TSD at the scene or at autopsy. To accurately estimate TSD, an investigator must first understand how a body changes postmortem, a process known as decomposition (Box 7.1). The ability to establish TSD depends on the materials recovered, the methodologies available, and an investigator's training or familiarity with those methods. Whether an anthropologist participates in TSD estimation depends on training and job responsibilities. Regardless, all anthropologists benefit from an understanding of the methods used to establish elapsed time.

Methods of Time-Since-Death Estimation

After death, a body passes through several stages of decomposition (Box 7.2). While it is tempting to impose a time frame onto each stage that would provide an estimate of TSD, to do so is ill-advised. It is important to differentiate between sequence, which is predictable, and rate, which is not. Although the decomposition of a body follows a known sequence, the rate

BOX 7.2
Stages of decomposition

1. *Fresh* The period from death until the first signs of bloating is characterized by a decrease in body temperature over time and a lack of odor. This stage terminates with the arrival of blow flies and **oviposition** (the deposition of eggs).

2. *Bloat* The bloated appearance of the body, with a distended abdomen as the result of gas accumulation, marks the beginning of this stage. It is characterized by discoloration or marbling of the skin, strong odor, extruded anus, and decreases in body mass. This stage terminates with the development of the maggot mass.

3. *Active decay* This stage is characterized by the presence of sizable maggot masses, strong odor, and a greasy appearance of the soft tissues. The stage terminates with the collapse of the thorax.

4. *Advanced decay* Diagnostic features of this phase include the disappearance of the maggot masses and a marked decrease in body mass. Soft tissue changes are extensive. This stage terminates with the disappearance of beetles and all nondesiccated soft tissue.

5. *Skeletal/dry remains* This is a terminal stage. Only bone, cartilage, and desiccated soft tissue remain. There is no odor or insect activity. Depending on environmental conditions, a body can remain in this stage indefinitely.

(Anderson and VanLaerhoven, 1996; Komar and Beattie, 1998a)

at which the different phases of the sequence occur is highly variable. Many intrinsic and extrinsic factors affect the rate of decay (Box 7.3). It is also difficult to quantify the morphological variation seen within and among stages. For example, how do you define degrees of bloating? Since extrinsic factors render estimates regionally or environmentally specific and intrinsic factors vary by individual, the direct translation of decompositional stages to accurate TSD estimates is not a readily attainable goal.

Three categories of TSD estimates are recognized: short-range, midrange, and long-range. **Short-range estimates** encompass the period from death to 72 hours postmortem. **Midrange estimates** cover the period from 3 days to 30 days postmortem. **Long-range estimates** begin 30 days postmortem and continue indefinitely (Gallois-Montbrun et al., 1988). A variety of methods are available for each category. Anthropologists primarily employ mid- and long-range methods.

Numerous methods of determining TSD have been studied, especially those dealing with the short-range period, while only a few less accurate methods have been proposed for the mid- and long-range periods. "As the

---

BOX 7.3

Intrinsic and extrinsic factors affecting decay rates

- Ambient temperature
- Depositional environment
- Humidity
- Sun versus shade exposure
- Presence of clothing or body covering
- Animal scavenging
- Insect colonization
- Size of the individual
- Presence and type of perimortem trauma

PMI [postmortem interval] increases, the accuracy of its determination necessarily decreases" (Schoenly et al., 1991: 1395). Early critics of TSD estimation methods felt that the estimates generated were flawed and wide-ranging enough to allow for "the farcical possibility of the person still being alive" (Camps, 1959: 78) and that establishing the PMI should be seen more as a matter of estimation rather than determination (Van Den Oever, 1976).

Short-Range Time-Since-Death Estimation

Methods proposed to establish short-range PMI include rectal temperature (Henssge, 1988; Knight, 1986; Marshall and Hoare, 1962), white cell counts (Babapulle and Jayasundera, 1993), vitreous humor creatine levels (Piette, 1989), electrical excitability of skeletal muscle (Madea, 1992), subcutaneous hemoglobin concentrations (Inoue et al., 1994), sweat gland morphology (Cingolani et al., 1994), cell composition of cerebrospinal fluid (Wyler et al., 1994), and free amino acid content (Mel'nikov and Alybaeva, 1995).

Of these, temperature remains the most commonly employed method, although its accuracy is questionable. Body temperature is recorded from probes inserted in the liver or rectum or thermal scanners. Attempts to use nasal or ear temperatures proved highly unsatisfactory (Nokes et al., 1992). Computer programs have been developed to estimate the TSD using decreasing body temperature in both humans (Lynnerup, 1993) and wildlife (Cox et al., 1994). These methods have been described by their authors as being accurate up to 18 hours.

Two formulas have been proposed to estimate TSD based on body temperature: Knight's (1986) and Henssge's (1988). **Knight's formula** allows for a 45-minute plateau after death in which body temperature remains constant, followed by a decrease of 1 degree per hour until ambient temperature is reached. **Henssge's formula** does not recognize a plateau. Temperature begins decreasing immediately after death at a rate of 1.5 degrees per hour until ambient temperature is reached. Both methods require multiple readings of body and ambient temperature over a period of time. Neither method has a high accuracy rate.

Decreasing body temperature, known as **algor mortis**, is considered part of the **triad of early postmortem change**, along with livor mortis and rigor mortis. **Livor mortis** is the purple discoloration of the skin of dependent parts of the body. The purple color results from unoxygenated hemoglobin in the blood. The pattern and distribution of livor mortis results from the settling of blood into the capillaries of the skin as they dilate after circulation ceases. Livor mortis is recognizable approximately 1 hour after death, is well developed after 4 hours and reaches a peak at 8 to 12 hours. After 12 hours, lividity "fixes" or becomes stable. The prefixation sequence can provide an approximate TSD estimate. Even in cases where PMI estimation is not required, field notes describing livor are necessary as it is used to identify changes in the position of the body after death (Spitz and Fisher, 1980).

Rigor mortis provides a less accurate estimate of elapsed time. **Rigor mortis** is the progressive hardening of the muscles as a result of chemical changes within the muscle fibers. Lactic acid and other by-products of tissue metabolism initiate the process. As acid accumulates in the muscles, the protoplasm gels, resulting in muscle rigidity. Rigor begins at death, manifesting in the smaller muscles of the face and hands within 2 to 4 hours, and reaches its peak in the large muscle groups at approximately 12 hours. Rigor normally recedes within 24 hours in the reverse order of its appearance. Rigor has been shown to be highly temperature dependent. The decedent's perimortem level of physical activity also has significant impact on the onset and rate of rigor development (DiMaio and DiMaio, 2001).

Midrange Time-Since-Death Estimation

Midrange postmortem intervals have been investigated by means of soft tissue biochemical studies in animal models (Gallois-Montbrun et al., 1988) but with limited success. Free amino acid content in liver and lung samples of cadavers with a PMI of 2 weeks, measured by using liquid chromatography (Mel'nikov and Alybaeva, 1995), produced estimates with 95% reliability. Mid- to long-range TSD estimates have been attempted in studies using putrefying muscle tissue and electron paramagnetic resonance (Zharov, 1996), volatile fatty acids in soil (Vass et al., 1992), and botanical evidence (Bock and Norris, 1997; Hall, 1997; Quatrehomme et al., 1997; Willey and Heilman, 1987) but with marginal success.

By far the most common midrange method is forensic entomology. **Forensic entomology** is the study of **necrophagous** insects and their patterns of colonization. Research on **carrion**-associated species has been conducted on every continent (Schoenly et al., 1991), but the method remains highly regionally specific and is of limited utility in cold climate regions (Komar, 1999a). Entomological TSD estimates are based on the observed life cycles of the major necrophages (blow flies and beetles) as well as the successive waves of colonization of different species on the body. Detailed examinations of forensic entomology, including scene collection protocols and methods of analysis, are provided in J.H. Byrd and Castner (2001), Catts and Goff (1992), Haskell et al. (1997), Kulshrestha and Satpathy (2001), and Tantawi and Greenburg (1993). The forensic use of insects extends beyond estimating PMI. Entomological evidence has been used to identify antemortem toxicology (Bourel et al., 2001; Gagliano-Candela and Aventaggiato, 2001; Introna et al., 2001), victim DNA (Campobasso et al., 2005; Linville et al., 2004), neglect of children and the elderly (Benecke and Lessig, 2001; Benecke et al., 2004), and perimortem trauma (Anderson, 1997).

Long-Range Time-Since-Death Estimation

Methods proposed exclusively to address extended PMIs include the quantification of animal scavenging (Willey and Snyder, 1989), which proved to be of limited accuracy. Invasive or destructive methods have also been studied, including DNA degradation (Perry et al., 1988), **luminol** studies on ground bone (Introna et al., 1999), and a histochemical study of the differential decomposition of **proteoglycans** and collagen in intervertebral disks (Komar, 1999a).

Unlike other disciplines within forensic science, those tasked with estimating elapsed time in forensic death investigations have generated no reports regarding the impact of the *Daubert* criteria (Chapter 3) on TSD methods and their admissibility in court. Byrd and Castner (2001) broadly discuss theories of admissibility of entomological evidence under the *Daubert* and *Frye* rulings. However, searches of relevant keywords in all major databases reveal no citations addressing the post-*Daubert* admissibility of TSD estimation in U.S. courts. Given the number of estimation methods and techniques, this lack of published reports may limit future attempts to introduce such evidence in court, in as much as many of the methods discussed here fail to meet the *Daubert* criteria. Having reviewed the first goal of forensic taphonomy, TSD estimation, we will now focus on the second, discerning human from nonhuman agents of bone modification.

Bone Modification

On scene, forensic anthropologists need to identify the types of scavengers interacting with remains. Accurate identification of necrophagous fauna and their patterns of behavior results in better search strategies and prevents the misinterpretation of natural defects as evidence of trauma. Classes of scavengers capable of modifying bone include canids, rodents, birds, mustelids (Bunch, 2006), and bear. Faunal interactions with remains fall into two categories: bone defects or modifications, and dispersal and disarticulation of remains.

Several features contribute to identifying bone defects: tooth morphology, bite force, dental formulas, and scavenging patterns. Canid scavenging has been examined by Haglund (1997a), Haglund et al. (1988, 1989), and Willey and Snyder (1989). Characteristic defects include canine puncture marks and scratches (Figure 7.1), as well as predilection for long bone epiphyses. Additional indicators of canid scavenging include the presence of paw prints or feces at the scene. Bear scavenging is characterized by large distinct puncture defects and extensive bone destruction (Figure 7.2). The skeletal manifestations of bear scavenging are detailed in Carson et al. (2000). Rodent scavenging produces distinct paired grooves or striae consistent with the morphology of the pronounced central incisors of these species (Figure 7.3; see also Haglund, 1997b). Rodent gnawing either is restricted to the margins of bones (Figure 7.3) or results in the progressive, layered destruction of cortical bone followed by cancellous bone (Figure 7.4). Bird scavenging has received less attention. However, birds have been shown to produce bone defects (Komar and Beattie, 1998b). Bird scavenging typically affects only the cortical and superficial cancellous layers. Distinctive conical defects reflecting beak morphology are characteristic (Figure 7.5).

Differentiating faunal modifications to bone from human action relies primarily on defect morphology. For example, Figure 7.6 illustrates both sharp force trauma and canid defects to rib bones. Although the defects are similar in location and gross appearance, closer examination reveals distinct differences in kerf morphology (Chapter 6). How each class of scavenger

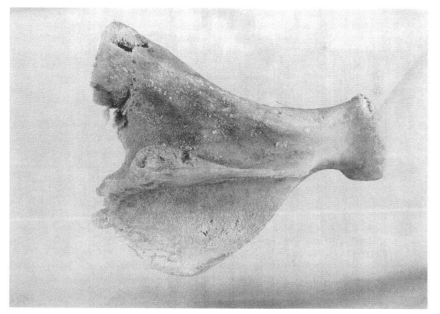

Figure 7.1 An example of domestic dog gnawing to a pig scapula. Note the destruction of the superior border and the characteristic puncture tooth defects along the spine. (*Photo—Debra Komar*)

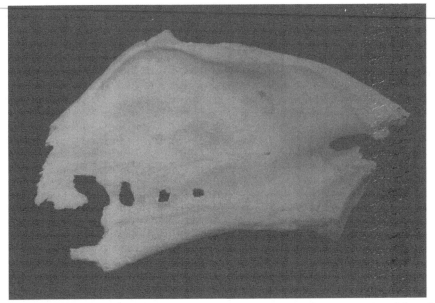

Figure 7.2 Human right scapula showing evidence of bear scavenging. Note the sequenced puncture defects to the lateral and inferior borders, indicating the animal's dental formula and tooth morphology. (*Photo courtesy of the New Mexico Office of the Medical Investigator*)

Figure 7.3 Proximal epiphysis of a juvenile tibia that has been gnawed by a rodent. The paired, grooved striae are indicative of tooth morphology. The gnawing is restricted to the outer margins of the bone. (*Photo by Debra Komar*)

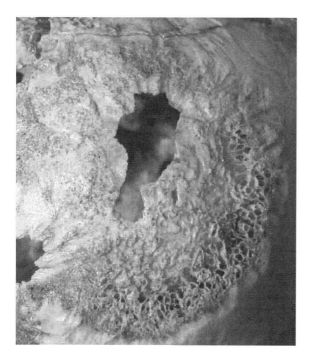

Figure 7.4 Rodent scavenging defects on a human cranial vault. This specimen shows both paired, grooved striae on the margins of the defect and progressive, layered destruction of the cortical and cancellous bone. (*Photo courtesy of the New Mexico Office of the Medical Investigator*)

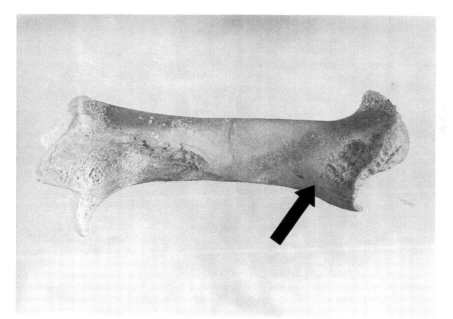

Figure 7.5 Juvenile pig humerus showing evidence of bony defects results from bird scavenging. The arrow indicates the defect, with characteristic conical impressions reflecting beak morphology. Bone destruction is limited to the thin cortical layers with some cancellous damage. Videotape of the scavenging event shows a black crow repeatedly pecking at the soft tissue attached to the bone. (*Photo by Debra Komar*)

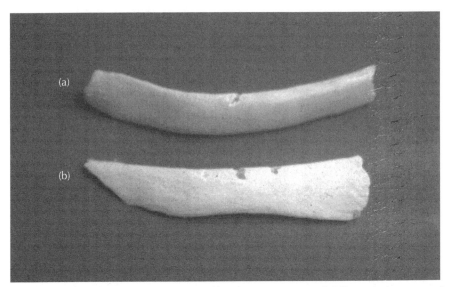

Figure 7.6 Two rib fragments, illustrating the importance of differentiating natural taphonomic change from human artifact: **(a)** Rib showing evidence of a sharp force wound from a knife; **(b)** Rib showing canine tooth defects. Accurate determination of the source of the defects relies on assessment of kerf morphology. (*Photo by Debra Komar*)

disarticulates and disperses remains will be addressed in the following section, which examines the third goal of forensic taphonomy, the selective transport of remains.

Mechanisms of Transport

The postmortem dispersal of remains presents a challenge for investigators. Transport of remains from the original deposition site occurs as the result of environmental, faunal, and human action. To what extent patterns of dispersal are predictable depends largely on the agent of transport. Among the least predictable is human agency. As criminal acts such as homicide are inherently not rational, the actions of those committing such crimes do not necessarily follow a logical course. Although other agents cannot be controlled, their influence can often be predicted.

Transportation of bodies in water has been extensively researched (Haglund, 1993; Nawrocki et al., 1997; O'Brien, 1997; Sorg et al., 1997). Models and simulations calculating drift trajectories allow investigators to estimate the location of water-deposited remains (Ebbesmeyer and Haglund, 1994). Reconstructing water flow patterns in a closed irrigation system has, for example, permitted the tracking of the remains of a single individual dispersed over 66 miles (Komar, 2004; see Box 8.8 in Chapter 8 for an expanded discussion of this case).

BOX 7.4
Mechanisms of transport

The following case study illustrates the need to understand the mechanisms of bone dispersal. In November 2002, a teenage boy was reported missing from his rural New Mexico home. Seven days later, a farmer discovered a defleshed human skull in one of his fields. Dental records revealed that the skull belonged to the missing boy. Tissue loss on the skull was consistent with canid scavenging, and a large population of coyotes and domestic dogs was reported in the area. An extensive search of the farmer's field revealed major postcranial elements dispersed over a 2-mile radius. The remains were skeletal, with all soft tissue removed as a result of canid gnawing, and no clothing was recovered. Law enforcement at the scene believed the distribution to be the result of canid scavenging. However, the dispersal pattern was unusual for canid activity in that both lower limbs (absent clothing and pelvis) were found together in one location, while portions of both arms were found in another, and the thorax and pelvis in a third location. All sites, including the one at which the skull was found, were separated from each other by more than a half mile.

Examination of the remains at autopsy revealed an unequivocal sharp force defect to a cervical vertebra, consistent with postmortem dismemberment. While the loss of soft tissue to the remains can be attributed to the action of the canids, the evidence of dismemberment and the unusual pattern of dispersal raises the possibility of the intentional distribution of body parts by the assailant following dismemberment. The case has not been resolved.

Terrestrial dispersal of remains often results from the action of faunal scavengers. Although insect scavenging alone can result in limited transport and modification of remains (Haskell et al., 1997; Komar and Beattie, 1998c), animal and avian interaction with remains can lead to widespread dispersal of skeletal elements (Box 7.4). As the distribution of faunal species is regionally dependent, anthropologists need to rely on experience and careful observation of local phenomena to understanding scavenging patterns in their jurisdictional areas. Dispersal patterns have been reported for canids (Haglund, 1997a; Haglund et al., 1989; Rothschild and Schneider, 1997), rodents (Haglund, 1997b), and birds (Komar and Beattie, 1998b). Evidence of canid and rodent scavenging should initiate a search of local dens and warrens for transported elements or evidence. Search strategies of scenes scavenged by birds must include a thorough search of adjacent nests (Figure 7.7).

Although animal scavenging often results in the recovery of incomplete remains, forensic anthropologists can encounter incomplete remains in a

Figure 7.7 Typical rib found in a downed magpie nest near surface-deposited remains in rural Alberta. Birds frequently act as mechanisms of transport, and search strategies should include all nests within the search area. (*Photo by Debra Komar*)

variety of scenarios and for a number of reasons other than animal scavenging, including deliberate and unintentional human action. Remains become incomplete as the result of intrinsic and extrinsic factors. These factors are discussed in the following section regarding the fourth goal of forensic taphonomy, identifying variables resulting in the selective preservation of remains.

Preservation and Collection Bias

Selective transport or differential preservation can result in incomplete remains. Selective transport and dispersal are extrinsic factors affecting element survivorship and recovery rates. Differential preservation of skeletal elements is intrinsic. Intrinsic factors affecting preservation include morphology, cortical thickness, age of the individual, perimortem trauma, and antemortem conditions affecting bone mineralization such as osteoporosis. Thin or fragile bones, such as the eye orbit or the scapular body, are more susceptible to postmortem destruction than more robust bones such as the femur. Elements of adult individuals generally survive at higher rates than those of juveniles and subadults. Poor bone mineralization and trauma, including fractures of bone, can result in the loss of elements and accelerated decomposition of remains.

BOX 7.5
Investigator bias

A field deputy medical investigator was called to a vehicle fire in an extremely isolated region of New Mexico. The vehicle was a camper that was parked diagonally, completely blocking a rural dirt road. The vehicle had been extensively damaged by fire. Firefighters at the scene reported the presence of what appeared to be two large masses of burned flesh and bone within the camper. After examining the material, the field deputy reported that the masses represented two burned dogs, "one large and one small." As the scene strongly suggested arson, the fire marshal requested an anthropological examination of the biological material to confirm that it was nonhuman. The forensic anthropologist called to the scene quickly determined that the remains were of an adult male, with the large mass representing the skull and thorax and the small mass representing the lower limbs. This case study reveals the importance of a thorough investigation of all potential biological material by a qualified observer. Had the police at scene accepted the initial opinion of the deputy investigator, the remains of a homicide victim would have been destroyed as nonmedicolegal animal remains.

Investigator bias also plays a role in the incomplete recovery of remains. Search, collection, recovery, and curatorial methods vary among anthropologists. Inexperienced investigators may fail to recognize certain types of biological evidence (Box 7.5), such as calcifications relating to pathological conditions or fire-modified remains (Box 7.6), potentially affecting the outcome of the investigation. Time and resource constraints or excavation techniques may result in the incomplete collection of fragmentary remains. Inadequate lighting during searches or reliance on improper equipment leads to collection bias. Curatorial methods, including those used to transport remains from the scene to the morgue, can result in additional fragmentation of remains or the continued decomposition of biological materials. Proper analysis and accurate reconstruction requires that remains be completely recovered. All methods and efforts that prevent information loss and increase recovery rates are worthwhile. Although little can be done to control completeness issues related to preservation, investigators can prevent collection bias.

Forensic anthropologists incorporating taphonomic analyses into their casework assist law enforcement and medicolegal investigators in addressing one of the most important aspects of scene investigation—the reconstruction of perimortem events. We will now consider this fifth and final goal of forensic taphonomy.

BOX 7.6
Fire modifications to bone

Fire is perhaps the most destructive force seen in forensic investigations. The capacity for heat to modify, destroy, or obscure remains is impressive; indicators of identity, pathological conditions, or trauma can all be lost to fire. Understanding the effects of thermal-induced change to human remains is a valuable skill to forensic anthropologists, as fire-related cases are frequently encountered.

Despite its destructive power, it is important to note that the total destruction of bone through fire is exceptionally rare, and some would argue impossible. Even professional crematoriums with furnace temperatures exceeding 2000° cannot completely destroy a body. Criminal attempts to destroy bodies through bonfires, vehicle fires, or building fires also fall far short of complete destruction of the remains, although vital information is lost nonetheless.

An excellent overall review of fire modification of bone is provided by Mayne Correia (1997), while Christensen (2002) examines the combustibility of the human body. Issues related to the identification of burned remains are examined by Bassed (2003). Recognizing perimortem trauma in fire-modified remains is reported by de Gruchy and Rogers (2002), Hausmann and Betz, (2002), and Pope and Smith (2004), while Bohnert et al. (1997) and Herrmann and Bennett (1999) describe the differentiation of traumatic and heat-induced fractures.

Interpreting thermal damage to human remains is a highly specialized aspect of forensic anthropology, requiring considerable experience and dedicated research and training. Forensic anthropologists called to testify in such cases should exercise caution and remain mindful of the important differences among observation, opinion, and speculation. Since much of our understanding of thermal-related change in human remains comes from direct observations and experience, testimony relating to burned remains would more aptly fall under the criteria outlined in *Kumho* regarding technical expert testimony, rather than *Daubert* (see Box 3.7).

Reconstructing Perimortem Events

The accurate reconstruction of perimortem events requires investigators to distinguish modifications produced by natural taphonomic processes from artifacts resulting from human action. These modifications fall into two categories: mimicry and obscuring processes. **Mimicry** involves changes resulting from natural actions that closely resemble specific artifacts of human action. For instance, defects in clothing similar to contact gunshot wounds can result from the penetration and subsequent decomposition of plant roots

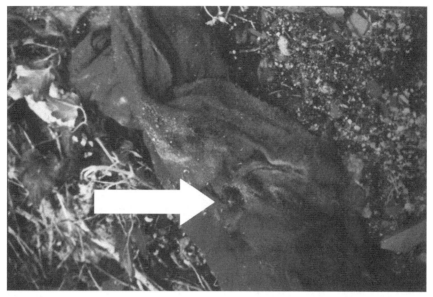

Figure 7.8 Mimicry: pseudo–contact gunshot defects in clothing. The circular defect (arrow) and blackened margins resulted from the penetration of the cloth by plant roots, which subsequently decomposed, causing the black discoloration. The clothed remains had been found near a creek bed following a postmortem interval greather than 4 years. (*Photo by Debra Komar*)

through clothing (Figure 7.8). Mimicry resulting from necrophagous insect activity is well documented. Maggot defects to skin can mimic the appearance of shotgun pellet defects. Pseudo–drag marks produced by the mass exodus of final instar maggots from a corpse is another example of mimicry (Haskell et al., 1997). Postmortem insect activity can produce clothing patterns similar to those seen in cases of perimortem sexual assault (Komar and Beattie, 1998a). Recognizing mimicry relies on experience and the careful examination of features.

Obscuring processes result in evidence loss or destruction due to natural processes, leading ultimately to the irrevocable loss of information over time. For example, insect and faunal scavenging patterns that favor bloody trauma or open wounds result in extensive soft tissue destruction that obliterates evidence of injury. Similarly, decomposition compromises the reliability of toxicological tests. Advanced decomposition ultimately renders such tests impossible.

Obscuring processes are progressive. However, all information is not lost immediately; rather, investigators must rely on alternative sources of evidence to reconstruct events. For example, Figure 7.9 illustrates the formation of a natural cast in a case involving a clandestine burial. At the time of discovery, the body had fully decomposed, and repeated scavenging of the

Figure 7.9 Example of alternative sources of information at the scene: a natural cast of a body, as found during investigation, of a homicide with clandestine burial. The decedent's head was to the left of the photo, legs to the right. Exposure of the cast resulted from repeated scavenging of the grave by canids. Cast reveals stature and body size, as well as impressions of clothing no longer present. (*Photo courtesy of the New Mexico Office of the Medical Investigator*)

shallow grave by animals had resulted in the extensive destruction and dispersal of skeletal remains. Fortunately, a natural cast revealed the original position of the body within the grave, the height and body mass of the individual as well as details of clothing no longer present. This case illustrates the significance of time in determining the survival of information. Had the body been discovered earlier, these observations could have been made directly from the remains. Alternatively, had the discovery of the grave been delayed, natural taphonomic processes would have compromised or destroyed the cast, resulting in a greater loss of information.

While the potential of obscuring modifications at scenes must be recognized, the loss of information inevitably reaches a threshold, beyond which there is nothing investigators can do to compensate for the loss. With examples of mimicry, however, the responsibility lies with investigators to recognize the potential for mimicry and to accurately assess the origin or source of the artifact.

CONCLUSION

The accurate interpretation of scenes, as well as the proper analysis of remains at autopsy, requires a thorough understanding of how decomposing remains interact with their environment. Differentiating natural from human-generated artifacts, estimating the postmortem interval, and recognizing the potential for natural agents such as scavengers to modify scenes or transport remains are just a few of the contributions anthropologists make to criminal investigations. Comprehensive training and experience are crucial in preparing forensic anthropologists to meet the challenges of our ever-expanding role in forensic investigations.

CHAPTER 8

Personal Identification

Putting names to the unidentified begins with the development of biological profiles (Chapter 5). An assessment of pathological conditions and trauma (Chapter 6), an estimate of the time-since-death, and a reconstruction of perimortem events (Chapter 7) also contribute to the developing profile of the decedent. Now the process of individualization begins, with the goal of scientifically verifying the person's identity.

BOX 8.1
Types and applications of radiological imaging

Traditional **diagnostic radiology** is the study of images of internal structures of the body for the purpose of identifying and evaluating disease. This use of imaging technology provides the antemortem standard for establishing identification. The exception is breast mammograms, which constitute 8% of all diagnostic radiographic studies and are of no use in identification (Brogdon, 1998b). **Forensic radiology** is the interpretation of radiographic images of interest to a court of law. In addition to the use of radiographs for ID purposes, forensic radiology also focuses on the evaluation of injury and pathology. Fractures, tumors, and other osseous injuries are evident on radiograph, as are projectiles and other foreign bodies. An excellent review of the forensic applications of radiology, including identification, can be found in Brogdon (1998a). Issues relating to the introduction of digital imaging in forensic investigations are discussed in Box 4.9 in Chapter 4. **Radiation oncology** is the use of radiant energy wavelengths and particles to treat malignant neoplastic disease. This application of radiology does not generate images, hence has little bearing on issues relating to identification.

The identification (ID) of unknown decedents is one of the most important responsibilities in forensic casework. Every unidentified body entering the medicolegal system represents a missing person with anxious family members. Delays in identification prolong the family's suffering. The resolution of legal issues following a death cannot be completed without identification of the decedent. Without a death certificate, next-of-kin cannot execute wills, obtain death benefits, or divide estates.

Methods used to establish identification include visual recognition, comparisons of medical (Box 8.1) or dental radiographs, fingerprints, and DNA testing. Determining which method to use depends on several factors—the decompositional state of the remains, the availability of comparative antemortem data, and the admissibility in court of the various methods within a jurisdiction.

In this chapter, we will examine how identification is established, as well as identification categories and the use of the term "Doe." We then introduce methods of establishing positive identification through individualizing features of the human skeleton, as well as DNA testing. Finally, the impact of the *Daubert* ruling on the ID process and investigative methods related to identity, such as facial reconstruction, will be discussed.

HOW IDENTIFICATION IS ESTABLISHED

Regardless of the method employed, all personal identification of deceased persons relies on pattern recognition. Visual identification by next-of-kin, fingerprints, radiographic comparisons, and DNA all involve the interpretation of patterns. In the case of visual recognition, patterns exist within memory or antemortem photographs. With all other systems, postmortem information patterns are compared against patterns derived from existing social data.

Identifications can be made anecdotally through visual recognition of the decedent, circumstantially through associated artifacts, or scientifically using verifiable evidence. Identifications by means of visual recognition or personal effects, although common, are not scientific and are subject to error. Scientific methods are valid and replicable when used by experienced practitioners (Bernstein, 1999).

Identifications are based on either the existence of a unique identifier or the collective and consistent agreement of multiple features. Identification can also be established as a statement of probability, based on the frequency of a suite of characters within a population. For example, likelihood ratios have been generated for genetic profiles within a specific geographic region. These likelihood ratios served as the basis for identifying individuals in a mass casualty situation (Gonzalez-Andrade and Sanchez, 2005). This method will be examined in more detail in Chapter 9. Similar probability statements form the basis of identifications based on dental formulas rather than direct x-ray comparisons and will be discussed in a subsequent section of this chapter.

Identification must also be considered a progressive process of inclusion and exclusion. The goal is to narrow the sample of potential missing persons down to a positive identification. Image a bull's-eye target. The largest outer circle contains all possible missing persons. Progression to the next smaller circle relies on inclusion or exclusion. A variable (such as sex or height) is introduced. All individuals who share the trait are included, all those who do not are excluded. For example, the body of an unknown female is discovered by police. At autopsy, she is estimated to be white, 20 to 30 years of age, and 5 foot 2 inches to 5 foot 5 inches in height. When comparing this individual to the missing persons list, investigators first exclude all males, then all females younger than 20 and older than 30 years. Of the remaining females on the list, they then exclude those outside the prescribed height range and who are not white. While investigators have not yet arrived at a positive ID, the pool of potential missing persons against which the unknown individual must be compared is dramatically reduced. The introduction of additional variables, such as the estimated time-since-death or weight, aids in reducing the pool to those most likely to match.

Responsibility for identifying unknown individuals varies by jurisdiction. In most regions, law enforcement is statutorily obligated to establish the identity of a decedent. Even so, most medicolegal offices perform a large number of identifications every year as part of their autopsy and investigative process (Murphy et al., 1980; Weedn, 1998). Statutes may also stipulate that the identification of unknown decedents is exclusively the responsibility of the medicolegal authority. For example, in cases of mass or natural disaster, the responsibility for identifying unknown individuals falls to the medicolegal authority with jurisdiction.

The introduction of DNA testing as a means of identification in 1985 raised new issues. DNA technology is more costly than traditional methods of identification such as dental or fingerprint comparisons. Whether law enforcement or the medicolegal authority bears the burden of this increased cost depends on which agency is statutorily obligated to establish identification. DNA testing also requires that comparative samples be obtained from living relatives. While such samples are often voluntarily provided by the next-of-kin, law enforcement has little recourse if relatives are unwilling to submit samples. Warrants can be sought to forcibly obtain biological samples from suspects but not from a decedent's family members. Collecting samples from decedent next-of-kin may also be outside the jurisdictional scope of the medicolegal authority, obligating police assistance to obtain the necessary comparative materials.

CATEGORIES OF IDENTIFICATION

Standard designations reveal the level of certainty attributed to identifications: tentative, presumptive, or positive. **Tentative** indicates that the identity of the decedent is suspected, based on circumstances or associated materials such as a wallet or driver's license. Typically, this designation is given to bodies entering the medicolegal system prior to establishing more substantial levels of identification. "Tentative" is synonymous with confirmatory or directed identifications. **Presumptive identifications** meet a higher standard than tentative but a lesser standard than positive. Circumstances or general characteristics provide investigators with a basis for identification, and there is no exclusionary evidence. For example, an elderly white male is found dead in his home. The door is locked from the inside; there are no signs of foul play or forced entry. The home owner was an 85-year-old white male known to live alone. However, no antemortem data are available to confirm the identification. The medicolegal authority can issue a death certificate based on a presumptive identification if satisfied with the basis of the ID. **Positive identifications** indicate that individualization has successfully been demonstrated. **Individualization**, as the term suggests, is based on features or characteristics that make each person unique—fingerprints, nuclear DNA sequences, medical and dental history. A direct comparison of AM and PM data results in the conclusive determination of identity.

Doe Designation

Individuals entering the medicolegal system without even a tentative identification are given the designation **Doe**. There are no circumstantial clues to provide identity, no missing persons report, and no basis for an a priori assumption of identification. The term "John Doe" is derived from a fourteenth-century British custom of identifying unknown parties in legal actions as John Doe or Richard Roe. The current legal definition of John Doe is "a party to legal proceedings whose true name is unknown or withheld." The female equivalent is Jane Doe or Mary Major. If the sex of the deceased can be determined, the appropriate designation of Jane or John Doe is given to the body. When sex cannot be determined, the designation is simply Doe. Infants are identified as Baby Doe or with a notation identifying sex such as Baby Boy Doe.

In these cases, anthropological analysis provides demographic information on the individual. A biological profile estimating the individual's sex, age, population affinity, and stature will be generated (see Chapter 5). Any indications of antemortem trauma or pathology are noted, as these features can also provide clues to the individual's identity. This profile is then compared against the local and national databases of missing persons. Any tentative matches will be subject to subsequent scientific testing to confirm the identification or exclude the individual from further consideration. This process continues until the body has been identified or all possible avenues have been exhausted.

Depending on protocols, each medicolegal authority must decide how long and in what form to retain Does or unidentified individuals. Because storage space is a limiting factor in most morgues, unidentified remains cannot be held indefinitely. State statutes stipulate what is to be done with unclaimed or unidentified bodies; frequently these remains are buried or cremated. If statutes permit, and the resources and facilities are available, the medicolegal authority can opt to render the remains to a skeletal or stable state and retain the body indefinitely pending identification. This provides the ME or coroner with the opportunity to continue testing the individual against any potential leads that develop, as well as the ability to return the remains of those ultimately identified to their next-of-kin. Medicolegal authorities without the resources to retain remains will collect all possible sources of identification, including harvesting DNA samples, as well as taking fingerprints and postmortem dental and medical radiographs prior to releasing the body for indigent burial or cremation. Should an identification subsequently be made, the decedent's next-of-kin can work with local or state authorities to recover the remains or ashes.

METHODS OF POSITIVE IDENTIFICATION

Depending on the jurisdiction, the following methods are commonly accepted as means of establishing positive identification: visual recognition, fingerprinting, DNA testing, dental formulas, and medical and dental radiographic comparison. Although visual recognition has always been employed to identify the dead, historically fingerprints were the primary means of scientifically establishing identification. In 1888, Sir Francis Galton, a British anthropologist and cousin of Charles Darwin, began his observations of fingerprints as a means of identification. In 1901 England and Wales introduced fingerprinting as a means of identifying criminals, with the United States following in 1902 (Moore, 2006). While the history of forensic odontology began in AD 49 with the earliest reported case of Lollia Paulina (whose teeth were examined in her decapitated head to confirm her identity), its widespread acceptance and development as a professional discipline began in the later part of the nineteenth century (Sansare, 1995). The use of DNA as a means of identification is of much more recent origin, with its potential uses in forensic contexts first described in 1985 (Gill et al., 1985). DNA profiling was first reported in 1985 by Alec Jeffreys, who found that certain regions of DNA contain sequences that repeat and that the number of repeated sections present in a sample differs from individual to individual. The regions are known as **variable number of tandem repeats (VNTR)**, while the technique used to examine them is called **restriction fragment length polymorphism (RFLP)** (Butler, 2005).

Visual Recognition

Visual recognition is considered the least reliable form of identification. Despite this, it remains the most common means of establishing ID, largely due to circumstance. When a person dies in a hospital or at home, accompanied by relatives or friends, personal identity is not in question. The majority of deaths are natural and in most instances expected, so identification is normally not an issue. Establishing personal identification becomes significant when an individual dies unexpectedly or away from home and is not known in the community where death occurs. In such cases, visual recognition based on associated photographs, such as a driver's license photo, may provide a source of tentative identification. However, a positive ID using a scientific method will still be required.

Fingerprinting

Dactyloscopy is the practice of using fingerprints to identify an individual. Fingerprinting is a system of biometric pattern analysis that uses different fingerprint characteristics (ridge patterns such as loops, whorls, and arches) to establish individualization. Fingerprints are believed to be unique. No two individuals, including identical twins, have the same prints. Prints are stable throughout life. Superficial injuries such as burns or cuts will not alter the ridge structure. New skin grows with the same pattern. **Latent prints** are those found at a crime scene. As latent prints are usually invisible to the human eye, numerous techniques are needed to render them visible, including powder, alternative light sources, glue fuming, and lasers.

The FBI currently stores over 250 million sets of fingerprint records in a computer database known as **IAFIS (Integrated Automated Fingerprint Identification System)**. This database contains both civil and criminal prints. Civil prints are fingerprint cards submitted by federal employees and applicants for federal jobs. Those considering careers in forensic science should know that employment will necessitate submitting a fingerprint sample to the FBI. Criminal prints are taken at the time of a suspect's arrest.

Despite some challenges (Box 8.2), fingerprint evidence continues to be an admissible and common form of identification. Fingerprint analysts are required to be certified (the FBI website www.fbi.gov offers more information). What constitutes an acceptable match depends on the jurisdiction. For example, South African courts accept seven concordant features as being "beyond reasonable doubt" for fingerprint ID (Phillips and Scheepers, 1990). Excellent reviews of fingerprinting method and practice are provided in the FBI publication *The Science of Fingerprints: Classification and Uses* (1985), as well as in Coppock (2001) and Maltoni et al. (2003).

DNA

Testing for **DNA** (deoxyribonucleic acid) in forensic contexts serves two purposes: (1) the positive identification of decedents and (2) the analysis of biological evidence, such as blood or semen, recovered at scene, and its attribution to specific suspects; other applications have been investigated (Box 8.3). This section will focus on the first goal, namely, the identification

BOX 8.2
Fingerprinting faces *Daubert* challenges

Despite the assumption of fingerprint uniqueness, dactyloscopy has faced some serious challenges as a method of identification. In the 2002 court proceeding *USA v. Plaza, Acosta and Rodriguez*, a senior judge declared fingerprint identifications inadmissible. The judge ruled that the method failed to satisfy the *Daubert* criteria for admissibility. Although the ruling was local, applying only to the Eastern District of Pennsylvania, and was subsequently **vacated** (canceled or rescinded), its implications were significant (Rogers and Allard, 2004). A more recent study (Dror et al., 2006) further challenged the validity of fingerprint IDs. The report indicated that the majority of fingerprint experts tested made different judgments on the same fingerprints based on the contextual information they were given, resulting in erroneous identifications. Another study raised the possibility that fingerprinting methods could result in the transfer of DNA, again leading to erroneous results through contaminated DNA tests (van Oorschot et al., 2005).

BOX 8.3
Genetic analysis to estimate sex

Estimating the sex of unidentified remains is an important step in developing a biological profile. In cases involving partial or fragmentary remains, without sufficient morphology to allow for sex estimation, researchers have investigated the use of DNA testing to differentiate males and females. Molecular techniques, in particular the typing for length variation in the X-Y homologous amelogenin gene (AMELX and AMELY), have been employed. Several commercial PCR multiplex kits are available for use, but high error rates during testing have raised significant concerns over their forensic application (Cadenas et al., 2006; Mitchell et al., 2006).

of remains. Two forms of DNA are found within each cell—nuclear DNA and mitochondrial DNA (mtDNA). Both are used to establish identification. However, significant difference in the structure of each DNA type necessitates different systems of analysis. These differences are outlined in Box 8.4.

The **Combined DNA Index System (CODIS)** was developed and is maintained by the FBI both to assist in identification and as a basis of comparison for biological evidence recovered at scenes. CODIS was introduced in the United States in 1998 and is organized into three tiers—local, state, and national. The **National DNA Index System (NDIS)** is the highest level

BOX 8.4

Nuclear versus mitochondrial DNA

Feature	Nuclear DNA	Mitochondrial DNA
Copies per cell	2	> 1,000
Number of base pairs	~3.2 billion	16,569
Chromosomal pairing	Diploid	Haploid
Individualization	Unique to individual or to identical twins	Not unique
Mode of inheritance	Combination of maternal and paternal inheritance	Maternal inheritance only
Method of analysis	Tetranucleotide STR loci	HVI and HVII analysis

After Butler, 2005

in the CODIS hierarchy, enabling law enforcement agencies across the country to access samples obtained from crime scenes as well as reference samples obtained from convicted offenders. As of January 2006, NDIS had 124,285 samples from crime scenes and 2,754,714 samples from convicted offenders. The FBI also compiles a mtDNA database, divided into laboratory forensic samples and published references. The forensic database contains 4,839 mtDNA profiles from 14 different populations, while the published database contains 6,106 profiles. More information is available at the FBI website http://www.fbi.gov/hq/lab/codis/program.htm.

Nuclear DNA

Nuclear DNA analysis for identification relies on **short tandem repeat (STR) typing**. **Tetranucleotide** (a compound of four nucleotides) STR loci are preferable. CODIS lists 13 tetranucleotide STR loci per specimen, allowing investigators to generate potential matches. STR (and mtDNA) analysis uses **polymerase chain reaction (PCR)**, an enzymatic process in which a specific region of DNA is replicated over and over to produce sufficient material for testing. PCR involves the heating and cooling of samples in precise thermal cycling patterns (typically over 30 cycles). This process results in a testable product known as **amplicons**. The amplicons are sequenced, and the STRs of interest are identified and compared against a known sample. The results allow investigators to include or exclude individuals as the source of the unknown sample.

DNA analysts working in forensic contexts contend with a number of difficulties.

- *Sample degradation* Because samples consist of biological materials subject to decomposition, the state of the material when found, the methods of collection and preservation, as well as the speed with which the sample is processed, all contribute to the viability of the specimen.

- *Contamination* **Contamination** is the presence of exogenous DNA in a sample. Contamination can occur at the scene, during collection or curation, as specimens are prepared, or during analysis. Stringent protocols are enacted to prevent contamination, and guidelines regarding the interpretation of test results were created to identify potentially contaminated samples. Protocols are outlined in Butler (2005: 152).

- *Mixed samples* Often, the nature of the crime results in the commingling of biological material; thus a sample from a rape victim may contain both assailant semen and victim blood. Researchers have proposed methods of dealing with such samples (see, for e.g., Clayton et al., 1998; Curran et al., 1999).

- *Quantity of the sample* The amount of biological material recovered at some scenes is minute. The ability to successfully extract and analyze DNA from such small samples has improved greatly in recent years.

- *PCR inhibition* Occasionally, PCR analysis is inhibited as the result of small sample size or specimen degradation. If the DNA cannot be copied, it cannot be sequenced and analyzed. Therefore, PCR inhibition has serious consequences for investigators (Butler, 2005).

Mitochondrial DNA

To overcome many of the problems inherent in working with degraded samples, investigators use mtDNA. Often, mtDNA's higher copy number per cell results in the successful extraction of this genetic material when nuclear DNA cannot be recovered. Successful extraction of mtDNA from teeth and bone, even after prolonged postmortem intervals, makes the use of mtDNA preferable in cases involving decomposed or skeletal remains. The use of mtDNA in biohistorical investigations will be discussed in Chapter 10.

Mitochondrial DNA is inherited along matrilineal lines. It is passed virtually unchanged from mother to offspring. Recombination in human mtDNA has been suggested (see, for e.g., Eyre-Walker and Awadalla, 2001), but the studies have been widely criticized (e.g., Innan and Nordborg, 2002). The use of mtDNA for identification purposes has both a significant benefit and a disadvantage as a result of its mode of inheritance. The benefit is that any surviving matrilineal relative of the decedent can provide a comparative sample. With nuclear DNA, only parents and offspring separated by one generation can serve as comparative relatives. The disadvantage is that mtDNA is not unique to an individual because it is inherited unchanged from the mother and thus all her children share the same profile. As with nuclear DNA, identifications based on mtDNA are therefore statements of probability, rather than absolutes, with mtDNA probability statements weaker than those of nuclear analysis.

Analysis of mtDNA for identification purposes focuses on the control region contained within the **D-loop**. The D-loop is a 1,122–base pair "control" region that contains the origin of replication for the mtDNA but does not code for any gene products (Butler, 2005). The sequencing of two portions of the region, known as **hypervariable region I (HVI)** and **hypervariable region II (HVII)**, allows comparisons. HVI is located between nucleotide positions 16,024 and 16,365, while HVII is located between 73 and 340. If needed, **hypervariable region III (HVIII)**, located between nucleotide positions 438 and 574, can help resolve equivocal samples (Bini et al., 2003).

The steps in the analysis of mtDNA are given in Figure 8.1. Samples are compared at all 610 nucleotide positions of HVI and HVII. Unknown and comparative samples are deemed concordant if every site matches. Because mtDNA is not unique to every individual, the results of mtDNA tests are not defined dichotomously as either inclusive or exclusive. Results are grouped into three categories: exclusion, inconclusive, or failure to exclude. *Exclusion* occurs when two or more nucleotides differ. Tests are considered *inconclusive* if there is a difference at one nucleotide position. Comparisons with no differences are designated as *failure to exclude* (SWGDAM, 2003). If failure to exclude is the conclusion, a statistical estimate of the match is needed to confirm the identification. This estimate determines if the **haplotype** (genetic sequence) is rare or common within a population. If the haplotype is common, it is possible the match occurred in unrelated individuals. An extremely rare haplotype supports the identification. Currently, the rarity of a mtDNA sequence is estimated by counting the number of times a particular haplotype occurs in a database. This method depends on the number of samples in the database. The larger the sample number, the more powerful the estimate (Box 8.5). Unfortunately, the population frequency for most mtDNA haplotypes is not known (Isenberg, 2004).

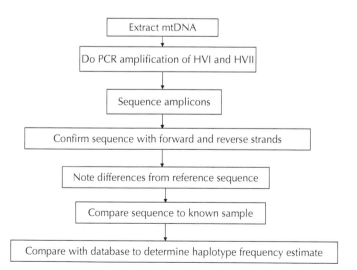

Figure 8.1 Flowchart illustrating the sequence of mtDNA analysis. (After Butler, 2005)

BOX 8.5
mtDNA haplotypes

To test the validity of mtDNA in the positive ID of individuals in forensic contexts, a team of Italian researchers (Grignani et al., 2005) sampled 271 unrelated Italian subjects. Their analysis revealed that 42% of the sample had a haplotype of H, the most common haplogroup in Caucasian European populations. This H haplotype subset was then screened for specific single nucleotide polymorphisms located in the coding region and categorized into H subclades H1–H15. A large portion of the mtDNA sequences could be assigned to subhaplogroups, with H1 and H5 being the most common. The study found that two-thirds of the individuals sharing common sequences could be subdivided and ascribed to specific H subhaplogroups. This study illustrates how subtyping within a common haplogroup can be extremely useful in forensic identification, particularly when many samples need to be analyzed and compared, as in cases of mass death (Grignani et al., 2005).

DNA and Forensic Anthropology

The introduction of forensic DNA testing for identification purposes led some to assume that anthropological analyses of skeletal remains would become obsolete (Saks and Koehler, 2005). However, despite the contributions of genetic analysis, osteological examination remains an integral part of the identification process (Kemkes-Grottenthaler, 2001). In both individual casework and investigations of mass death scenes such as the World Trade Center, putative matches resulting from DNA testing must be confirmed by a review of the individual's biological profile in relationship to his or her AM demographic data. Should discrepancies emerge, additional DNA testing will be required to confirm the identification (Mundorff, 2003).

Anthropologists are frequently called upon to harvest DNA samples from decomposed or skeletal remains. A few issues warrant attention. The possibility of subsequent DNA testing should be considered prior to using any skeletal preparation or maceration techniques to deflesh remains, as many techniques strongly impact genetic analysis (Rennick et al., 2005). Steadman et al. (2006b) found that maceration techniques using bleach, hydrogen peroxide, ethylenediaminetetraacetic acid, room temperature water, and sodium carbonate followed by degreasing resulted in low DNA concentrations and the inability to generate nuclear PCR products. The team recommended using high temperatures (90° C or higher) for short durations. Contamination resulting from sampling protocols is also problematic. Kemp and Smith (2005) recommend the use of bleach to eliminate contaminant DNA from bone and tooth samples. They found that, as DNA adsorbs to hydroxyapatite in bone, endogenous DNA remained stable and that immersion of the sample in various concentrations of sodium hypochlorite removed surface contaminants.

A number of researchers have examined the effects of decomposition, taphonomic processes, and depositional environment on DNA preservation in bone. A recent study (von Wurmb-Schwark et al., 2005) found that DNA could be extracted from commercially cremated remains, although the origin of the DNA (endogenous to the victim or contamination from other sources) could not be elucidated. Iwamura et al. (2005) reported the extraction and analysis of DNA from femoral cortical bone of surface-deposited and buried remains in various stages of decomposition. Staiti et al. (2004) and Ye et al. (2004) reported successful extraction from burned bone as well as from remains that were fully decomposed, or had been buried or submerged in water.

This section was intended to introduce the basic principles of DNA testing in forensic contexts. For more information, excellent reviews of the forensic applications of DNA are found in Butler (2005) and Kobilinsky et al. (2005).

Forensic Odontology

Forensic odontology or dentistry uses antemortem and postmortem dental radiograph and record comparisons to confirm identification. Statements of probability based on comparisons of decedent dental traits (called dental formulas) with databases of trait frequencies within populations also allow forensic dentists to establish ID. Analysis of bite marks recovered at scene or autopsy, either on victims or from evidence such as food, is another important contribution made by odontologists.

Identifications based on radiographic comparisons are most commonly employed. However, charts and notes detailing antemortem dental condition, including missing, filled, and unrestored teeth, can be used in lieu of x-rays. Large databases and the OdontoSearch computer program (www. jpac.pacom.mil/CIL/OdontoUse.htm) allow North American odontologists to calculate objective frequency information regarding the incidence of specific dental traits in the general population. In North America, there is no definitive number of points of agreement needed to establish identity; rather, each case is assessed individually (Adams, 2003a, 2003b). In South Africa, where this method has been accepted for some time, a minimum of 12 concordant characteristics must be shown to establish identification (Phillips and Scheepers, 1990).

In the United States forensic odontologists are certified by their governing board, the American Board of Forensic Odontologists. Information on membership and certification can be found on their website http://www. abfo.org. Students interested in learning more about forensic odontology should review Bernstein (1999), Clark (1992), and Willems (2000).

Radiologic Identification

Radiographic identification of human remains depends on matching features or characteristics evident on both AM and PM radiographs (Box 8.6). Sometimes a suite of common anatomical changes can be used to establish

---

BOX 8.6
Radiology: an overview

Understanding fundamental concepts in radiology allows anthropologists and odontologists to read radiographic films, interpret medical records including radiology notes, and communicate effectively with technicians when requesting radiographs for ID purposes or trauma assessment. Extracting the maximum information from imaging techniques begins with an appreciation for how such images are formed.

Radiological imaging relies on x-rays and gamma rays. Both have very short wavelengths and are high-frequency, high-energy radiation. Both gamma and x-rays are forms of electromagnetic radiation, consisting of vibrations in the electric and magnetic fields. Electromagnetic radiation can be visualized as waves described by the following characteristics:

- **Velocity** is the rate of travel of the waves.
- **Frequency** is the number of wave cycles per second, measured as hertz (Hz).
- **Period** is the length of time it takes to complete one wave cycle.
- **Wavelength** is the distance between the peaks or valleys of the waves.
- **Amplitude** is the maximum height of the wave.
- **Intensity** is understood through the inverse square law. All radiation travels at the speed of light and diverges from its source. The **inverse square law** states that as the distance increases, the intensity decreases. Doubling the distance decreases the intensity to one-quarter of the original. Reducing the distance to one-half increases the intensity by four times.

Knowing how x-rays travel provides a basis for understanding how x-rays interact with the objects they strike. X-rays entering tissue can be absorbed, scattered, or transmitted. When X-rays are absorbed, all their energy is transferred into the tissue. When they are scattered, they change direction and lose energy, which reduces the image contrast produced on the film. When transmitted, x-rays pass through the tissue without interaction. Most diagnostic x-rays are scattered or absorbed.

Attenuation is the partial or complete loss of x-ray energy in the tissue. The degree of attenuation depends on the following factors.

- *X-ray energy* X-rays with higher energies, meaning shorter wavelengths, have greater penetration and lower attenuation rates.
- *The thickness of the tissue* As tissue thickness increases, rates of attenuation increase. More x-rays are lost in 10 cm of tissue than in 7 cm of tissue.
- *The density of the tissue* **Density** refers to how tightly packed the atoms of a substance are; attenuation increases as density increases. In the body, air or gas has the lowest density, muscle is denser than fat, and bone is denser than muscle.
- *The material composition or atomic number of the tissue* Materials with higher atomic numbers have higher attenuation rates. Air, iodine, and barium are contrast materials introduced to improve image quality. They are effective contrast media because their atomic numbers differ from those of surrounding tissues. Substances that are highly attenuating are called **radiopaque**. Substances with low attenuating values are **radiolucent**.

(Bontrager, 2002; Bushong, 1988; Callaway, 2002)

identification, while other cases rely solely on a single unique characteristic. Identification methods using radiographic comparison fall into three categories: classification of morphological features, unique feature confirmation, and collective characteristic concordance.

Classification systems of morphological features assign a class to morphological variants that collectively describe a feature. Classification systems have been proposed for frontal sinus pattern (Cameriere et al., 2005; Christensen, 2005), bone trabecular pattern (Mann, 1998), ectocranial suture patterns (Chandra Sekharan, 1985; Rogers and Allard, 2004), and palatal rugae patterns (Limson and Julian, 2004; Muthusubramanian et al., 2005). Frontal sinus pattern analysis is examined in greater detail in Box 8.7 and illustrated in Figure 8.2.

Unique feature confirmation relies on the presence of a stable, individualizing character evident on antemortem and postmortem radiographs. Figure 8.3 shows a femoral head replacement confirmed on AM and PM films. Permanent orthopedic appliances are inscribed with a serial or identification number that can be used as a source of positive identification (Figure 8.4), provided the surgeon noted the number in the medical records.

Collective characteristic concordance uses a suite of consistent features, evident in single or multiple radiographic views, to establish identity (Figure 8.5). Evaluation of anatomical variants resulting from pathological or degenerative change serves as the basis for ID (Hulewicz and Wilcher, 2003). A minimum of eight characteristics, with no evidence of exclusionary or inconsistent features, provides a reliable basis for identification (Box 8.8).

BOX 8.7
Frontal sinus pattern analysis

The frontal sinuses are anatomical structures located at glabella and in the supraorbital region. Like fingerprints, frontal sinuses are believed to be unique to each individual, even twins. The sinuses begin to develop at 6 months to 2 years of age, are recognizable radiographically by 5 years of age, and continue to develop through puberty. Although anatomically stable once developed, the sinuses are subject to modification by pathological and physiological processes (Brogdon, 1998b; Cameriere et al., 2005). Normally bilateral, sinuses can develop unilaterally and are totally absent in 5% of the population (Brogdon, 1998b). Radiographic comparisons of frontal sinuses are conducted using AP or preferably PA views of the skull.

In the past, forensic radiologists argued that simple visual evaluation sufficed in comparisons of AM and PM radiographs (Brogdon, 1998b). The use of tissue paper or acetate overlays was also common. The margins of the sinus were traced from the AM radiograph and then superimposed over the PM radiograph to confirm or refute the ID. This required precise matching of the position, projection, and scale of the AM and PM radiographic films. With the introduction of the *Daubert* criteria, classification systems were adopted that quantify or codify the frontal sinus using morphological characteristics such as area size, bilateral asymmetry, outline of superior margins, partial septa, and supraorbital cells (Cameriere et al., 2005; Yoshino et al., 1987). This system permits the calculation of error rates and testing, thereby meeting the *Daubert* criteria and rendering frontal sinus ID admissible in court.

When working with **radiographic films**, general guidelines for identifications include the following.

1. *Review the patient information blaze* All AM radiographs have patient information recorded on the film, **blazed** or imprinted into an identification bar. Faulty hospital record keeping results in misfiled films. Carefully examine each film's ID blaze to ensure that all films represent the same person. Discrepancies in birth date, social security number, or patient number should alert the examiner to the possibility of a misfiled film. A different name could reflect changes in marital status. If in doubt, confirm such changes with other sources of AM data.

2. *Consider the age of the film* Except for fractures and surgical interventions, morphological changes to bone occur relatively slowly. Despite this, growth or degenerative changes do accumulate over time. Anthropologists reviewing radiographs taken more than two years prior should expect

(a)
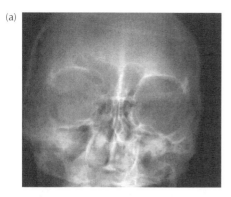

Figure 8.2 Frontal sinus pattern comparison using (**a**) antemortem and (**b**) postmortem radiographs. Note the bilateral asymmetry, the outline of the superior margins of the supraorbital cells, and the configuration of the septum. (*Photo courtesy of the New Mexico Office of the Medical Investigator*)

(b)
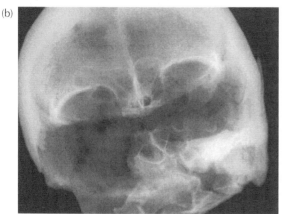

some age-related change. This is especially true in films of juveniles and subadults. Consider the age of the film relative to the date of the decedent's disappearance, rather than recovery, in cases involving prolonged postmortem interval.

3. *Match AM view* Once AM films have been located, review the films to establish the anatomical regions depicted as well as the projection. Even subtle changes in position or projection (Box 8.9) can produce significant changes. Matching position (Box 8.10) and projection (Box 8.11) of the AM film is crucial to establishing a valid identification. Similarity of image quality between the films also aids in accurate assessment (Box 8.12).

4. *Consider the anatomical region* Studies indicate that identification rates vary depending on the anatomical region evaluated. Hogge et al. (1994) found that the skull and cervical spine are most likely to be identified correctly. AP views of the chest are also identified with a high degree of accuracy (91–98%), while AP views of the lower leg score poorly (79%) (Hogge et al., 1994). Murphy et al. (1980) found chest, skull, and abdominal radiographs the most useful. Although always limited by the AM x-rays

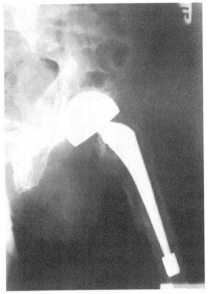
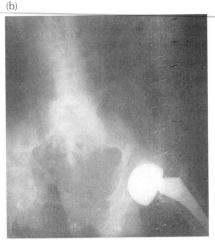

Figure 8.3 Unique feature confirmation by radiography: (**a**) antemortem and (**b**) postmortem radiographs of a right hip, showing an orthopedic appliance. The serial number on the device further supports the identification. (*Photo courtesy of the New Mexico Office of the Medical Investigator*)

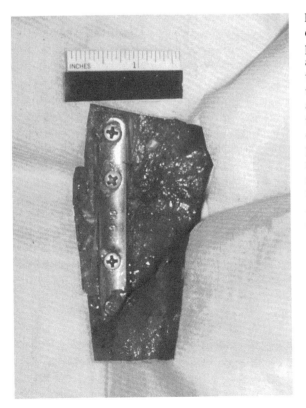

Figure 8.4 A surgical repair of the tibia, with stabilizing plate and screws, provides an example of unique feature confirmation. The identification number of the appliance, visible between the two middle screws, could be used to establish positive ID because the number had been recorded in the surgeon's operative notes. (*Photo courtesy of the New Mexico Office of the Medical Investigator*)

(a)   (b)

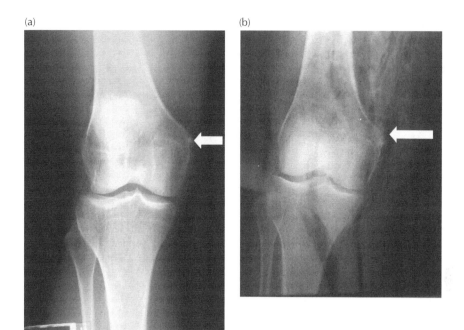

Figure 8.5 Collective concordance of features: arrows indicate the same osteological variant on the antemortem (**a**) and postmortem (**b**) films. At least seven additional features would be needed to positively establish the ID. (*Photo courtesy of the New Mexico Office of the Medical Investigator*)

BOX 8.8

Radiographic identification: the tale of Simon and Steven

On July 10, 2002, Steven, a 55-year-old Native American male, was released from the Santa Fe County Detention Center in northern New Mexico. He met with Simon, a 31-year-old Native American male. Both men were residents of the San Felipe Pueblo. The men lost contact with family and friends and were reported missing shortly thereafter.

On October 12, 2002, a decomposed human torso was discovered in an irrigation ditch in Bernalillo County. Subsequent searches revealed isolated bones scattered over six miles of the irrigation system. Problems in communication among law enforcement agencies excluded Steven and Simon from the missing persons list initially provided to the medical examiner. The remains were retained, unidentified, by the ME.

On November 29, 2002, a maintenance man working at a sewage treatment plant in Santa Fe County found a defleshed human skull. The skull was identified as belonging to Steven by means of dental radiographs. The skull was recovered 25 miles from the detention center (the last confirmed sighting of Steven) and more than 50 miles from the San Felipe Pueblo (where the two men were reported missing).

On December 19, 2002, an employee monitoring the irrigation system in Bernalillo County discovered a human skull trapped below one of the control grates. Searches of the canals and surrounding area produced no additional remains. Using radiographic frontal sinus pattern comparison, the skull was identified as that of Simon. Simon's skull was recovered 57 miles from Steven's skull and 12 miles from the Pueblo where the men were last seen.

A medical examiner's deputy investigator recalled the headless torso recovered from the same irrigation system two months earlier and suggested that the postcranial remains might also belong to Simon. Comparisons of AM x-ray films excluded Simon, but examination of multiple views of the chest and legs confirmed that the remains belonged to Steven. Although no unique features were evident, over 15 concordant characteristics were noted between the AM and PM radiographs. The distance between the recovery sites for Steven's skull and postcranial remains was more than 66 miles, with the irrigation system serving as the mechanism of transport. Using a map of the system and known water release dates, it was possible to reconstruct the path the remains traveled through the system. The cause and manner of death for both men remain undetermined (Komar, 2004).

This case required the use of three forms of radiographic identification—dental, frontal sinus and concordant characteristics—in order to establish the positive identification of both men. This case also highlights the difficulties inherent in identifying partial remains separated by time and distance.

BOX 8.9

Position and projection

Radiology technicians use standardized terminology to describe the placement of the patient or specimen during imaging. To improve communication, anthropologists should familiarize themselves with such terminology. **Position** refers to the physical orientation of the patient. Standard positions include dorsal recumbent and prone. A list of standard radiologic positions is given in Box 8.10. **Projection** refers to the direction of travel of the x-ray beam. Standard projections include anteroposterior (AP) or lateral. A list of projections is provided in Box 8.11.

---

available, anthropologists should remain mindful of the overall utility of each region, particularly in cases presented in court.

5. *Time's arrow* The introduction of new features on PM radiographs is not automatic grounds for exclusion, provided there is sufficient time between

BOX 8.10
Radiographic positions

Radiographic positions describe the position of the body while it is being radiographed.

Supine lying on the back, face up

Prone lying on the abdomen, face down

Lateral lying on one's left or right side; designation of right or left lateral is based on which side is placed against the film

Oblique the body or body part is rotated to a predetermined degree between the prone and supine positions

Right posterior oblique (RPO) the body is rotated, placing the right posterior side next to the film

Left posterior oblique (LPO) the body is rotated, placing the left posterior side next to the film

Right anterior oblique (RAO) rotation of the body to place the right anterior side toward the film

Left anterior oblique (LAO) rotation of the body to place the left anterior side toward the film

Decubitus the patient is lying down and the x-ray beam is parallel to the floor

Dorsal decubitus the body is supine, with x-rays entering from the right or left; may also be called a cross-table lateral projection

Ventral decubitus the body is prone, and the x-rays enter from the right or left side

Lateral decubitus the body lies on the right or left side, and the x-rays enter from the anterior or posterior aspect

AM and PM films to explain the discrepancy. The AM film could represent a preoperative radiograph or an earlier, unrelated examination. Features that appear on AM films but are absent on PM films are reason for exclusion, however, provided the feature represents a permanent morphological alteration. For example, amputations, congenital anomalies, or orthopedic hardware evident on AM radiographs should be discernible in PM films. The absence of such features on PM films that cannot be explained through medical records rules out the putative identification.

Alternative Medical Imaging

The widespread use of new imaging technologies (Box 8.13), including computed tomography and magnetic resonance imaging, has complicated

BOX 8.11
Radiographic projections

Radiographic projection indicates the path of the x-ray beam as it passes through the body.

Anteroposterior (AP) x-ray beam enters the anterior aspect and exits the posterior aspect

Posteroanterior (PA) x-ray beam enters the posterior aspect and exits the anterior aspect

Lateral x-ray beam enters the right side of the body, exiting on the left or the inverse

Oblique x-ray beam strikes the body at an angle, between lateral and AP or PA

Axial a longitudinal angulation of the x-ray beam along the long axis of the body

BOX 8.12
Radiographic image formation

Optical density describes the degree of blackness of a radiographic image. Darker images have higher optical densities. The optical density of a film is controlled through manipulation of three variables: milliamperes, exposure time, and source–image distance. **Millamperes (mAs)** control the quantity of x-rays by controlling the electrical current flowing through the x-ray tube during exposure. Thicker body parts require higher mAs to achieve the proper optical density. Like mAs, **exposure time (s)** measured in fractions of a second (s), controls the quantity of x-rays. The longer the exposure time, the greater the quantity of x-rays striking the tissue. **Source–image distance (SID)** is the distance between the film and the x-ray source. The inverse square law dictates that the intensity of the x-rays depends on the SID. Increasing the SID decreases the number of x-rays striking the film. Despite this, the SID is normally not adjusted to change the optical density of a radiograph. SID is standard or fixed; the mAs and exposure time are manipulated in response to variations in target density.

    Contrast is the difference in density between two areas of an image. Radiographic contrast is comprised of subject contrast and film contrast. The energy of the x-rays, expressed as **kilovoltage (kVp)**, controls the contrast. Manipulating the mAs or SID does not affect contrast. **Detail** is the sharpness of the image. The greater the detail, the more clearly fracture lines will be visualized. Detail is synonymous with definition and sharpness.

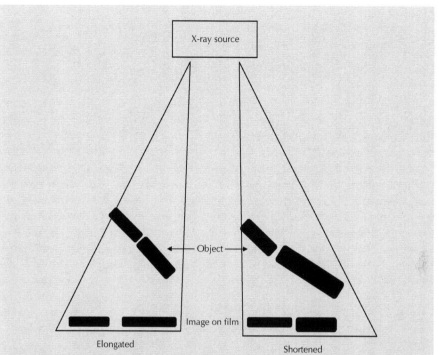

Figure 8.6 Schematic illustration showing how differing alignments among the x-ray source, object, and film produce distortion of the final image

---

REFRESHER

mAs = quantity of x-rays based on the electrical current flowing through the x-ray tube

s   = exposure time, controlling the quantity of x-rays in fractions of a second

kVp = quality of x-rays, based on energy and relating to penetration capabilities

SID = distance from x-ray source to film

Image distortion is inherent in all two-dimensional images of three-dimensional objects. The quality and appearance of the final image reflect the delicate balance of mAs, exposure time, kVp, SID, and the orientation of the x-ray source to the object. **Distortion** is the increase or decrease of image size in comparison to true object size. This is described as the magnification factor. A **magnification factor** of 1 means the image is the same size as the object. There is no distortion along the central ray of the x-ray beam. Shape distortion seen in a film results from the alignment of the x-ray source, the object, and the film (Figure 8.6).

(Bushong, 1988; Callaway, 2002; Devos, 1995)

BOX 8.13
Other medical imaging techniques

In addition to traditional radiology, antemortem images generated by other medical imaging techniques serve as a basis of comparison for iden-tification purposes. Computed tomography (CT or CAT scan) uses an x-ray tub and radiation detectors to produce a series of cross-sectional images (Figure 8.7). "Tomography" is derived from the Greek words *tomos* meaning "slice" and *graphia* meaning "describing". CT scans are used to image the brain, thorax, abdomen, spine, and extremities. CT

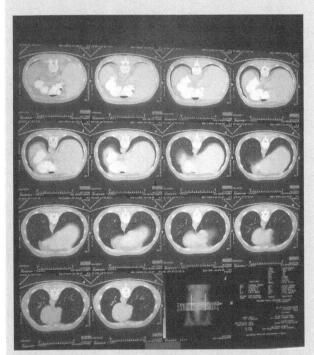

Figure 8.7 Computed tomography (CT) scan of the lower thorax and upper abdomen. Adjacent to the patient information blaze in the bottom right-hand cor-ner is a small reference AP view. (*Photo courtesy of the New Mexico Office of the Medical Investigator*)

generates a series of axial images that can be manipulated, using a process known as **windowing**, to recreate the image in different planes. **Digital geometry processing** produces a three-dimensional image of internal structures from a series of two-dimensional x-ray images.

X-ray slice images are generated using an x-ray source that rotates around the specimen. X-ray sensors are positioned opposite the source. Many images are progressively taken as the specimen passes through the sensing gantry. The individual images are then converted, by the use of a mathematical procedure known as tomographic reconstruction, into the final detailed representation. Intravenous, oral, and rectal contrast media, such as intravenous iodinated contrast, may be introduced to enhance image quality. Although recent improvements in CT technology have reduced radiation dosages and scan times, the radiation exposure from CT scans is several times higher than that acquired from conventional x-ray.

Magnetic resonance imaging (MRI) does not use ionizing radiation but rather strong magnetic fields, pulsed radio waves, and the nuclei of hydrogen atoms in the body to produce images. MRI generates cross-sectional images similar to CT (Figure 8.8). MRI produces good

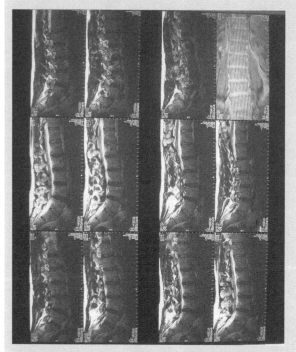

Figure 8.8 Magnetic resonance image (MRI) scan of the lateral lower spine. Patient information is blazed on each individual image; reference image is in the upper right-hand corner. (*Photo courtesy of the New Mexico Office of the Medical Investigator*)

contrast between tissues, particularly muscles, blood vessels, ligaments, and tendons, making MRI the preferred imaging technique of specialists in sports medicine and physical therapists.

In MRI, radiofrequency waves are directed at protons, the nuclei of hydrogen atoms, in a strong magnetic field. The protons are excited then relaxed, emitting radio signals that are detected and processed digitally to form an image. Because protons are abundant in the hydrogen atoms of water, the MRI shows differences in water content and distribution in various tissues. MRI is limited in its ability to image bone, which has little water content, although bone marrow can be visualized.

Two other imaging techniques offer potential but highly limited sources of AM data: nuclear medicine and ultrasound. **Nuclear medicine** produces images with poor resolution compared with other techniques but is used to assess the function of various anatomical systems including the skeletal system. Radiation-emitting materials called **radionuclides** are administered orally, intravenously, or through inhalation. Based on the type of tracer used, the radionuclide localizes to a specific area or tissue type. A radiation-detecting gamma camera measures the amount of radiation coming from the patient. Currently, because of the poor resolution, nuclear bone scans are not a viable basis for identification. **Diagnostic sonography**, or **ultrasound**, imaging relies on sound waves rather than radiation. Sonography can differentiate between solid and fluid structures in the body. Sound waves emitted from a handheld source pass through body fluids and reflect back off solid tissues such as bone. Air and bone act as acoustic barriers, obscuring structures beneath. Ultrasound images are readily available, and the technique is commonly employed in emergency rooms, and by obstetric gynecologists to image fetal development. However, the resolution of the images and the transience of the body's fluid structures constrain the use of ultrasound as a source of personal identification postmortem.

(Haacke et al., 1999; Seeram, 2001)

identification. These techniques generate a source of AM comparative data that could potentially be used in identification. However, MRI and CT display anatomical structures in planes different from those of conventional radiographs. Obtaining a suitable postmortem image for comparison against an antemortem CT or MRI image is problematic for most medicolegal offices unequipped with such technologies. Despite this, Rogers and Allard (2004) contend that cranial CT scans are a more useful modality for identification than conventional x-ray, due to better resolution and greater availability. Smith et al. (2002) used bone morphology evident on serial CT scans as the basis for identification.

The small reference radiographic image (as in, e.g., Figure 8.7) that accompanies MRI or CT films is potentially useful when such AM data are available. Multislice images require a standard AP or lateral view showing the anatomical distribution of the slice image sequence. Although these conventional radiographic views are small, computer enlargement renders them useful for identification purposes. Also, the MRI or CT films provided by hospitals are copies of computer-generated images. Depending on the elapsed time between the radiographic study and death, it may be possible to access the original computer images. Hospitals and other institutions retain these images in varying formats and for different time periods. Students wishing more information on imaging techniques should consult Bontrager (2002), Bushong (1988, 2000), Buxton (2002), Callaway (2002), Carlton and Adler (1992), Devos (1995), Fauber (2004), Haacke et al. (1999), and Seeram (2001).

Having reviewed the methods of establishing positive identification, we now focus on their use in court. Although the availability of antemortem comparative materials limits the methods an investigator can use to establish ID, practitioners should also remain mindful of the admissibility of various methods within their jurisdictions.

IMPACT OF *DAUBERT* ON POSITIVE IDENTIFICATION

With the introduction of the *Daubert* criteria to U.S. federal courts and the *Mohan* ruling in Canada, standards changed regarding admissibility of scientific evidence in court (see Chapter 3). Methods of establishing personal identification must conform to these criteria. In light of these new rules, anthropologists and odontologists have reexamined many ID methods. Christensen (2004a) noted that the use of frontal sinus patterns to establish identity failed to meet the new criteria. The method lacked the empirical testing, known error rates, standards of operation, and objective evaluation criteria necessary for a technique to be admissible. In response, Christensen (2005) and Cameriere et al. (2005) published studies to correct these limitations. By improving on existing classification systems and establishing error rates, both studies provided anthropologists with the necessary data to allow for the use of frontal sinus patterns as a means of identification.

In addition to developing more stringent methodologies, researchers have been reviewing the accuracy of practitioners as well. For instance, Koot et al. (2005) compared the accuracy rates in individualization of 12 examiners from the forensic science community at large to those of 10 forensic anthropologists. The researchers provided each set of examiners with simulated AM and PM radiographs of the hand. While the overall accuracy of the forensic examiners was 95%, the anthropologists correctly identified all potential matches. Hogge et al. (1994) found disparity in the accuracy rates of radiographic identifications based on the experience level of the examiners and argued for the use of trained radiographic interpreters in cases of forensic identification. Pretty et al. (2003) tested the accuracy rates of odontologists who used digitized radiographs to perform identifications. A statistically significant increase (from 85% to 91%) in accuracy was evident when the IDs were made by experienced practitioners.

Anthropologists whose duties include establishing personal ID should review the standards for the admission of scientific evidence in their jurisdictions, as well as the admissibility of the methods they plan to employ. Practitioners also benefit from conducting intraobserver tests using their methods of choice, as this information is now frequently requested in court. Knowing before you take the stand both your own error rate and the known error rate of the method will facilitate the process of declaring the identification in question to be admissible.

Fingerprints, DNA, and dental and medical radiographs are court-accepted methods of establishing ID. We will now consider other techniques related to the identification process that have not yet attained acceptance.

PHOTOGRAPHIC SUPERIMPOSITION AND FACIAL RECONSTRUCTION

A number of proposed identification-related techniques have captured the public's attention and have been the subject of countless documentaries and crime shows. Although popular in the media, photographic superimposition and facial reconstruction fail to meet the *Daubert* criteria and are inadmissible in a U.S. court of law. It is important to recall that such methods are used for investigative purposes, such as generating leads, rather than to establish positive identification.

Photographic superimposition uses computer technology to layer a photographic image over an unidentified skull. Concordance between soft tissue landmarks and bony features supports the inclusion of the individual for further testing, while inconsistencies result in exclusion (Figure 8.9). The method requires that the skull be oriented to the exact position of the face in the photograph. Originally called video superimposition, the method relied on mixing a live video image of the skull with a computer image of the photograph. For a historical overview of the technique, see Austin (1999). A less expensive method, using a photograph of the skull and an acetate overlay of the photo, has also been reported (Shahrom et al., 1996).

Superimposition is not admissible in North American courts as a method of establishing positive identification. The method lacks the necessary rigorous scientific testing, known error rates, and acceptance. However, much as with polygraphs or lie-detector tests, investigators continue to use superimposition as a preliminary investigative tool to include or exclude tentative identifications. Publications have reported a more accepted role for superimposition in Hungary (Angyal and Derczy, 1998) and India (Jayaprakash et al., 2001).

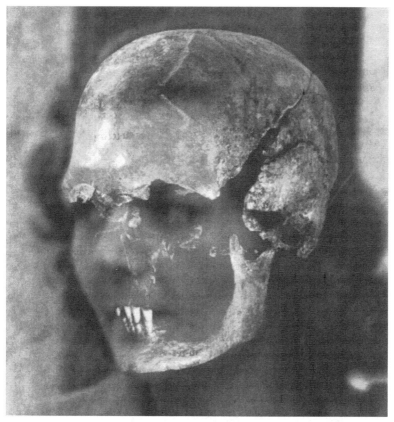

Figure 8.9 Photo superimposition, using an antemortem photo of the suspected individual and the cranial remains recovered. Note the concordant alignment of skin and bone along the jaw line and supraorbital torus. Although useful in including or excluding individuals from further testing, superimposition alone cannot be used to positively establish identification. (*Photo courtesy of the New Mexico Office of the Medical Investigator*)

Facial reconstruction is the artistic reproduction of the soft tissue features of an individual. Published tissue depth standards serve as a basis for determining muscle and skin thickness. Spacers corresponding to the known tissue depth are applied directly to the skull or cast at prescribed anatomical landmarks. Clay is then molded to the skull and the artist renders an approximation of the soft tissues and features of the face (Figure 8.10). Hair and accessories such as glasses or facial hair can also be applied.

Facial reconstruction goes by many names, including 3-D craniofacial reconstruction (DeGreef and Willems, 2005), facial approximation (Hayes et al., 2005; Stephan, 2003), and forensic reconstruction (Aulsebrook et al., 1995 Tyrrell et al., 1997). Rhine (1990a) suggested "facial reproduction" as the

(a)

(b)

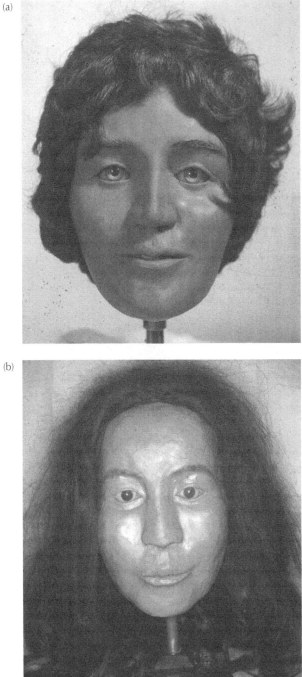

Figure 8.10 (a) and **(b)** Examples of three-dimensional clay facial approximations. Features such as wrinkle patterns, lip and nose shape, and hair style are entirely at the discretion of the artist. (*Photos courtesy of the New Mexico Office of the Medical Investigator*)

precise and preferred term but with little effect. Aulsebrook et al. (1995) stated that facial reconstruction was divided into four categories: (1) replacing or repositioning damaged or distorted soft tissues onto a skull, (2) the use of photographic transparencies and drawing, (3) photographic superimposition, and (4) three-dimensional reconstruction of a face over a skull using clay. Most other researchers recognize a distinction between categories 1 through 3 and the fourth category, identifying clay reconstruction as facial reconstruction.

Although clay is the traditional medium of reconstruction, the use of computer imaging programs or two-dimensional drawing has also been proposed (De Greef and Willem, 2005; Shahrom et al., 1996; Tyrrell et al., 1997). A purely mathematical model designed to create a "face space" by applying principal component analysis to an unknown skull has also been reported (Turner et al., 2005). Hayes et al. (2005) suggested a combination of clay modeling to create an approximation and computer graphics to enhance features and allow for easy modification as new information is received.

Facial reconstruction (FR) fails to meet the standard of scientific acceptance and validity required for admissibility in court. Stephan (2003) contends that the term itself implies everything the method is not—scientific, technical, exact, or credible. Stephan argues that marketing and political ploys common within the FR community have misled the public through impressive but unjustified claims of success and validity. Other researchers find similar fault with the method's lack of scientific rigor. For example, early standards for tissue depth were established using needle probes on cadavers with dehydrated tissues. Recent studies have revised the depths using data from ultrasonography, CT scans, and radiographs (DeGreef and Willems, 2005; Phillips and Smuts, 1996; Tyrrell et al., 1997). Sample size has also proven problematic. Phillips and Smuts (1996) argued that the tissue depth measurements from 32 individuals were "representative of the variation in facial tissue thickness of South Africans of mixed racial origins." In no other area of physical anthropology would such a small sample size be considered representative. Swan and Stephan (2005) attempted to develop predictive models to estimate eyeball protrusion based on osteometric variables such as stature. The models were described as weak, generating considerable error in 65% of test cases.

The fundamental problem with facial approximation is that it ignores the paradigm of pattern recognition that is the basis of all other acceptable forms of identification. While tissue depth standards may allow for a reasonable estimation of facial structure, there are no credible ways to estimate nose, ear, or lip shape, wrinkle patterns, eye color, hair style, or facial hair patterns. These features are what family members recognize and respond to. What law enforcement and the general public fail to realize is the extent to which "forensic" reconstructions are an act of artistic expression rather than scientific endeavors. Although successful reproductions are often publicized, unsuccessful attempts receive little attention. Stephan and Henneberg (2006) offer a rare report of a case in which facial reproduction was not successful in a forensic setting. They found that a strong resemblance of the facial reconstruction to the target individual did not equate to recognizability.

The problem forensic reconstruction seeks to address, namely, identification of unknown skeletal remains, continues to challenge the forensic community. While the problem is real, the answers offered by facial approximation are not. The research question to be answered is whether FR can ever be rendered sufficiently credible, or whether it will remain an artistic rather than scientific exercise. At present, the highly speculative nature of reconstructions suggests that investigational time and resources are best spent elsewhere.

CONCLUSION

The category of identification reflects the degree of certainty, with "Doe" designation revealing no basis for the ID, while positive identifications indicate verifiable evidence of identity. The methods used to establish positive ID rely on pattern recognition. The introduction of the *Daubert* criteria has challenged some methods of ID and has resulted in increased empirical testing. Methods that fail to meet these standards, such as facial reconstruction and video superimposition, continue to play a limited role in the resolution of unidentified persons, although they cannot serve as the sole basis for an identification.

All deaths, whether resulting from natural or unnatural causes, require that decedents be positively identified. Establishing identity is vital to issuing death certificates, prosecuting criminal acts, and providing answers to next-of-kin. Anthropologists play a key role in the identification process. Despite the introduction of DNA testing, the majority of identifications rely on dental records, radiographic comparisons, and fingerprint analysis. As such, the responsibility of identifying the dead will continue to fall to forensic anthropologists and odontologists.

CHAPTER 9

Mass Death and International Investigations of Human Rights Violations

Forensic anthropologists have become increasingly engaged in mass death investigation (see, e.g., Andelinovic et al., 2005; Hunter et al., 2001; Komar, 2003a; Skinner et al., 2003; Steadman and Haglund, 2005; Vanezis, 1999). Mass death results from catastrophes due to natural disasters such as flood and hurricanes or to mass disasters such as plane crashes, as well as conflict, which includes acts of war, terrorism, and genocide. While extensive loss of life unites such disasters and conflicts, the distinctive needs and goals of each prompt different investigative approaches and protocols. Medicolegal teams responding to **natural disasters** focus on body recovery and personal identification of victims. As these events do not represent criminal activity, protocols for death investigations in natural disasters closely resemble those developed for accidental or natural deaths. The significant difference is scale, in terms of both the number of individuals involved and the geographic area affected. **Mass disasters** result from human action or inaction. Such disasters may result in court proceedings; however, these are often civil trials involving liability claims, wrongful death, or negligence. Protocols invoked for mass death vary, depending on the cause and nature of the event. With mass disasters, the medicolegal response centering on body recovery and identification is coupled with perimortem event reconstruction, which focuses on the event as a whole. Deaths resulting from **armed conflicts** are homicides. Investigations may encompass individual deaths, temporally or spatially isolated events, or patterns of violence representing a totality of war crimes. Protocols must include victim recovery and identification, as well as trauma assessment and reconstruction of perimortem circumstances.

While many of the subjects treated in earlier chapters concerning crime scene investigation, body identification, perimortem circumstances, and postmortem interval are relevant to situations of mass deaths, there are other topics that are specific to such contexts. Therefore, in addition to addressing issues relating to war crimes charges, including genocide, this chapter will examine establishing group identity and personal identification. We will also consider the differences between humanitarian and medicolegal responses to mass death, how jurisdiction is established in postconflict regions, and the responsibilities of various governmental and nongovernmental agencies. Finally, issues relating to mass graves, commingling, and investigator objectivity will be discussed.

WAR CRIMES

Acts of war are episodes of aggression that terminate the normal international law of peacetime and activate the international law of war. **War crimes** are those violations of the laws of war or international humanitarian law that incur individual criminal responsibility (Ratner, 1999). More information on the laws of war and international humanitarian law is available in Detter (2000), Gutman (1999), and Roberts and Guieff (2000). The first true war crimes trial was that of Peter von Hagenbach (Box 9.1), who was tried in Austria in 1474 and sentenced to death for wartime atrocities (Ratner, 1999).

Acts defined as war crimes (Detter, 2000; Gutman, 1999) include the following:

- Genocide.
- The blocking of humanitarian aid.

BOX 9.1

The first international war crimes tribunal

Peter von Hagenbach was beheaded in 1474 following his conviction by the first international criminal tribunal. The tribunal consisted of 28 judges drawn from throughout the Holy Roman Empire. By appointing the court to adjudicate von Hagenbach's case, the precedent was set for future international courts, including the Nuremberg trials and the UN criminal tribunals of Rwanda and the former Yugoslavia.

Peter von Hagenbach had been appointed by the Duke of Burgundy to maintain order in Austria's territories. He ultimately used the appointment to terrorize the population. Charged with violations of "the laws of God and man," including murder, rape, and perjury, von Hagenbach claimed he was merely following the duke's orders. The "just following orders" defense has since been raised by defendants at Nuremberg and war crimes trials for the former Yugoslavia, but it is a defense rejected in international law (Grant, 2006).

- The use of prohibited means of warfare, including poison or biological weapons designed to cause unnecessary suffering.

- Perfidy and treachery, two closely linked but distinct concepts. **Perfidy** is causing an individual to believe a falsehood; persons feigning surrender under a flag of truce or combatants pretending to be civilians are acting perfidiously. **Treachery** involves acts of betrayal that actually cause harm, such as killing those who have surrendered. Although some acts of deception are permitted under international law (such as the use of decoys and camouflage, and the dissemination of false information), acts of perfidy and treachery are punishable as war crimes.

- **Pillage**, which is the unlawful acquisition or destruction of material goods. Pillage must be differentiated from **requisitioning**, which is the legal taking of necessities such as food, fuel, and medicine from civilian populations for the use of an army of occupation. Requisitioned goods must be paid for, when possible, or receipt must be acknowledged.

- **Terrorism**, defined as violence with a political or social purpose, designed to intimidate or incite fear. Terrorism requires three parties: the terrorist, the victim, and a third party to be intimidated by what happened to the victim.

- Unjustified destruction of property, including seizure or willful destruction of cultural properties.

- Targeting of civilians or other protected persons.

Of all defined war crimes, none carries the severity of penalty or the emotional impact of genocide.

Genocide

The crime of **genocide** is legally defined in Articles II and III of the 1948 Convention on the Prevention and Punishment of Genocide (Resolution 260[III] A of the United Nations General Assembly). Article II describes two elements of the crime: the *mental element*, including the "intent to destroy, in whole or in part, a national, ethnical, racial or religious group" and the *physical element*. The physical element encompasses five crimes punishable as genocide: conspiracy to commit, incitement to commit, attempt to commit, complicity (Box 9.2), and successful acts of genocide.

In addition to these five crimes, Resolution 260 identifies five acts considered evidence of the intent to commit genocide: (1) killing members of the group; (2) causing serious bodily or mental harm through widespread torture, rape, or mutilation; (3) deliberately inflicting conditions of life calculated to destroy a group, including deprivation of resources, detention in camps, or forced relocation; (4) prevention of births; and (5) forcible transfer of children.

BOX 9.2

Complicity

Complicity is legally defined as the union with others in an ill design; one who is an associate, confederate, or accomplice in the ill design is said to be in complicity with it. Being in complicity is acting knowingly, voluntarily, and with intent shared by the principal offender. Guilt of complicity implies that one was present and aiding in the crime or was absent at the time of the commission of the crime but had previously advised or encouraged the act. Regardless of how heinous the crime, the following are not legally considered acts of complicity:

- Silence or acquiescence
- Mere presence during the act
- Falsely denying having knowledge of the crime

(Garner, 2001)

The law affords protection to four defined groups: national, ethnic, racial, and religious. Their respective identities are defined as follows.

- A **national group** is defined by common country of nationality or origin.
- An **ethnic group** shares common cultural traditions, language, or heritage.
- A **racial group** is defined by physical characteristics.
- A **religious group** shares common religious creeds, beliefs, doctrines, practices, or rituals.

Successful prosecution of genocide requires two evidentiary components—intent and action. Intent is revealed through purposeful action. Intent can be demonstrated by orders or statements but is most often inferred from patterns of action, such as forcible detention and transfer of victims or clandestine mass burial. Intent differs from motive. Motive is the perceived justification for committing an act (see Chapter 2 for an extended discussion of intent and motive). Examples of genocidal motives may include national security, land appropriation, ancient hatred, or maintaining racial purity. If, however, the actions of the accused reveal intent, the motive for the killing is immaterial under the law.

Scale of action has no relevance to the identification and prosecution of genocide. It is not the number of victims that defines an act as genocide but the intent. The definition of genocide reads "in whole or in part." A perpetrator may be guilty of genocide after killing only one individual, provided the intent was a larger plan to destroy the group as defined.

Group Identity versus Personal Identification

As the legal definition includes the notion of group membership, prosecuting acts of genocide requires establishing the social identity of the victims. Cultural, ethnic, religious, or national identity extends beyond traditional forensic notions of identification, in which the decedent is named, to concepts of group membership, self-identification, perceived membership, and systems of recognition. The crime of genocide requires that the victims share the defining class characteristics of the targeted group. Determining identity can prove more difficult than the forensic act of establishing identification.

Ethnic identity is derived from formal criteria, such as country of birth or religion, or from self-designation. Given the ascriptive nature of ethnicity, defining ethnic groups in terms of descent or ancestry can be problematic. Referencing the origin of ancestors, for example, rather than the use of contemporary sources of identification excludes individuals who adopt their ethnic identity based on marriage or relocation and includes those who have renounced the culture of their ancestors. Definitions of ethnicity may also include issues of status and prevalence. Ethnic affiliation includes individuals of a numerical minority who constitute a social, political, or economically subordinate group. In this sense, ethnic groups have been defined as those who cannot identify with the ruling class (Wilkinson, 1999). This sense of marginalization, perpetuated both within and outside the affected ethnic group, is a common precursor to acts of genocide, which usually erupt in areas with long-term, institutionalized differential access to wealth and power based on group affinity.

Ethnic identity in the context of genocide literally becomes a matter of life and death, raising issues of self-identification and systems of recognition. Self-categorizing represents an individual's desire to claim membership within an ethnic group. Self-identification is flexible. Ethnicity is founded on shared culture, which is learned and changeable. Individuals can adopt cultural codes and acquire ethnic traits (Lindholm, 2001). Self-identification allows individuals to renounce their membership within an ethnic group. When one's group is under genocidal attack, the motivation to do so can be considerable.

The malleable nature of self-identification conflicts with social systems of recognition and perception. How an individual choses to self-identify may not coincide with how others perceive that individual. While ancestry may be reflected in physical features that are undeniable clues to racial categorization, ethnic identity is truly revealed only in cultural practices, such as language, rituals, food choice, and ceremonies. How outside groups, including those perpetrating the genocide, recognize or identify members of the targeted group can differ from how members of the oppressed group recognize themselves.

In many ways, investigators tasked with establishing the ethnic identity of the victims of genocide must rely on the same system of recognition as the perpetrators of the crime. Reliance on external indicators, such as clothing, is often overemphasized. Such physical characteristics survive and are recognizable postmortem. In situations such as late twentieth-century

Bosnia, in which there was remarkable uniformity of material culture among all three ethnic groups (Box 9.3), artifact and clothing styles provided little evidentiary value. More difficult to acquire, and yet more powerful and relevant, is evidence of cultural components such as language and ritual. Such behavioral aspects of identity are not always captured in the archaeological record.

Further confounding investigations is the highly variable nature of culture expression within a group. Physical isolation, as well as decades or centuries of institutionalized segregation, can result in relatively uniform culture expression. However, as personal expression of cultural affinity remains variable, establishing ethnic identity postmortem is often more a statement of statistical probability or likelihood, rather than an absolute. Most recent genocides have involved conflicts between ethnic groups in a single nation-state, for example, Rwanda, the former Yugoslavia, and Iraq.

The definition of genocide also includes the targeting of groups based on **racial identity**. Definitions of race vary (Chapter 5). Some coincide with the biological definition adopted by the United Nations (see, e.g., Rhine, 1990b). Others (e.g., Graves, 2001) argue that race has no absolute content. Lindholm (2001) contends that genotypical analysis of clusters of shared traits may be used to trace ancient migration patterns but that clusters do not coincide with culturally formed concepts of race. Wilkinson (1999) argues that while race is an anatomical designation based on biological criteria, it is also a group definition based on how individuals are perceived and treated by others in response to these defining external characteristics.

Investigators tasked with establishing the racial identity of genocide victims may chose to focus on the biological aspects of race contained in the UN definition of genocide. Biological differences between groups can be identified

---

BOX 9.3
Ethnic groups in Bosnia

The Bosnian conflict involved three distinct ethnic groups or "constituent peoples," as they are formally known: the Bosniaks, the Serbs, and the Croats. The term "Bosniak" has replaced the former designation, "Muslim", in part to avoid confusion with the strictly religious connotations of the word. As of 2000, approximately five years postconflict, the distribution of the groups was 48% Bosniak, 37% Serb, and 14% Croat (www.cia.gov—Field Listing of Ethnic Groups, 2000).

The Bosniaks are descendants of indigenous converts to Islam during Bosnia's Ottoman period. They predominantly adhere to Islam and share a common language, Bosnian. The majority of Serbs belong to the Serbian Orthodox Church, although some are atheists. Their language is called Serbian. Croats are predominantly Roman Catholic and speak Croatian.

through measures of distance. Methods include metric and nonmetric trait assessment (see Chapter 5). DNA testing and mitochondrial haplotypes also provide sources of identification (see Chapter 8). Investigators embracing more culturally defined notions of race may forgo biological evidence in favor of evidence derived from material culture or mortuary practice. The same criteria may also be used when one is investigating genocide of religious groups.

Genocide of a specific national group may be the most difficult to address. In life, nationality is flexible, altered by marriage, migration, or choice. However, in death the nationality of the victim cannot be assumed merely from the location of the death or burial. Individuals not holding citizenship can be found within a nation's borders, and those of the targeted nationality may have temporarily or permanently relocated to another country. Establishing national identity to a legal standard of proof requires the positive identification of the victim, thereby providing access to supporting social data such as birth certificates or passports.

Definitions of genocide, we re-emphasize, do not focus on notions of scale. It is the "intent to destroy" a specifically designated group, not the scope of the destruction, by which investigators identify genocide. Anthropologists should not rely solely on the presence of any one feature—large-scale inhumations, violent trauma, or clandestine burials—to identify possible acts of genocide; rather, they should focus on any symbolic, documented, or implied evidence of the cohesive identity of the victims and the intent of the perpetrators.

PERSONAL IDENTIFICATION

In each mass death scenario, regardless of scale, the question is raised: To what extent is identification of the victims possible or practical? While the positive identification of every individual is the ethical, professional, and humanitarian ideal, the ability to realize this ideal depends on two crucial factors. The first is resources. In addition to the enormous financial costs of such endeavors, large-scale identification efforts require significant commitments of personnel, equipment, facilities, and support staff. The second consideration is the availability of the antemortem social data necessary to establish ID. Without known sources against which to compare the unidentified remains, there can be no scientific basis for establishing identification. To illustrate the evolution of large-scale identification efforts, we will examine recent examples from Rwanda, Bosnia, and Kosovo.

In 1994, the ethnically motivated genocide of approximately 800,000 Tutsis by Hutu extremists known as the Interahamwe was carried out in 100 days. Although some guns were used, the predominant weapon was the machete. Minimal identification efforts were mounted by the United Nations International Criminal Tribunal for Rwanda (ICTR). Although financial and manpower resources were limited, in this case the primary constraint was a lack of antemortem data. DNA testing in forensic applications was still in its infancy in 1994, and testing on such a massive scale was inconceivable. Poverty, purposeful destruction of records, and limited preconflict social infrastructure also confounded identification efforts. Traditional sources of data, such as fingerprints and dental or medical records, were nonexistent. In an effort to generate identifications, investigators held "clothing shows." The clothing and personal effects from unidentified victims were cleaned and laid out, grouped by individual. Survivors were invited to view the clothing from victims recovered from their area. If the next-of-kin recognized an article of clothing, the family was interviewed and asked to describe the physical characteristics of their missing loved one. Anthropologists compared these traits to the physical characteristics of the decedent. If the individual could not be excluded, a presumptive match was made and the body was returned to the family. Despite the best efforts of investigators, fewer than 20 identifications were made using this method (Physicians of Human Rights, Rwanda Project, 1996).

Anthropologists and other members of the forensic community encountered similar problems when responding to the "ethnic cleansing" of Bosniak Muslims in Bosnia. From 1992 through 1995, hundreds of thousands of Muslim Bosniaks were killed by Serbian forces (Figure 9.1). Again, the United Nations International Criminal Tribunal for the Former Yugoslavia (ICTY) headed the investigation. Identification of the victims was given lower priority than evidence collection, description of injuries, and cause of death determination. In 1996 Physicians for Human Rights (PHR), an American-based organization, developed and implemented an identification effort in Bosnia. As in Rwanda, limited access to health care and the destruction of existing social data during the conflict hindered traditional identification methods in Bosnia. Building on the idea of the clothing shows used in Rwanda, PHR produced "the Book of the Dead," containing photographs of the clothing and personal effects of victims that was shown to survivors. PHR also attempted to compensate for the lack of social data by conducting extensive interviews with survivors who had reported missing family members. The next-of-kin responded to a standardized questionnaire to describe physical attributes of their loved ones—height, weight, hair color—as well as any known medical history such as broken bones or past surgeries. This information was then compared against the biological profiles of decedents generated at autopsy. In cases of agreement between the antemortem and postmortem characteristics, a presumptive identification was made. In extremely rare instances, DNA testing was also used to confirm identity. Again, as in Rwanda, these efforts generated few (<100) identifications (Komar, 2003a). In 2000 the International Commission on Missing Persons (ICMP) introduced the widespread use of mtDNA as the predominant method of identification in Bosnia, and considerable success was reported (Huffine et al., 2001). As of 2004, more than 2,500 DNA-supported identifications had been completed (Klonowski et al., 2004). As of January 2007, the number had risen to 10,989 (www.ic-mp.org).

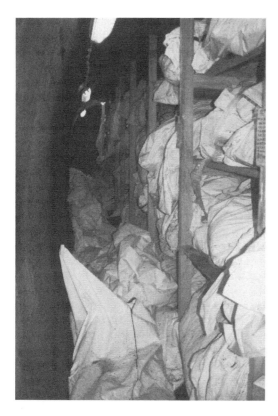

Figure 9.1 Temporary body storage facility in Tuzla, Bosnia. Remains of individuals recovered from the Tuzla area (including the massacre site of Srebrenica) were temporarily stored in a series of tunnels under a factory while proper autopsy and storage facilities were being constructed. The tunnels of Tuzla became a symbolic meeting place for mothers and widows seeking to learn the fate of their missing loved ones. *(Photo by Debra Komar)*

In 1995 intervention by the international community resulted in the end of the conflict in Bosnia. Unfortunately, the "ethnic cleansing" of Muslims by Serbs then began in another region of the former Yugoslavia—Kosovo. In 2000 UN peacekeeping forces managed to end the violence in Kosovo, and the ICTY began investigations into potential charges of genocide in the province. This time, the ICTY mandate prioritized personal identification of victims. Kosovo is a predominantly rural, poor region with little access to health services; again, very limited dental and medical records were available as potential sources of antemortem data. Although initial efforts at identification relied on the clothing shows (Figure 9.2) and third-party reported data methods used in Rwanda and Bosnia, resources for the large-scale collection (Figure 9.3) and testing of DNA were ultimately made available, and DNA became the primary method for establishing identification.

In the aftermath of investigations in the former Yugoslavia and on the site of the World Trade Center, the large-scale use of DNA has become the accepted form of identification in mass death scenarios. Even so, anthropologists continue to play a critical role. Although mtDNA can be extracted from decomposed remains, it is important to recall that mtDNA is not unique to an individual (see Chapter 8 for more details on the use of mtDNA

Figure 9.2 A "clothing show" in Djakova, Kosovo, in 2000. Clothing and personal effects from unidentified victims recovered from mass graves in the area were cleaned and displayed, grouped by individual. Surviving members of the community were invited to view the clothing. If they recognized any articles, they would meet with an investigator, who recorded the family's account of physical characteristics of the missing person (such as height, weight, and prior medical history). This information was then given to an anthropologist, who compared it against the biological profile of the body associated with the clothing. If the body could not be excluded based on morphological characteristics, DNA testing was ordered to confirm the identity. (*Photo by Debra Komar*)

Figure 9.3 DNA sample collection point in Djakova in 2000. Surviving matrilineal relatives were encouraged to provide blood samples to the International Commission on Missing Persons. DNA obtained from the blood samples allowed technicians to develop mtDNA profiles, database of which were subsequently compared against a mtDNA profiles of unidentified decedents. (*Photo by Debra Komar*)

in forensic identifications). In areas such as Bosnia, many of the survivors providing DNA samples as known comparatives are missing several family members, further confounding ID efforts. While mtDNA may successfully identify family groups, it cannot differentiate closely related individuals, such as cousins or brothers. Furthermore, the possibility remains that unrelated individuals may share mtDNA sequences, resulting in false positive tests. It remains standard for anthropologists to confirm mtDNA matches by reviewing the physical characteristics of all individuals against whatever antemortem data are available.

The use of nuclear DNA as the basis for identification in such scenarios is rarely a viable option. First, the decomposed and sometimes heat-altered nature of remains prevents the successful extraction of nuclear DNA. Second, comparative standards are overwhelmingly not available. For nuclear DNA testing to be useful either the individual must have provided a DNA sample to a database during life or the appropriate surviving immediate family members (i.e., parents) must provide comparative samples. While nuclear DNA provides a more specific finding, it also requires a more specific basis for comparison.

HUMANITARIAN VERSUS MEDICOLEGAL RESPONSE

In postconflict regions, as well as in the aftermath of natural disasters, the nature of the response by the forensic community falls into two categories: humanitarian efforts and medicolegal investigation. **Humanitarian efforts** concentrate primarily on the needs of the living, even when dealing with the dead. Humanitarian response focuses on the recovery and identification of the deceased. Bodies are removed from affected areas prior to the return of surviving inhabitants and transported to permanent or temporary morgue facilities. There, the remains are examined, identified, and returned to their next-of-kin. Humanitarian efforts are the primary response to natural disasters where deaths do not represent any form of criminal act. A humanitarian response can also be enacted in postconflict regions, provided it does not interfere with or undermine medicolegal investigations.

A **medicolegal response** by the forensic community to a mass death event contains all the elements seen in traditional death investigations. Scene, autopsy, and investigative protocols must be established and maintained to ensure sufficient quality and rigor, despite the scale of the event or the number of victims. Investigators recognize that each death represents potential criminal activity, as in acts of genocide or possible war crimes against civilians. Thus, to meet the needs of any subsequent prosecution, there must be full documentation of the scene, complete autopsy to establish the cause of death and collect evidence, as well as the identification of individual decedents. Medicolegal response requires all participants to have adequate training, certification, and expertise to fulfill their assigned roles.

A primary concern in medicolegal responses is the issuance of death certificates. If the affected region had an existing social tracking system that included death certification, certificates will be required for all decedents. This allows next-of-kin to resolve legal issues. Positive identification of all victims will be required or must be attempted. Investigation and autopsy to establish the cause and manner of death are also necessary to allow certification. Prevailing local standards or laws dictate the professional qualifications of those who can certify death and sign or issue death certificates.

If the region lacked a formal medicolegal system prior to the conflict, it is not necessary or even valid to issue death certificates solely for decedents of the conflict. When the death certificate was not a required, recognized legal or social document, and the government had not actively tracked births or citizenship, the issuance of a death certificate provides no social or political benefit. Should the perpetrators of the violence be subject to prosecution, formal autopsy reports will suffice in identifying individuals (if possible) and in documenting their injuries or circumstances of death.

Establishing Jurisdiction

The participation of international forensic teams in response to mass disasters or conflict raises an important question: What gives foreign nationals the right or authority to enter a country, conduct investigations, collect evidence, or issue death certificates? Board certification or licensing afforded to professionals in their home countries will not be valid in foreign lands. Sworn police officers have no authority outside their own geographic jurisdictions. So how are such investigations carried out, and what circumstances must exist to render such investigations valid or legal?

As in all other venues of forensic inquiry, establishing authority in a mass death situation is a matter of jurisdiction. While certain aspects of a humanitarian response (such as body collection or burial) can be carried out by local citizens, family members, or volunteers, a medicolegal response requires the participation of law enforcement and forensic professionals. However, the qualifications of the participants alone are not sufficient to constitute medicolegal authority. Legal jurisdiction must be established.

Establishing jurisdiction in postconflict areas can prove problematic. If a local governmental structure is functionally intact, that government may invite foreign teams to assist or head investigations. However, if the conflict involved a civil war, genocide, or political coup, the area may be without a formal government. In the absence of governmental structure, functional court systems, and valid laws, it is impossible to establish jurisdiction locally (see Chapter 2 for a review of how jurisdiction is established). Many regions of the world have no medicolegal system. Even if a functional government remains following the violence, the infrastructure needed to mount an effective response locally is usually nonexistent.

Establishing medicolegal jurisdiction in postconflict areas without a functional government requires a number of steps.

- First, the end of the conflict must be recognized, and the power or authority of all prior governmental bodies must be dissolved.

- Second, an interim government must be established. This interim government must include a judicial branch. The temporary government can be local or international. For example, following the end of the Bosnian conflict in 1995, the United Nations established the Office of the High Representative to serve as the interim authority while the local people reestablished self-governance. The International Criminal Court (ICC) in The Hague serves as the judicial branch of the United Nations.

- Third, the interim government must have the authority to issue mandates to organizations or agencies to fulfill specific tasks. Jurisdictional authority is derived from these mandates, according to the time frame and responsibilities outlined in the mandate. It is through these mandates that temporary medicolegal authority can be granted to foreign investigative teams.

Agencies

Two types of agencies are recognized in large-scale international human rights investigations: governmental agencies and nongovernmental organizations. **Governmental agencies** are organizations with established personnel, protocols, and mandates. These agencies are capable of expanding their normative operations to include additional geographic jurisdictions or circumstantial responsibilities. Governmental agencies derive their authority and financial support directly from the government that creates them. Examples of forensic or medicolegal governmental organizations in the United States include DMORT (Disaster Mortuary Operational Response Team, www.dmort.org), JPAC (Joint POW/MIA Accounting Command, www.jpac.pacom.mil), and EMAC (Emergency Management Assistance Compact, www.emacweb.org). International examples include the International Criminal Tribunal for the Former Yugoslavia (www.un.org/icty), the branch of the United Nations tasked with forensic investigations of war crimes in Bosnia and Kosovo.

Nongovernmental organizations (NGOs) are independent agencies created by individuals without direct governmental supervision. Typically nonprofit, such agencies rely on funds obtained privately or through governmental grants or subsidies. The jurisdictional authority of NGOs is derived from the mandates issued to them by interim or permanent governments. These mandates are valid for a prescribed period of time, and the role and responsibilities of the agency are clearly defined. Examples of NGOs include the Argentine Forensic Anthropology Team (www.eaaf.org), Physicians for Human Rights (www.phrusa.org), and the International Commission on Missing Persons (www.ic-mp.org). Once a mandate has been issued, the NGO is responsible for hiring the appropriately qualified personnel to fulfill the mandate. The jurisdictional authority of those hired is derived solely from the NGO's mandate, mooting the issue of local certification or licensing for professionals. For example, in certain instances, foreign pathologists employed by an NGO with a medicolegal mandate can issue death certificates.

Even when jurisdiction can be clearly established, large-scale investigations continue to present unique challenges to anthropologists. Primary among these is mass graves.

MASS GRAVES

Mass graves are complex deposits in the archaeological record. Proper excavation, documentation, and interpretation of mass graves require patience, experience, special training, and equipment. Widespread acceptance of what constitutes a mass grave has not been reached; specifically, no minimum number of individuals has been established that classifies a grave as "mass." An attempt to define grave type based on taphonomic variables found significant differences in decay rates and grave dynamics among single, multiple (2 to 5 individuals) and mass (6 or more individuals) graves (Komar, 2001). Other researchers recognize only the distinction between single and multiple graves (Vanezis, 1999).

The challenges posed by mass graves begin with attempts to locate them. Despite their size, mass graves are formed illicitly and with the intent to avoid detection. Locating mass graves relies primarily on witness statements. Survivors, those involved in the burials, or even the perpetrators lead investigators to the site. Because the scale of the graves renders them visible from space, satellite photography is also useful in detecting grave locations. Tests were conducted in Bosnia in 1999 using cadaver dogs to locate graves but with little success. Moreover, since most graves are created by means of heavy equipment and are substantially deeper than conventional graves, the usefulness of handheld probes and metal detectors may be limited, although ground-penetrating radar has proved beneficial (Miller, 2003; Schultz, 2006).

Once a mass grave has been located, excavation requires the use of heavy equipment such as backhoes or excavators. The use of such equipment should be limited to the removal of the overburden and peripheral trenching to expose profiles. The entire grave should be defined and exposed prior to the removal of any remains. This can be a lengthy process, and therefore a long-term commitment for site security, as well as protection from the elements through tents or tarpaulins, must be part of the investigative plan.

Mapping a mass grave requires more than conventional archaeological methods involving levels and tape measures. The use of global information systems (GIS) technology, and sophisticated survey equipment, commonly known as "total stations," is standard. The use of these laser- and computer-based systems allows archaeologists to capture the site in three dimensions. Programming the computer-supported total station prior to data collection facilitates the mapping process but requires considerable effort. Through the use of associated computer programs, the data are translated into three-dimensional maps that can be manipulated and viewed from multiple perspectives. In addition to indicating body placement, this system allows for the mapping of evidence, including shell casings and ligatures.

Archaeologists working on mass graves should anticipate encountering ephemeral evidence, including tire tracks and blade gouges from the heavy equipment used to form the grave. Mapping and documenting such features can provide powerful evidence for prosecutors or allow investigators to link graves associated with a particular group of perpetrators.

In addition to mapping, photography (both video and still) and the taking of extensive field notes are crucial to the documentation process. The excavation of a single large grave can take weeks to complete, and so changes in personnel are possible. Investigators from different countries or training programs may be present on scene. Finally, those excavating the grave may not be responsible for preparing the final written report. Problems arising from these conditions can be overcome by establishing protocols, including standardized field recording forms, that ensure uniformity of data collection and allow the integration of an enormous amount of information into a single report.

An important variable in mass grave interpretation is body position. Two types of body orientation are recognized: orderly placement and erratic commingling. **Orderly placement** of multiple bodies a grave (Figure 9.4), such that all individuals are afforded sufficient space, bodies are placed supine with purposeful arrangement of the extremities and all heads are

Figure 9.4 Schematic illustration of orderly placement in a multiple grave. Individuals were deposited in a supine position, affording sufficient space to each; all heads are oriented in the same direction, and the placement of limbs is purposeful. Orderly placement implies respect or care for the decedents by the agents carrying out the burial.

oriented in the same direction, represents an act of burial more respectful than **erratic commingling** (Figure 9.5). In cases of erratic comingling, bodies are literally dumped into the grave manually or deposited en mass from heavy equipment such as dump trucks. The allusion to "bodies as refuse" is appropriate, as erratic commingling is strong evidence of disrespect and violation.

Commingling

Although not exclusive to modern mass graves, commingling is frequently encountered in mass interments and mass disasters, such as those involving aircraft. **Commingling** is the mixing of body parts or skeletal elements from two or more individuals. Commingling can be the result of purposeful human action, as seen in secondary graves and **ossuaries**, or natural taphonomic processes such as **bioturbation**. Progressive loss of soft tissue through decomposition results in the overall decrease of the biomass of the grave and the settling of skeletal elements within the grave. This alone can disarticulate individuals and produce commingling.

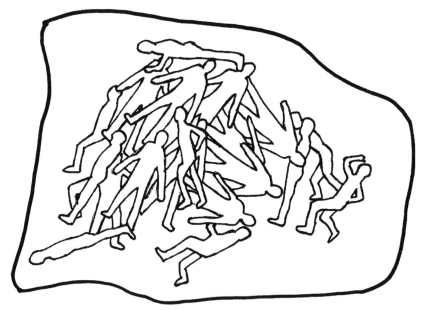

Figure 9.5 Erratic commingling in a mass grave. In this schematic illustration, individuals are deposited in the grave in any position, with heads oriented in all cardinal directions and with variable arrangement of the limbs. Bodies overlap and, as the remains decompose, extensive commingling of elements results. Erratic commingling reflects a level of indifference to or disrespect for the dead on the part of those performing the burial.

Issues relating to commingling must be addressed both at the scene and at autopsy. Proper scene processing requires the recognition of commingling prior to mapping or body recovery. Because bodies are evidence, protocols need to be established for the assignment of evidence numbers in cases of commingling. If anatomical regions or skeletal elements cannot be definitively assigned to specific individual it is best to assign a separate evidence number to each set. Careful mapping, noting the position of the body parts and all nearby or associated individuals, offers the best chance of correct reassociation during analysis.

Reassociating separated elements is best attempted at autopsy rather than in the field. Achieving consistency in sex, age, and size of elements is the first step. Proper inventory ensures that there is no duplication of elements per individual. Methods of resolving small-scale commingling include visual pair-matching, articulation, process of elimination, osteometric comparison, and taphonomy (Adams and Byrd, 2006; Byrd and Adams, 2003). Large-scale commingling, as seen in secondary graves in Bosnia or following the collapse of the World Trade Center towers (Box 9.4), has been addressed by means of DNA testing (see also Byrd et al., 2003).

The amount of time and resources dedicated to reassociation varies depending on the project, its goals, the number of victims, and the degree of commingling present. While small-scale commingling, involving a few individuals, can be resolved, large-scale events, involving hundreds of highly fragmentary victims, can quickly exceed the capacity of investigators. Reasonable efforts are expected and are required as part of establishing identification and calculating MNI. When it comes to asserting their abilities to identify and reassociate every fragment of biological material, however, it is important for anthropologists to be realistic with family members, the media, and themselves. Often, a compromise is reached in which the next-of-kin receive material positively attributed to their loved one, while the families

---

BOX 9.4

Separating commingled remains using mtDNA

On September 11, 2001, two commercial aircraft crashed into the twin towers of the World Trade Center, resulting in the deaths of 2,749 individuals and the largest single forensic identification effort in U.S. history. Of these, 1,565 had been identified as of 2005. Recovery efforts generated almost 20,000 fragments of human remains that required reassociation and identification. The extensive use of mtDNA, in combination with traditional anthropological methods of analysis, both identified and reassociated these remains. To date, the highest number of fragments reassociated to a single individual is 209 (Mundorff et al., 2005).

BOX 9.5
Genocide memorials

In response to the need to acknowledge and venerate victims of genocide, as well as the need to address the practical requirement for body disposition, many communities have opted to create communal memorial sites. One of the largest is the Kigali Memorial Center in Rwanda. Kigali provides open public space, allowing families and communities to inter remains. Currently, over 250,000 victims of the Rwandan genocide are buried at the site. In addition, permanent exhibitions displaying clothing, photographs, and commemorative objects are housed in a museum on the site.

To commemorate the 5,000 residents killed in a poison gas attack in March 1988, the Kurdish community of Iraq has erected a memorial at Halabja. The display includes photographs of individuals killed in the attack. Kosovo, a province in the former Yugoslavia, also plans a communal public memorial to accommodate the unidentified and fragmentary remains from the 1999–2000 genocide of the ethnic Albanians (Kosovar Research and Documentation Institute, 2004).

---

affected by the disaster collectively agree to the disposition of unidentifiable fragments (Box 9.5).

OBJECTIVITY

In the wake of a disaster such as Hurricane Katrina or a conflict such as the attacks on the World Trade Center, forensic professionals are not immune to strong emotional responses and personal motivation to assist the victims. However, good intentions are not enough. Proper death investigation, identification, and certification require that legal and professional standards be maintained despite the circumstances. Evidence collected without the authority to do so is inadmissible in any later legal proceeding. Jurisdiction and the authority to investigate must be established prior to implementing a response.

Those experiencing their first mass disaster deployment are particularly vulnerable. The desire to help is overwhelming, but limited experience and an incomplete understanding of the complex issues governing such events can result in violations of protocol or even illegal interference with ongoing investigations. A good rule of thumb is this—if you did not have the authority to do something before the disaster, you do not have it during or after the event. Investigators and relief agencies cannot allow the scale of the event to compromise standards. The willingness of a volunteer to participate should not outweigh the need for legally binding investigations. When the emotion of the event passes, downstream court proceedings will result in the discarding of any evidence not acquired in accordance with professional standards. New participants are also susceptible to solicitations from organizations that lack the authority to carry out their proposed missions. First-time participants should endeavor to work with well-established organizations.

CONCLUSION

Forensic anthropologists wishing to engage in large-scale investigations of conflict and human rights violations must begin with an understanding of international law and criminal charges specific to such inquiries, including war crimes and genocide. Familiarity with jurisdictional concerns encountered in such scenarios allows anthropologists to ensure that both their actions and the investigations themselves meet legal and professional standards. Mounting effective and appropriate investigations to mass death begins with recognizing the differing goals and parameters of medicolegal and humanitarian responses, as well as the variable authority of governmental and nongovernmental agencies.

Mass death resulting from environmental hazards, acts of aggression, or human-generated disasters pose uncommon challenges to investigators: mass graves, commingling, and large-scale identification efforts are only a few. Forensic anthropologists are uniquely qualified to meet such challenges, providing expertise in recovery, analysis, and identification of human remains. However, methods commonly used in individual casework may not translate to a mass death environment. The time, resources, and attention afforded to individuals in isolated cases cannot always be allocated when death occurs on a greater scale. All forensic investigators, including anthropologists, must remain flexible, prioritizing and maintaining professional and ethic standards when faced with the tragic but all too common events that result in mass death.

CHAPTER 10

Biohistory

Historical Questions, Methods, and Ethics

In Chapter 8 we examined methods now used to scientifically confirm identity to a legal standard. In this chapter, we will consider the important distinction between the questions asked and methods used on the one hand, and the legal ramifications of the results on the other. While the medicolegal investigation of recent deaths is relevant to the court, the legal implications of similar methods or inquiries applied to historical decedents are less clear.

Biohistorical approaches use methods of bioanalysis in the study of historical personages. **Bioanalysis** is any method of scientific inquiry (e.g., DNA, histology, radiography) in which the subject material is of biological origin. Depending on the nature of the case, anthropological techniques may be primary, though the full suite of scientific analytical procedures is potentially available. Issues commonly addressed include confirming identification, establishing cause of death, defining ancestral ties, and exploring health status. In very few cases, however, are these studies done to a legal standard, nor are the results of medicolegal or forensic significance.

Biohistorical investigations can be undertaken within forensic contexts, as the featured case study here will demonstrate. Typically, however, biohistorical studies are conducted by scholars or hobbyists pursuing questions not relevant to the court. Forensic anthropologists considering engaging in biohistorical inquires must be mindful of the jurisdictional constraints on nonmedicolegal inquires. Even cases involving the personal identification of a prominent historical figure are not inherently forensic or of legal consequence.

In biohistorical studies, ethical issues frequently emerge. When does a legitimate research question or request from descendants cease to be

BOX 10.1

Who "owns" the past?

Determining who controls access to and authority over the remains of the dead can be complicated. For the recently deceased, each state has statutes dictating who can give consent for autopsy, agree to organ or body donation, and determine the disposition of the remains. For example, according to New Mexico state law, the descending order of priority of next-of-kin or authorized grantor is as follows:

(1) the decedent during his or her lifetime through a living will,

(2) a spouse,

(3) an adult child over 18 years of age,

(4) a parent,

(5) an adult sibling,

(6) any other relative, or

(7) the entity with custody of the body (i.e., the medicolegal authority or funeral home).

For deaths that occurred in the more distant past, any descendant or heir to a historic personage can claim authority over the disposition of the remains. If there are no direct descendants (as identified in points 2 through 5), all heirs have equal rights under the law. For example, should researchers wish to exhume a famous historic figure and should living descendants be found, those descendants may grant or deny permission for the exhumation. However, if more than one relative exists and there is disagreement between the parties over the exhumation request, a court will be called on to render a decision. Further complicating the situation is the burden of proof required to establish ancestry. Prior to DNA testing, an individual had merely to offer a genealogy demonstrating relationship to the decedent. With the advent of DNA testing, it is the court with jurisdiction that determines how ancestry must be demonstrated when a challenge is presented.

For descendants claiming tribal or group affiliation rather than direct ancestry, federal laws apply. Details on ancestral rights granted to Native Americans, for example, are detailed in the section on NAGPRA in Chapter 5.

compelling in the face of media attention, economic advantage, and lack of respect for deceased people and their heirs? Legally, who has the right to engage in biohistorical study? Many of the same issues that have been raised in relationship to the archaeological disinterment of more ancient materials pertain here. Just who owns the past? (Box 10.1).

In this chapter, we will examine the types of questions posed in biohistorical analyses, as well as the methods used. Both invasive (destructive) and noninvasive methods will be discussed. To illustrate some of the ethical issues raised in biohistorical studies, as well as the potential to blur the line between forensic investigations and commercial endeavors, a case study involving the proposed exhumation of a well-known historical figure, the outlaw Billy the Kid, will be examined in detail.

BIOHISTORY: PAST AND PRESENT

Although overwhelmingly conducted for academic or personal interest, biohistorical investigations have addressed matters of forensic significance. One of the earliest medicolegal examples was the identification of a Nazi war criminal. In 1985 skeletal remains believed to be those of Dr. Josef Mengele were exhumed in southern Brazil (for an account of the involvement of anthropologist Clyde Snow, see Joyce and Stover, 1991). Mengele, known as the Auschwitz "Angel of Death", had evaded capture following the end of World War II and was reported to have escaped to Brazil. In 1979 an individual known locally as Wolfgang Gerhard drowned in a swimming accident and was buried there. Six years later, an international team of forensic pathologists was assembled to examine the exhumed remains of Gerhard/Mengele and establish identity.

Standard forensic dentistry anchored initial attempts to link the remains with Mengele. In one of the earliest applications of photo superimposition as a means of presumptive identification (Helmer, 1987), investigators also performed a detailed comparison of skeletal features with known photographs of Mengele. The results of these tests led investigators to conclude, beyond reasonable doubt, that the remains were those of Josef Mengele (W.J. Curran, 1986; Eckhert and Teixeira, 1985; Helmer, 1987). However, doubts raised by Israeli police and an Israeli pathologist challenged this conclusion, and additional testing was ordered by German authorities.

Although the use of mitochondrial DNA (mtDNA) to establish identity had already been introduced to American courts, investigators chose to compare nuclear DNA markers from the skeletal remains with samples from Mengele's living son and wife to establish the paternity of the son (Box 10.2). This posthumous paternity test was then used to confirm Mengele's identity. Since the remains had been interred for six years prior to exhumation, successful extraction of nuclear DNA proved difficult. Results published by Jeffreys et al. (1992) argued that the test strongly supported the identification of the remains, but since 0.1% of unrelated Caucasians could not be excluded, questions persisted regarding whether the test afford enough specificity to establish identification. In 1993 Rogev examined the remains as well as reports of previous examinations and found that the dental, photo superimposition, and osteobiographical evidence presented earlier as means of positive identification were inadequate or incorrect. Rogev, however,

BOX 10.2
Paternity testing

Paternity testing establishes the identity of a child's father. Testing involves comparing the child's genotype to those of both the mother and the alleged father. Failure to match results in exclusion, indicating that the man could not have fathered the child. If a match (inclusion) results, a comparison of the DNA profile is made to a population database containing the genotype of unrelated individuals. Since many alleles will be shared when samples are drawn from related individuals, this statistical refinement are necessary. Paternity tests typically use the same short tandem repeat (STR) standards as forensic testing.

Reverse parentage testing is used to identify the remains of a missing person, particularly in mass disaster victim identification or in biohistorical analyses. In paternity testing, the question posed is "Who are the parents?" In reverse testing, the question is "Who is the unknown individual?" (Butler, 2005).

conceded that Jeffreys' DNA results established the identity of the remains as Mengele beyond reasonable doubt.

Mengele's case is the rare exception in biohistorical analysis, in that the results had legal implications, such as the altering of a death certificate. More traditional inquiries conducted purely for scholarly interest have been carried out for centuries. Consider, for example, the case of German poet and dramatist Friedrich von Schiller, who died in 1805. Some believed that Schiller's remains, originally buried in Kassengewolbe, had been disinterred in 1826 and reburied the following year in Furstengruft. By 1883, prominent German anatomists expressed doubts that the skull in Furstengruft was that of Schiller. In 1911 an anatomist named von Froriep excavated the site of Kassengewolbe, uncovering over 70 crania. Attempts at facial approximation of the Kassengewolbe and Furstengruft skulls were compared with all known busts and portraits of Schiller, and von Froriep finally declared one skull as belonging to Schiller with absolute certainty. Countless other German anatomists remained unconvinced, and the debate continued until the early 1960s, when a dental analysis of the suspect skull excluded it from further consideration (Henschen, 1965). Since then, no skull has been found that definitely can be attributed to the famed poet.

Identity has also been established to a nonlegal standard through the comparison of historical accounts of injuries suffered by a prominent figure and the perimortem wounds evident on the purported remains. For example, Francisco Pizarro, conqueror and governor of Peru, was killed by assassins in 1541. Attacked while eating lunch, Pizarro received a wound to the throat and was then stabbed repeatedly with rapiers as he lay on the floor.

According to historic accounts, Pizarro was buried behind a cathedral in Lima but was exhumed four years later. His bones were then placed in a wooden box in the crypt under the main altar of the Lima cathedral.

In 1891 a well-preserved mummy believed to be Pizarro was removed from the crypt and placed in a ceremonial sarcophagus. In 1977 workmen discovered a wooden box containing human remains in the crypt under the altar. In 1984 anthropologists Robert Benfer and William Maples examined both the mummy and the remains from the box. Sharp force defects to the neck, thorax, and arms of the remains from the box matched historic accounts of the assassination of Pizarro. An exhaustive examination of the mummified remains revealed no indications of perimortem injury. Based on these results, the anthropologists determined that the remains in the box were those of Pizarro (Maples et al., 1989).

Of all the prominent figures investigated biohistorically, none have received more attention or caused more debate than the Romanovs. The bodies of seven members of Russia's ill-fated royal family, shot by Bolsheviks in July 1918, quickly disappeared. In 1991 the remains of nine individuals were unearthed from a shallow grave in central Russia. DNA testing during the 1990s suggested that the bodies were five of the seven royal family members, (Tsar Nicholas II, his wife, and their five children), plus their doctor and three servants. In 1998 a special Russian government panel confirmed that the remains were those of the Romanovs and their attendants. However, other researchers have challenged the results, and the debate over the identity of the nine individuals continues to this day (see, e.g.,Gill and Hagelberg, 2004; Knight et al., 2004; R. Stone, 2004; Zhivotovsky, 1999). The focus of these challenges is either the validity and integrity of the original DNA analysis, which some believe was tainted, or the interpretations of the subsequent mitochondrial DNA test. The original DNA study used amplified short tandem repeats (see Chapter 8 for more information on DNA testing) to establish the familial relationship of the five royal decedents. Subsequent samples of mtDNA extracted from these five sets of remains were then compared to samples provided by the Duke of Edinburgh, Prince Philip, and two living descendants of the tsar's maternal grandmother. The samples from the bodies of the tsarina and the children matched the specimen provided by Prince Philip, but the purported body of the tsar did not. Despite this discrepancy, investigators claimed the test established the identity of all five Romanovs (Stone, 2004), although what legal implications (if any) such findings have is unclear.

The use of DNA testing, particularly mitochondrial DNA (Box 10.3), has allowed researchers to address a variety of questions pertaining to historic figures. Comparisons of mtDNA extracted from skeletal material with samples donated by living descendants provided the means for identifying the remains of Nazi Martin Bormann, a high-ranking member of the Third Reich whose postwar life had long been a subject of debate (Anslinger et al., 2001). Samples from living maternal relatives of Marie Antoinette were used to

BOX 10.3

Why mitochondrial DNA?

Mitochondrial DNA (mtDNA) offers certain advantages over nuclear DNA in the identification of historic human remains. For example, mtDNA has a high copy number; it is found in very large numbers within each cell, increasing the likelihood of successful extraction. Each cell carries only two copies of nuclear DNA. The rapid rate of evolution seen in mtDNA results in a high probability of exclusion, should test results vary between samples. Both these features make mtDNA preferable to nuclear DNA when the remains in question are historic. Although mtDNA sequences are not unique to every individual, the rate of evolution allows researchers to differentiate those who are closely related from those who merely share a common ancestral origin.

Most importantly, when living descendants are available to provide reference DNA samples in cases of historical remains, the use of mtDNA allows for identification in situations where establishing identity based on nuclear DNA is virtually impossible. As a result of recombination and segregation, nuclear DNA is unique to each individual (excluding twins). However, this extensive genetic reshuffling means that very little of the nuclear DNA sequence is shared by relatives separated by even a few generations. In contrast, mtDNA has a haploid and maternal mode of inheritance—it passes largely unaltered from mother to children. As a result, any living maternal relative of the historical figure in question could provide a sample, since that individual should have an identical mtDNA type. It is for these reasons that mtDNA is the genetic material of choice in biohistorical investigations.

(Butler, 2005)

challenge the putative identity of Carl Wilhelm Naundorff, who had claimed to be the son of King Louis XVI of France and Marie Antoinette (Jehaes et al., 1998). Mitochondrial DNA tests on the possible remains of Prince Branciforte Barresi, a member of the Italian royal family in the sixteenth century, as well as two of his children, his brother, and another family member, were conducted in 2001 after five bodies were discovered buried in a church in Sicily. Although morphological analysis and historical reports had already tentatively identified the remains, mtDNA tests confirmed that the five skeletons were related to each other. Genetic analysis supported the tentative identifications, although they did not positively identify the remains as those of the prince and his family (Rickards et al., 2001).

Occasionally, DNA testing raises more questions than it answers. In 2001 a group of Italian scientists extracted mtDNA from the body attributed to the evangelist Luke. A pivotal figure in Christianity, Luke was believed to have been born in Syria and died in Greece. His body was first taken to Constantinople and then finally to Padua, Italy. Some scholars believe that the Padua body is not that of Luke but an illicit replacement, a person from Greece or Turkey. Although no modern or historic sample exists that could confirm the identity of the Padua body, the Italian investigators attempted to ascertain whether these remains represented an indvidual of Greek or Syrian descent. The researchers compared the Padua body against typed modern samples from Syria and Greece and published samples from Anatolia. The sequence derived from the Padua body was an atypical variant of alleles common in the Mediterranean region. Based on these results, the researchers rejected the notion that the body belonged to a Greek substitute, but could not rule out a replacement from Turkey or Syria. The evidence does not establish the identity of the body as Luke, nor does it rule out the possibility that the body came from Syria. It also does not rule out the possibility of a substitute from Constantinople (Vernesi et al., 2001). Although the body is potentially of great historic and religious significance, the genetic investigation did little to address the mystery.

Anyone planning to engage in biohistorical inquiry must appreciate the inconclusive nature of many DNA test results. Costs incurred in the extraction and analysis of genetic material, as well as the travel or transport expenses, exhumation costs, and legal fees associated with such investigations can be considerable. Published reports seldom indicate that funding was obtained from traditional sources such as governmental research grants. Potential conflicts of interest can develop when private funding is obtained to facilitate the investigation. Funding provided by putative or established descendants or heirs of historical figures, as well as private citizens or organizations having a vested interest in the outcome of such testing, should be accepted with caution. Support provided by media, educational, or entertainment organizations, particularly in exchange for broadcast rights or exclusive control over the dissemination of results, also invites concern over intellectual property rights.

Biohistorical questions regarding identity have also been raised in the absence of biological material. Researchers have relied on published materials, photographs, and prior reports to answer questions or challenge previous findings (Box 10.4). For example, in 2005 Marchetti and others conducted an exhaustive review of all available documentation relating to the death of Adolf Hitler to address questions regarding the cause of his death, as well as the identification of his purported remains. Included in the review was Hitler's autopsy report, as well as documentation from previous reviews by Russian and English investigators. The team found that conclusions regarding cause of death were not possible because toxicology testing was not conducted at the time of autopsy. Issues relating to positive identification were also inconclusive. Marchetti called for mtDNA testing of Hitler's purported skull fragments against living maternal relatives of the German dictator to confirm identity, although to date this has not taken

BOX 10.4
Establishing identity without examining the body

In 2005 N.L. Rogers et al. used a review of published documents to revisit the contentious efforts to identify of the remains of John Paul Jones, the eighteenth-century naval hero famous for having "not yet begun to fight." The return of his remains had not been requested by the U.S. government following the French Revolution, and the exact location of his grave was unknown, although putative remains had been discovered. Following what the researchers termed a "substantive analysis" of all available evidence, including comparisons of written descriptions and artworks depicting John Paul Jones with photographs of the putative remains, the team agreed with previous findings and supported the identification of the contested remains as those of John Paul Jones.

place. This inquiry, despite being conducted with scientific rigor, is outside the jurisdictional authority of the medicolegal system, and the findings have little bearing outside of academic interest.

Other Biohistorical Questions

While establishing individual identification is by far the most common biohistorical question, other life history issues can be addressed. For example, questions remain as to the exact cause of death of Senator Huey P. Long of Louisiana, who on September 8, 1935 suffered a single gunshot wound to the abdomen. Long died 30 hours after he was shot, reportedly by physician Carl Austin Weiss (for a fascinating biohistorical review of Weiss' remains, see Ubelaker, 1996a). Mrs. Long refused to grant permission for an autopsy despite repeated requests. Given the nature of the wound, there was speculation that Long could have survived the injury but that during his treatment, the attending surgeons had made a serious error. In 1994 Garcelon and O'Leary examined the original hospital records, including the surgeons' notes, to determine the exact cause of death. These authors, both surgeons themselves, believe that an unrepaired kidney injury and retroperitoneal hemorrhage were the cause of Long's demise, indicating that attending physicians had in fact fatally erred (Garcelon and O'Leary, 1994). While interesting, these result again have no legal bearing on the resolution of Long's death.

In addition to cause of death, the antemortem health states of historical figures have also been the subject of biohistorical inquiry. For example, recent examinations of a lock of hair from Ludwig van Beethoven employed a variety of analytical tools to examine the health state of the famed composer. The hair, historically attributed to Beethoven, was recovered from a locket with established provenience. Although less informative than assays of other tissues such as blood, spectrometry analysis of hair is commonly used in forensic science to examine drug or toxin levels (Shen et al., 2006), mineral toxicity (Peterson et al., 2006), or nutritional status (Lachat et al., 2006). A **radioimmunoassay** (Box 10.5) on twenty of the hairs from the locket was conducted to determine whether Beethoven had taken opiate pain medication during the last months of his life. The assay proved negative, indicating that Beethoven had declined pain-relieving medication during his last days. Scanning electron microscope energy dispersion spectrometry (Box 10.5) was used to determine lead levels. The test revealed Beethoven's lead levels were 42 times higher than three control samples taken from living donors. The effects of lead toxicity include personality changes and likely contributed to Beethoven's chronic illnesses and his death. Another team used nondestructive synchrotron radiation x-ray analysis to examine lead levels in six of Beethoven's hairs (Box 10.5). This test showed that Beethoven's hair had lead levels 100 times greater than modern hair samples. It has been suggested that such high lead levels can produce the variety of conditions from which Beethoven reportedly suffered, including his deafness. The same research teams performed tests to determine mercury levels in Beethoven's hair. Both reported that mercury, if present, was below the detection levels of their instruments. These findings lent support to the opinion among Beethoven scholars that the composer did not suffer from syphilis. Since this venereal disease was commonly treated with mercury during the 1820s, the absence of evidence seen in the mercury level tests has been argued to represent evidence of absence (Martin, 2000).

Investigators are not yet finished with Beethoven. In 2005 geneticists at the University of Münster began DNA testing of several fragments of skull reportedly removed from Beethoven's body following its exhumation in 1863. Samples extracted from the skull fragments will be compared against those from the known hair samples. Researchers are also hoping to determine whether Beethoven inherited a gene that may have resulted in his hearing loss as a young man. Regardless of the outcome, historical figures such as Beethoven will continue to interest scientists and historians alike for as long as their physical remains are available.

Biohistorians interested in assessing the health states of prominent figures are not restricted to examination of biological material. Documentation provided the basis for a recent attempt at a posthumous psychological autopsy (Box 10.6) of Vincent van Gogh. The famed but troubled artist committed suicide in Paris at the age of 37, following a life of heavy drinking as well as psychiatric and health problems. Hughes (2005) contends that these factors explain van Gogh's erratic behavior and periodic losses of consciousness, rather than pathological conditions such as Geschwind syndrome or temporal lobe epilepsy.

BOX 10.5
Methods of bioanalysis

RADIOIMMUNOASSAY

Radioimmunoassay uses materials such as radioactive iodine as a detection mechanism to evaluate the levels of chemical components, such as hormones, in a biological sample. Of growing interest in anthropology, protein radioimmunoassay (pRIA) has been used to identify morphologically unidentifiable bone fragments as either human or nonhuman (Ubelaker et al., 2006).

SCANNING ELECTRON MICROSCOPIC ENERGY DISPERSION SPECTROMETRY

Spectrometry is used to identify atoms and isotopes and to determine the chemical composition of a sample. Scanning electron microscopy (SEM) uses a beam of electrons to scan an object. The electrons are then captured by detectors that produce an enlarged image on a cathode-ray tube. SEM energy dispersion spectrometry uses detectors that capture and analyze the object's scattered particles to determine its chemical composition. A pilot study in which dental and osseous material was distinguished from other fragmentary evidence by means of SEM (Ubelaker et al., 2002b) found that, while the proportions and amounts of calcium and phosphorus are informative in differentiating teeth and bone from other substances, the analytical tool was unable to exclude ivory, mineral apatite, and some types of coral.

SYNCHROTRON RADIATION X-RAY IMAGING

Commonly known as "supermicroscopy," synchrotron x-ray imaging combines x-ray beam features with new-generation detectors and computers to provide imaging techniques including absorption, phase contrast, diffraction, and fluorescence. Its uses include spectroscopy, x-ray lithography, and traditional crystallography. It has widespread applications throughout the medical and physical sciences as a means of understanding and quantifying the smallest components of all forms of matter. Recent studies include investigations of alveolar support tissue and the remodeling process (Dalstra et al., 2006), *in vivo* bone deposition dynamics (Komlev et al., 2006), and a cadaveric study of the mechanical strength of os calcis cancellous bone (Follet et al., 2005).

BOX 10.6

Forensic psychiatry

Forensic psychiatry is the application of theories, methods, and principles of psychiatry and mental health to proceedings in a court of law. Forensic psychiatrists evaluate victims and suspects of crime, addressing issues such as competency to stand trial, psychological trauma, mental illness, and risk of **recidivism** (repeating the criminal act). While such evaluations are possible with living subjects, forensic psychiatrists also perform **psychological autopsies** to reconstruct the state of mind of a decedent. A forensic psychiatric autopsy synthesizes information from multiple informants and sources.

Psychological autopsies involve collecting all available information on the deceased by means of structured interviews with family members, friends, and attending health care personnel. Information is also drawn from health and psychiatric care records, other documents (such as letters or journals), and forensic examination. For example, a psychiatric study of Jeffrey Dahmer, a notorious serial killer, was conducted using photographic scene documentation and autopsy findings of the victims. The team found that the basis for Dahmer's behavior was unconscious feelings of hate and ambivalent feelings about his homosexuality (Jentzen et al., 1994).

NONINVASIVE ANALYSES

While documentary searches may lack the specificity of technological approaches, they have two significant advantages. First, they are noninvasive and nondestructive. Second, such searches do not challenge the ethical and professional boundaries often confronted in attempts to use destructive analysis. Proposed biohistorical research designs involving noninvasive methods of analysis encounter less resistance from authorities safeguarding historical remains, as they do not require destruction or excessive disturbance of the remains.

The most promising nondestructive avenue of research is the use of virtual autopsy techniques on archaeological material. A discussion of virtual autopsy was provided in Chapter 2. Recently, Hughes et al. (2005) used CT scans and a computer reconstruction program to examine the skull of an Egyptian priestess dating from 770 BC. The mummified remains of the priestess were encased within an anthropoid coffin. CT imaging allowed the capture of information without removing the remains. Twenty-seven reconstructed skull measurements were entered into the CRANID morphometric analysis software package and compared against 64 groups from around the world. Results indicated a 53% probability that the mummy was an

Egyptian female and an 81% combined probability that the remains represented an African female. The findings were consistent with the inscription on the coffin.

High-definition imaging techniques such as computer tomography and magnetic resonance imaging offer significant advantages as well as disadvantages in biohistorical research. Advanced imaging techniques have the potential to visualize internal structures without invasive procedures. However, such techniques have limitations. Imaging methods allow for the examination of form, but offer little in terms of understanding function or chemical composition. Such techniques are also expensive and require the transport of remains to medical centers housing the necessary equipment.

Other noninvasive methods involving biological materials have been used to establish the identity of historic remains. Among the least exacting methods has been the comparison of skeletal remains to paintings or other artistic images of the individual in question. The putative skull of Duchess Eleonara Gongaza della Rovere, a prominent figure of the Italian Renaissance, was superimposed over a portrait of della Rovere by the artist Titian. Discrepancies between skull and portrait in the length of the nose did not exclude a presumptive identification but prompted the researchers to hypothesize that the artist had altered the dimensions of the duchess' nose to conform to notions of beauty prevalent at the time (Rollo et al., 2005).

Two examples illustrate the process taken a step further. A facial reconstruction (see Chapter 8 for more details) was performed using the purported skull of Antal Simon, a well-known priest and teacher in nineteenth-century Hungary. The reconstruction was then compared to a portrait of Simon. Kustar (2004) reports results indicating that 62% of the reconstructed features showed great resemblance to the portrait, while 35% showed close resemblance and 3% were approximate.

A similar academic study involves a facial reconstruction using the remains of the Red Queen, a Classic Maya ruler discovered in Temple XIII at Palenque, Mexico. The reconstruction was compared with the Red Queen's funerary mask and local paintings believed to depict the wife of Janaab' Pakal, a famous Maya ruler of the Classic period. The perceived similarity between the reconstruction and available artwork led investigators to argue for a presumptive identification (Tiesler et al., 2004).

These studies, while creative, do not achieve the standards necessary to provide positive or even presumptive identification. They are not "forensic" exercises, as they are not introduced into a court of law. Nor are the methods forensic, as neither facial reconstruction nor superimposition meets the standards of low known error rate and general acceptance within the scientific community. These examples succeed, however, in capturing the attention of the media and general public, who often confuse these academic studies with forensic investigations.

There are currently no accepted professional standards governing biohistorical analysis. Andrews and others (2004) reviewed guidelines from professional organisations overseeing the scientific, forensic, and historic communities and proposed some ethical constraints on the use of modern technologies to address questions in the historical record. Chief among the concerns raised was the fear that biohistorical analysis was often conducted for commercial gain or media sensationalism.

To demonstrate why such fears are justified, we will examine the media-driven, politically motivated attempts to exhume the remains of the famous U.S. outlaw known as Billy the Kid (BTK). This case illustrates the types of questions addressed through bioanalysis of historical material, the methods commonly used, and the ethical issues raised by such investigations. It also serves as a cautionary tale to those who would misuse science for personal gain.

A CASE STUDY IN BIOHISTORICAL AND FORENSIC INVESTIGATION: BILLY THE KID

Billy the Kid is not the first infamous outlaw to receive such attention. In 1995 the presumptive remains of Jesse James were exhumed and subjected to mtDNA testing. The results, while not conclusive, supported the identification of the exhumed remains as those of Jesse James (Stone et al., 2001). Publicity generated by the exhumation of Jesse James was among the factors that prompted investigators to turn to Billy the Kid.

A Brief History of Billy the Kid

Although Billy the Kid died before his twenty-first birthday (Figure 10.1), his short life and violent death have become legendary in the American Southwest. Born in the 1860s, Billy was arrested for stealing laundry in 1875; his life of crime soon escalated to murders in Arizona and New Mexico.

On the evening of July 14, 1881, Lincoln County sheriff Pat Garrett shot Billy in the chest after tracking him to Pete Maxwell's house in Fort Sumner, New Mexico. The following day, a coroner's inquest ruled Billy's death "justifiable homicide." That afternoon, Billy was buried in the Fort Sumner cemetery, alongside his friends Charlie Bowdre and Tom O'Folliard, marked by a stone engraved "The Pals." This headstone was repeatedly stolen and replaced throughout the years, until cemetery officials encased the entire grave site within a metal cage (Figure 10.2).

In early 1882, Pat Garrett published *The Authentic Life of Billy the Kid*. Although now considered a work of fiction, the book helped cement the Kid's legend. Rumors that Garrett had not killed the Kid in the shootout at the Maxwell house began almost almost immediately after Billy's death, and over time, various individuals have stepped forward claiming to be the real Billy the Kid. Among the more notorious contenders were "Brushy" Bill Roberts of Texas and John Miller of Arizona.

Figure 10.1 The only authenticated photograph of William Bonney, a.k.a. Billy the Kid. This tintype was taken in 1880, approximately a year prior to his death.

Figure 10.2 The purported grave site of "The Pals," William Bonney, Tom O'Folliard, and Charles Bowdre, in the Fort Sumner cemetery, New Mexico, circa 2003. The cage was erected decades earlier to prevent vandalism. *(Photo by Debra Komar)*

The (Almost) Exhumation of Billy the Kid

Sometimes historical analysis is undertaken for commercial consideration or mere sensationalism —Andrews et al., 2004: 215.

On June 10, 2003, New Mexico governor Bill Richardson held a press conference to announce his support for the reopening of the investigation into the events surrounding the death of Billy the Kid. The goal of the investigation was to determine whether the outlaw was buried in the New Mexico site that was a well-known tourist attraction or, as had been claimed by some, was interred in Texas or Arizona. Heading the investigation was a group, referred to here as the "BTK team," comprising local sheriffs, the mayor of Capitan, a county attorney, and a University of New Mexico history professor. Conspicuously absent were the forensic scientists necessary to achieve the goal of the investigation as stated in the press release: "to put modern forensic science to the test to answer the questions surrounding those days in New Mexico history." The press conference ignited a flurry of media attention. Reporters from around the world interviewed the investigative team and carried stories of the upcoming exhumation.

Following the press conference, the New Mexico Office of the Medical Investigator (OMI) decided that the high-profile nature of the case warranted the office's involvement, as New Mexico statutes (1978 NMSA 24–14–23D) detail the role of the medical examiner in all exhumations. To lend credibility to the investigation, and to clarify jurisdictional issues, the BTK team had already declared the case a criminal investigation, going so far as to open an official homicide file three days before the press conference (Lincoln County Sheriff's Office case no. 03–06–136–01). Despite repeated requests by the OMI for a copy of the file during the early stages of the investigation, and throughout the legal battle that ultimately ensued, no official documentation was ever provided.

At the very least, investigators should disclose . . . the investigative question posed —Andrews et al., 2004: 216.

During preliminary meetings with the OMI, the BTK investigators outlined the goals of the criminal investigation: (1) to determine whether Sheriff Pat Garrett had shot William Bonney (a.k.a. Billy the Kid) in Fort Sumner on July 14, 1881; (2) to exhume the remains in the designated grave for the purposes of DNA testing; and (3) to exhume the remains of William Bonney's mother, Catharine Antrim, to serve as the comparative standard for mitochondrial DNA tests. Initial conversations with the BTK team also revealed the professional and financial involvement of multiple documentary film companies, including the "deputizing" of one documentary cinematographer who would have unrestricted access to film the exhumation and would retain the rights to the footage. With the mid-July anniversary of BTK's death a month away, the investigation was given top priority, and a seemingly unlimited budget of "private funds" was secured through, among other sources, financing from the documentary film companies.

Given the goals of the investigation, Komar (2006) researched the exact location of the graves of BTK and his mother; she also collected information that could assist in establishing the identity of either individual. Data came from seven major archives in New Mexico and Arizona, as well as Fort Sumner and Silver City. The results, as presented to the BTK investigators, are summarized in Box 10.7.

BOX 10.7

Why Billy the Kid's remains are not likely to be found

- The exact location of Billy the Kid's grave in the Fort Sumner cemetery is not known, despite the presence in the cemetery of a marker.

- The four individuals who claimed to have buried Billy could not agree on the location of the grave. Prior to their deaths, historians persuaded the four men to return to the cemetery and point to the exact location of the grave. In a moment of pure comic timing, the four men entered the cemetery and proceeded to point in four completely different directions.

- Virtually every historic account describing the grave's location referenced the gate in the north wall of the cemetery. A recent geographic survey indicates that the north wall of the cemetery is now approximately 30 feet further north than it was in an 1872 survey. All indications of the original gate have been obliterated.

- In 1904 the Fort Sumner cemetery flooded. All grave markers, walls, and gates were washed away. The current placements of these features are approximations only, based on the memory of local citizens.

- In 1906 Will E. Griffin was hired to move all the bodies of soldiers from the Fort Sumner cemetery to the National Cemetery in Santa Fe. Griffin removed what he believed to be the body of a soldier shot for deserting. This individual is historically listed as the body buried directly beside those of Billy, Tom O'Folliard, and Charles Bowdre. According to written reports, only two bodies remained when Mr. Griffin was finished. It is possible BTK's body was removed accidentally at this time.

- In 1924 a *New York Times* report indicated that a local rancher had attempted to exhume Billy's body but no remains had been found. The cemetery flooded again the following year, and all markers were destroyed.

- On July 21, 1881, the Las Vegas newspaper reported that BTK's remains were exhumed by a local physician, Dr. Demaris. The skull was cleaned and sold for an undisclosed amount. According to the report, Dr. Demaris allowed the remaining postcranial elements to decompose, wired them together, and displayed them as an anatomical specimen in Las Vegas. The report also indicated that Demaris removed Billy's trigger finger, set it in alcohol and sold it to a woman on the East Coast for $150.00. In 1908

Jim Carlin of Albuquerque claimed to have the skull. Since that time, the skull's location remains unknown.

- In response to a prior request to exhume Billy's remains in 1962, a New Mexico court ruled that BTK's grave could not be exhumed because "the search would inevitably lead to disturbing the remains of other persons." This prior petition was brought by Lois Telfor, who claimed to be the last surviving relative of BTK and wanted to stop what she felt was the overcommercialization of Billy's grave.

Based on the results, Komar offered the opinion that the remains of Billy the Kid were not likely to be found. Although reports were conflicting, evidence suggested that the remains were likely no longer in the cemetery. Even if the body were still interred in Fort Sumner, its exact location was not known, and any attempt at exhumation would undoubtedly disturb the other remains in the cemetery. Komar also noted that a preliminary search for information on Catharine Antrim indicated that the location of her grave appeared better established than that of BTK. Upon informing the BTK investigators of these findings, it became immediately clear that the focus of the investigation would be the exhumation of Catharine Antrim, as the general location of her grave was believed known.

Often, investigators fail to pose an investigative question capable of resolution by genetic testing   —Andrews et al., 2004: 216.

When asked what purpose Antrim's exhumation would serve to the criminal investigation, without the exhumation of BTK, the investigators indicated that an individual claiming to be a direct heir of Billy had stepped forward. The BTK team proposed testing the claimant's DNA against that of Catharine Antrim. Despite attempts to explain how mtDNA is inherited, the investigators remained resolved to carry out this plan. Concerns that such a test was scientifically invalid (not to mention pointless in the context of the criminal investigation) were ignored. It became clear that the increasing participation of documentary film companies at this stage dictated that someone, anyone, of historical significance needed to be exhumed and tested, regardless of the validity of the outcome.

The investigators also indicated their intent to test Antrim's DNA against that of "Brushy" Bill Roberts, who had claimed to be Billy the Kid. Further, the BTK team intended to test Antrim's sample against genetic material from John Miller, an individual from Arizona who also had claimed to be Billy the Kid. Since both men were deceased, this line of inquiry would necessitate their exhumations as well.

At this point, it was obvious a long-standing controversy involving tourism was driving the investigation. Arizona, New Mexico, and Texas all claim, through historical markers and tourist information, to have the grave site of Billy the Kid. The BTK investigators argued that their work would resolve the mystery and allow the "winning" state (presumably New Mexico) to definitively claim the final resting place of Billy the Kid. Amid growing concerns as to the scientific validity of the inquiry, the medical examiner's office declined to issue a permit for any exhumation and stated that the team would need either a letter from Governor Richardson or a court order to proceed.

Under these circumstances, Komar was tasked with researching the possibility of exhuming the remains believed to be those of Catharine Antrim. The results are summarized in Box 10.8. In the end, the research showed that neither Billy the Kid nor Catharine Antrim was likely to be successfully

---

BOX 10.8

Why the remains of Catharine Antrim are not likely to be found and other issues relating to her proposed exhumation

- The location of the body of Catharine Antrim is not known to a reasonable degree of scientific certainty, despite the presence of a grave marker bearing her name in the Silver City cemetery called Memory Lane.

- Antrim was buried in 1874 in a cemetery within the Silver City limits. In 1877 the cemetery flooded. Records indicate that the floodwaters disturbed the graves and markers within the cemetery.

- In 1882, in response to a change in city ordinance requiring burials outside the city limits, Antrim's body was reportedly moved to a new site in the Memory Lane cemetery. Antrim's exhumation was one of 19 conducted in a single day by unskilled laborers. Only Antrim's new grave was eventually marked, months after the transfer. It is not certain that the body exhumed and moved in 1882 was that of Catharine Antrim.

- Any attempt to exhume the putative body of Catharine Antrim would likely result in the disturbance of other individuals. The burial site marked with Antrim's name is D-27 in Memory Lane cemetery (Figure 10.3). This plot is shared by 12 other individuals of known identity (Figure 10.4). In addition to the known individuals, cemetery records list 276 persons buried in the cemetery whose exact location is "unknown." Similarly, cemetery records indicate that there are at least 455 additional unmarked graves in the cemetery. It is important to bear in mind that Memory Lane is not a large cemetery, roughly the size of a city block.

- Silver City and the cemetery flooded 10 times between 1877 and 1915. Newspaper and written reports detail the disturbance of grave markers throughout the cemetery following each flood and note the presence of human remains along the fence line.

Figure 10.3 The purported current location of the grave of Catharine Antrim, plot D-27 of Memory Lane Cemetery in Silver City, New Mexico. The headstone in the center of the photograph commemorates Catharine Antrim. The remains of at 12 other individuals are known to share the plot. *(Photo by Debra Komar)*

- If the purpose of exhuming the remains of Catharine Antrim is to compare her DNA to living individuals claiming to be descendants of Billy the Kid, such tests would be invalid. Given the length of interment (in excess of 130 years), only mtDNA would be successfully extracted from any recovered remains. This test provides proof of matrilineal lineage only—it passes from mother to child, not father to child (Box 11.3 and Chapter 8). Billy carried his mother's mtDNA; however, his biological children received their mtDNA from their mother, not Billy.

- If the purpose of exhuming the putative remains of Antrim is to compare her mtDNA against samples extracted from Brushy Bill Roberts of Texas, such a comparison would also be misrepresentative. Roberts never claimed to be the biological child of Catharine Antrim, merely a relative sent to live with her. Genealogical records provided by Bill Roberts indicate that he does share a matrilineal relative with Catharine Antrim (she was his great aunt, the daughter of Roberts' grandmother). If Roberts' family history is correct, and mtDNA could successfully be extracted from both individuals, in all likelihood the test would prove positive. The media attention and political agenda of the investigation, coupled with the lack of scientific background evident in the BTK team,

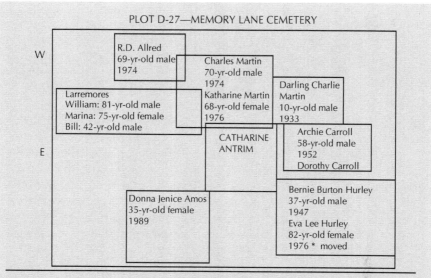

PLOT D-27—MEMORY LANE CEMETERY

W

R.D. Allred
69-yr-old male
1974

Charles Martin
70-yr-old male
1974

Larremores
William: 81-yr-old male
Marina: 75-yr-old female
Bill: 42-yr-old male

Katharine Martin
68-yr-old female
1976

Darling Charlie
Martin
10-yr-old male
1933

CATHARINE
ANTRIM

Archie Carroll
58-yr-old male
1952
Dorothy Carroll

E

Donna Jenice Amos
35-yr-old female
1989

Bernie Burton Hurley
37-yr-old male
1947
Eva Lee Hurley
82-yr-old female
1976 * moved

Cypress Road

Figure 10.4 Schematic map indicating the locations of known individuals sharing plot D-27 of Memory Lane cemetery with Catharine Antrim. The close proximity of those known to be buried in this plot, as well as the large number of individuals recorded as buried throughout the cemetery in unknown locations or in unmarked graves, indicates the strong possibility that an attempted exhumation of Catharine would result in the disturbance of the remains of others and recovery of remains other than hers.

might result in a misinterpretation of the test. The test would not prove that Brushy Bill Roberts was, in fact, Billy the Kid. It would merely confirm a matrilineal relationship already known to exist.

exhumed. The proposed use of DNA extracted from the putative remains of Antrim as a known standard violates scientific and ethical constructs because independent means of establishing Antrim's identity to a reasonable scientific standard were unavailable (Box 10.9). The benefit of Antrim's exhumation alone to the criminal investigation was negligible.

It is interesting to note that despite his initial enthusiasm for the project, Governor Richardson had begun to distance himself from the investigation. The negative outcome of Komar's research had altered the medical examiner's perception of the investigation and this message had been conveyed to the governor's office. Governor Richardson declined to comment publicly on the investigation, and even after repeated requests by the BTK investigators, he failed to issue a written request to the medical examiner to perform the exhumation.

BOX 10.9

Independently establishing the identity of Catharine Antrim

The possibility of exhuming Catharine Antrim raises an important research question: Could the remains removed from the grave be identified? If the exhumation of Antrim was intended to provide a definitive, known sample against which all other tests would be compared, how are investigators to know the remains exhumed are those of Catharine Antrim?

One method relies on basic biological profile (Chapter 5). According to historical records, Catharine Antrim was of Irish descent. She was described as being of medium build and height with straight and graceful features, light blue eyes, and golden hair. She is known to have died of tuberculosis, although her age at death remains uncertain. The current headstone indicates Catharine was born in 1829 and died 1874, suggesting that she died at age 45. However, the original headstone (upon which the current headstone is based) featured a misspelling of her name and an incorrect date of death, raising the possibility of an incorrect date of birth. Further, several historic records indicate Catharine's year of birth as either 1831 or 1840, making her as young as 34 when she died. The first step in confirming the identity of Catharine Antrim would be to recover remains consistent with what is known about Catharine's physical appearance: a female of European descent, of medium height, with a history of tuberculosis and an age-at-death of 34 to 45 years.

Since Catharine shares a burial plot with at least 12 other known individuals and the exhumation could potentially encounter the remains of any or all of them, it is necessary to obtain demographic information on any individual who might be mistaken for Catharine. Of the known dozen, five are females, all of whom are of European descent. Three of the females, however, are older than Catharine, with ages at death of 68, 75, and 82 years. The two remaining females are believed to be buried directly adjacent to Catharine. The female to Catharine's right died at age 35. The age of the female to Catharine's left is unknown (Figure 10.4). These circumstances offer the possibility that the wrong body could be exhumed, confirming the need for an independent means of establishing identity.

A recovered skull raises the possibility of photo superimposition. While photo superimposition is not acceptable in a court of law as a means of identification, it continues to be used as a means of including or excluding tentative identifications. Although the use of superimposition in typical biohistorical investigations falls outside the jurisdiction of court-established standards for acceptable methodology, this case is considered criminal, and all methods are subject to judicial review. While such circumstances would normally restrict the use of photo superimposition, in this instance the question is moot. Although several photos purporting to depict Catharine Antrim, exist, their provenience has been challenged. None of the photographs have been authenticated, and thus the use of photo superimposition in this case would not produce meaningful results, regardless of the issue of admissibility.

The final avenue of inquiry available to investigators is to find alternative known sources of mtDNA, such as living descendants. This is ironic, for if such established samples were available, there would be no need to exhume Catharine Antrim. To date, no living descendants of Catharine Antrim have been identified.

One briefly promising possibility came to light during the archival records search. In addition to Billy, Catharine had one other son, Joseph. Joseph died in Colorado, of "apoplexy," on November 25, 1930, at age 76. His body was donated to the Colorado medical school. If retained, Joseph's remains could have provided a source of mtDNA to compare against the putative remains of Billy the Kid, thereby rendering the exhumation of Catharine unnecessary. Investigators urgently sought to determine whether any portion of Joseph's body had been retained by the medical school. Unfortunately, all bodies donated to the school during that period were cremated following their dissection. No specimens attributable to Joseph could be found.

This case illustrates the dangers inherent in using the remains of one historical individual to establish the identity of another set of historic remains. For DNA testing to be valid, comparisons of an unknown must be made against a known sample. In this instance, there was no scientifically credible way to establish the identity of Catharine Antrim. Without an established known, all subsequent tests would be purely speculative and inconclusive.

Although the scientific reasons for not proceeding mounted, the BTK team remained undaunted. In November 2003, the investigators petitioned for a court order, directing the medical examiner to exhume the body of Catharine Antrim for the purpose of obtaining a DNA sample. In response, both Komar and the chief medical examiner, Dr. Ross Zumwalt, issued affidavits to the court outlining the results of the research thus far and the scientific grounds for objecting to the request. Upon reviewing the affidavits, the judge ruled against the petition, citing that the time was not "ripe" for the exhumation and that, by itself, the exhumation served no purpose relevant to the investigation. Breathing a collective sigh of relief, the OMI believed the matter to be resolved.

In October 2005, one of the BTK investigators, Tom Sullivan, wrote to the *Ruidoso News*. In his letter Sullivan blasted critics who had described the investigation as a "stunt" and had called the investigators "stupid cowboys," "slick varmints," and "unsavory characters." He also indicated that the team had located the bench on which Billy's body was believed to have rested after the shooting. The bench bore a stain the investigators argued was blood. The team had also exhumed the remains of John Miller for the purposes of obtaining a DNA sample (Box 10.10), presumably to be compared against the stain on the bench. Local newspapers ran a photograph of one of the investigators standing proudly, holding John Miller's freshly exhumed skull.

The introduction of new evidence, specifically the possibility of obtaining a DNA sample from the bench and the mtDNA sample to be extracted from John Miller, provides the BTK investigators with grounds to re-petition the court and potentially obtain an order for the exhumation of Catharine Antrim. In spite of the strong scientific rationale for denying such an exhumation, it remains to be seen whether the desired order will be issued. At the time of this writing (June 2006), the final outcome of the case is unknown.

This example is presented to illustrate the legal and ethical issues encountered in the course of biohistorical investigations. One of the most significant is the question of control of access to biological materials in the absence of living descendants. Without next-of-kin consent for exhumation or sample extraction, the decision falls to the courts (as in the case of Jesse James) or to the individuals who have physical possession of the remains (as was the case with Beethoven's hair and skull fragments). Depending on state statutes, the authority to permit exhumations may lie with the local medicolegal authority, as evidenced in the case presented here. Without professional guidelines or regulations, individual outcomes rely on the ethical standards of those empowered to grant or deny access to the biological material. Such decisions may be tempered by the possibility of political or financial gain.

BOX 10.10

BTK investigators face possible criminal charges

The *Albuquerque Tribune* of May 14, 2006 reported that two members of the BTK investigative team faced possible criminal charges in relation to the exhumation of John Miller's body in Arizona. The report states that the BTK team had exhumed the body without either a court order or a permit from the local medicolegal authority. The BTK team could not argue that they did not know such measures were necessary, having gone through a court battle in New Mexico attempting to secure such permission to exhume BTK. The investigators contend that they were not guilty of a criminal act because there was no intent to commit a crime on their part. It now falls to the court in Arizona to explain to these former lawmen that intent relates to the decision to commit the act (in this case, illegally exhuming human remains), not what they intended to do with the body afterward.

CONCLUSION

At present, biohistorical questions can be raised and investigations conducted by scientists or hobbyists. As seen here, cases initiated without carefully considering scientific issues may lack the scientific basis and integrity needed to produce meaningful results. Professional codes of conduct in the related scientific communities show concern for the outcomes of research, in terms of social impact and the interpretation and dissemination of results (Andrews et al., 2004). This challenges scientists and forensic professionals to behave responsibly in biohistorical investigations. It is not enough to simply distance oneself from an unethical or invalid inquiry; it is our professional obligation to establish standards and enforce them. Until our professional organizations provide guidelines and regulations for such investigations, it falls to individual scientists undertaking this research to maintain ethical standards and thus be accountable.

GLOSSARY

²¹⁰Pb: Symbol for the radioisotope lead 210

²¹⁰Po: Symbol for the radioisotope polonium 210

a-: prefix meaning without or not

AAAS: American Association for the Advancement of Science

AAFS: American Academy of Forensic Sciences

ABFA: American Board of Forensic Anthropology

abrasion: a disruption of the superficial layer of skin

academy: a place of learning or study

accidental deaths: death resulting from trauma or elevated toxicity but without intent

accuracy: the degree to which a method is correct and a true representation of the variable being tested

acquired: describing a pathological condition developed after birth as the result of external factors

acts of war: episodes of aggression that negate the normal international law of peacetime and activate the international law of war

actualistic studies: research that studies and describes a process through observations of recovered materials or controlled field studies

actus reus: the physical element of a crime; the prohibited act

acute: a single episode of a pathological condition

advocacy: assisting on behalf of another; working as an invested participant

age of attainment: the relationship between chronological age and the age of achievement or development of a morphological trait

ALAF: Latin American Forensic Anthropology Association

algor mortis: one of the triad of early postmortem changes, characterized by decreasing body temperature; see also *livor mortis* and *rigor mortis*

amelogenin gene: gene that codes for proteins found in tooth enamel

amino acid racemization: partial conversion of one enantiomorph into another (such as an l-amino acid to a corresponding d-amino acid) so that the specific optical rotation is decreased or reduced to zero in the resulting mixture (Pugh, 2000:1501)

amplicons: the product yielded by the polymerase chain reaction (see *PCR*) during DNA testing

amplitude: the maximum height of a wave of electromagnetic radiation

anatomical specimens: cleaned dry bone or tissue prepared for teaching or research purposes

ancient remains: a type of nonmedicolegal remains synonymous with prehistoric remains, involving individuals whose deaths predate the historic period of a region

anomaly: deviation from the norm that is nonlethal or nondisruptive to functioning

antemortem period: from birth to the beginning of the perimortem period

appointment: a formal, ongoing working relationship between an anthropologist and a medicolegal authority

Archaeological Resources Protection Act: 16 U.S. Code 470cc, enacted in 1979, which regulates the excavation of archaeological sites and the issuance of permits

armed conflicts: acts of aggression between opposing parties involving weaponry

arteria: Terminologia anatomica meaning artery

articulatio: Terminologia anatomica meaning joint

arthro-: prefix referring to joint

as-found: contextual designation indicating that accidental or intentional human interaction with the remains has occurred

asphyxia: impaired or absent exchange of oxygen and carbon dioxide on a ventilatory basis (Pugh, 2000)

associated materials: nonbiological goods or artifacts recovered at scene, including clothing, jewelry, or personal effects

association: theory of evidence recognition based on a direct link between the crime and a piece of physical evidence

attenuation: the partial or complete loss of x-ray energy in tissue

authority: derived from laws and statutes granting specific powers to specific individuals, with boundaries and circumstances of authority defined by jurisdiction

autolysis: *self-digestion*, occurring through the action of digestive enzymes that break down complex proteins and carbohydrates. The same mechanism that aids in digesting food in life begins to digest the gastrointestinal tract after death. Considered a minor component of decomposition.

autopsy: *to see for one's self*; the examination of a dead body to determine cause of death

axial: longitudinal angulation of the x-ray beam along the long axis of the body

background checks: comparisons of employment records, transcripts, and other forms of social data to confirm the qualifications claimed by job applicants

ballistics: the study of the travel of a projectile

Bayesian methods: used in the calculation of probabilities

bias: an inclination or prejudice, a sampling or testing error caused by systematically favoring some outcomes over others

bioanalysis: collective name for a group of physical sciences, such as DNA analysis, serology, and toxicology, traditionally employed in the forensic sciences in which the subject material is of biological origin

bioarchaeology: the study of human remains from archaeological settings

biohistory: the use of bioanalysis to investigate historical figures and their diseases, causes of death, behavior, and lineages

biological profile: synonymous with osteobiography, representing the demographic information (age-at-death, sex, stature, biological affinity, and health state) of a decedent

bioturbation: the mixing or disturbance of archaeological features (including remains) as the result of natural biological taphonomic processes such as burrowing rodents

black powder: a type of firearm propellant formed by mixing potassium nitrate, sulfur, and charcoal

blade progress: the plane of advancement from contact through the termination of a cut in sharp force trauma

blade stroke: the direction of the individual strokes or cutting action in sharp force trauma

blazed: describing patient information encoded onto a radiographic image

body orientation: the position of bodies in a grave, relative to each other and to the cardinal directions

BTK: William Bonney; a.k.a, Billy the Kid

burning rate: the speed at which burning gunpowder changes to gas

"but for": an informal rule applied in death investigations to determine the causative agent—"But for" what would the individual still be alive?

cadaver dogs: canines trained to detect the air scent of human remains and alert handlers

caliber: a system of firearm designation, relating to the diameter of the bore of a gun barrel or the bullet diameter and case length of a cartridge

CAMI: Civil Aeromedical Institute

canalis: Terminologia anatomica meaning canal

canonical analysis of the principal coordinates (CAP): in statistics, a regression method that quantifies a relationship between a predictor variable and a criterion variable by the coefficient of correlation r, the coefficient of determination r^2, and the standard regression coefficient β. Multiple regression analysis expresses a relationship between a set of predictor variables and a single criterion variable.

carbon 14 (^{14}C): a beta-emitting radioisotope having a half-life of 5715 years, used as a tracer in studying various aspects of metabolism. Carbon, the major bioelement, is a nonmetallic tetravalent element with two natural isotopes (^{12}C and ^{13}C) and two artificial, radioactive isotopes of interest (^{11}C and ^{14}C). Naturally occurring ^{14}C results from cosmic ray bombardment;its presence can be used to date materials containing natural carbonaceous materials (Pugh, 2000)

cardio-: Terminologia anatomica referring to the heart

carrion: remains that host necrophagous organisms

cartilago-: Terminologia anatomic referring to cartilage

cartridges: ammunition used in firearms; one hand gun or rifle cartridge, consisting of a casing, a primer, propellant, and a bullet

cause of death: the disease or injury that initiates a sequence of events that result in death

cavitation: temporary expansion of tissue seen in gunshot wounds as a result of the transfer of kinetic energy from the bullet to the tissue

cemental annuli: incremental lines in dental cementum representing sequential deposition, believed to be associated with the age of the individual

chain of custody: synonymous with *evidence continuity* and *chain of evidence*: investigators must maintain the integrity and security of evidence at all times, from collection through analysis to presentation at court

chattering: a form of sharp force trauma consisting of a series of discontinuous abrasions occurring when a serrated edge fails to fully engage across the tissue

choke: the constriction of the barrel at the muzzle of a shotgun

chondro-: Terminologia anatomica referring to cartilage

chopping: a form of sharp force trauma, similar to stab wound, in which the force is delivered along the long axis of a sharp object

chronic: describing a persistent pathological state

CILHI: Central Identification Laboratory in Hawaii, a U.S. military facility

circumstantial evidence: evidence based on inference, rather than personal knowledge or observation

civil court: the court with jurisdiction over noncriminal cases such as divorce and insurance claims

clandestine burial: an unmarked grave, not in a cemetery, created with the intent of evading detection (excluding archaeological interments)

class evidence: group affiliation based on common characteristics

clinical terms: terminology used in a clinical setting, typically the English version of Terminologia anatomica or the surname of the individual who first described the element or condition

coefficient of variation: the coefficient of variation (CV) is reported as a percentage and calculated from the average and standard deviation as follows:

$$CV = 100 \times \frac{\text{standard deviation}}{\text{average}}$$

Combined DNA Index System (CODIS): a computer database, developed and maintained by the FBI, that allows investigators to compare unknown DNA samples against known specimens.

commingling: the mixing of body parts or skeletal remains from two or more individuals

comminuted fracture: a crush or complex fracture with multiple extensions

common names: nonmedical terminology used to describe phenomena in anatomy or pathology

compel: to cause or bring about by force or overwhelming pressure

complicity: the union with others in an ill design or one who is an associate, confederate, or accomplice, acting knowingly, voluntarily, and with intent shared by the principal offender (Garner, 2001)

compound fracture: a form of blunt force trauma involving direct communication with the external environment, also known as open fracture

computed tomography (CT): a medical imaging technique using serial radiographic images

concentric fractures: fractures produced from force transmitted in waves encircling the point of impact in blunt force trauma

concordance analysis: examination of two samples to quantify congruent points of comparison

congenital: describing a pathological condition existing at birth

consultancy: a case-by-case working relationship between an anthropologist and a medicolegal authority or police agency

constructive intent: evident when the outcome of an act can be reasonably expected, such as the intent of the individual playing "Russian roulette"

contact gunshot wound: wound produced when the muzzle of the gun is in full contact with the skin at the time of discharge

contained scenes: scenes in which the remains are isolated in a single location or defined area

contamination: the presence in a sample of exogenous materials

context: the physical environment, associated materials, and location of human remains

contrast: the difference in density between two areas of a radiographic image, comprising subject contrast and film contrast

core: a soil sample extracted with the use of a hollow probe

coroner: an elected or appointed official responsible for determining the manner of death of persons in his or her jurisdiction

crime: a social harm that the law makes punishable, or a breach of a legal duty treated as the subject of a criminal proceeding such as medical malpractice or police misconduct (Garner, 2001)

criminal court: the court with jurisdiction over criminal cases involving felonies and misdemeanors

CSFS: Canadian Society of Forensic Science

cultural affiliation: group membership within a particular cultural group, determined by class characteristics

DABFA: diplomate of the American Board of Forensic Anthropology

dactyloscopy: the practice of using fingerprints to identify an individual

datum: a fixed point established at a scene to which all other evidence or objects are referenced

***Daubert* criteria:** federal standards specified in *Daubert v. Merrell Dow Pharmaceuticals* (92–102, US 579, 1993). The ruling outlines four criteria governing the introduction of scientific evidence at trial: (1) whether the theory or technique has been or can be tested, (2) whether the method has been subject to peer review and publication, (3) the known or potential error rate of the technique, and (4) the degree of acceptance of the theory or technique in the relevant scientific community.

DDS: doctor of dental surgery

decomposition: the systematic destruction of biological material through the processes of putrefaction and autolysis

decubitus: radiological position in which the patient is lying down and the x-ray beam is parallel to the floor

deduction: reasoning that moves from general to specific, arguments based on law or widely accepted principles are said to be deductive

defect: an imperfection, failure, or absence of tissue

density: a statement of how tightly packed the atoms of a substance are

depositional environment: the physical surroundings of a body postmortem, including surface deposits, burials, and water

derma: Terminologia anatomica meaning skin

detail: the sharpness of a radiographic image, synonymous with definition and sharpness

diagnostic radiology: the study of images of internal structures of the body for the purpose of identifying and evaluating disease

diagnostic sonography: ultrasound, a medical imaging technique in which sound waves are pulsed from a handheld transmitter

digital geometry processing: the computer program that converts two-dimensional computed tomography images into a three-dimensional representation of internal organs or structures

diploid: the state of a cell containing two haploid sets of chromosomes derived from the mother and father, the normal chromosome complement of somatic cells (Pugh, 2000)

direction: the path of travel of a projectile, sharp force weapon, or blunt forces

disclosure: the revealing of facts or the making known of previously unknown information

discovery: compelled disclosure of information that relates to a litigation at a lawyer's request

discriminant function analysis: used to determine which variables distinguish between two or more naturally occurring groups

disease: a pathological condition with recognized etiologic agent(s), an identified group of symptoms, and consistent anatomical alterations

disorder: a disturbance of function, structure, or both

distance consultations: occasions on which digital photos and information are sent to an anthropologist from scene investigators via e-mail or cell phone, allowing visual assessment of the scene without travel

distortion: a discrepancy between radiographic image size and the true object size

D-loop: the 1,122 base-pair control region that contains the origin of replication for one of the mitochondrial DNA strands but does not code for any gene products (Butler, 2005)

DNA: deoxyribonucleic acid

Doe: designation of an unknown individual

double jeopardy: being charged twice for the same offense; disallowed in the United States under the Fifth Amendment

duration: the length of time energy is applied to tissue

dys-: prefix meaning bad or difficult

EAAF: Argentine Forensic Anthropology Team

EAFG: Guatemalan Forensic Anthropology Team

endogenous: originating inside an organism

en bloc: autopsy technique involving removal of a complete system, such as the cardiopulmonary or gastrointestinal

en masse: autopsy technique involving removal of all thoracic and abdominal contents as a whole

ephemeral evidence: temporary physical evidence or feature at scene that cannot be collected in a traditional manner, such as fingerprints, and footprints, and tool marks

erratic commingling: description of body position within a mass grave in which bodies are deposited in a disorderly manner

ethnic: describing group identity derived from country of birth, cultural heritage, or self-designation

etiology: the cause or source of a pathological condition, as well as the study of cause of disease

evidence: testimony, documents, and tangible objects presented in court

exculpatory evidence: evidence that establishes a defendant's innocence

exogenous: originating outside an organism

expert witness: an individual with specific technical or scientific qualifications who offers testimony focusing on a defined area of expertise; may be asked to render an opinion in addition to contributing to the factual record

exposure time: the length of time, in fractions of a second, that an x-ray beam emits radiation

expression: the physical presentation or manifestation of a disease

exsanguination: blood loss to the external environment

external: originating outside an organism

external ballistics: the study of a projectile's flight from the gun muzzle to a target

external validity: characteristic of a model that can be generalized to different populations, useful as a global reference

extrinsic: originating outside a part or organism

FAC: the Forensic Anthropology Center at the University of Tennessee

facial reconstruction: the use of skin-depth standards to reconstruct the soft tissues of a decedent's face. Originally proposed as a source of identification, the method is not admissible as scientific evidence in U.S. courts of law. Synonymous with facial reproduction and facial approximation.

fact: stipulated evidence

fact witness: witness providing court testimony drawn from within the law enforcement or medicolegal systems, including witnesses who attend scenes or process evidence

falsifiability: characteristic of an empirical statement that is amenable to testing through the presentation of counterarguments

FBI: Federal Bureau of Investigation

FDB: the Forensic Data Bank at the University of Tennessee

feature: any nonportable remnant of human activity, such as a grave

Federal Rule of Evidence 702: source of the five criteria governing the admissibility of scientific evidence in court: (1) validation of the expertise of the witness; (2) the responsibility of evaluating experts falls to the judge; (3) testimony must be based on facts or data; (4) testimony must be the product of reliable theories and methods; and (5) the witness must apply the theory or method reliably to the facts of the case

felony: a serious crime punishable by imprisonment or possibly death

felony murder: homicide committed during the course of another dangerous felony such as rape or kidnapping

firearm: a gun that uses a spark to ignite a powder charge to create gas to propel a projectile

first-degree murder: homicide that is willful, deliberate, and premeditated

fissura: Terminologia anatomica meaning fissure

forensic: used in a court of law

forensic computing: the use of electronic media, images, and data in legal contexts

forensic entomology: the study of necrophagous insects and their patterns of colonization to establish time-since-death and to identify antemortem toxicology, victim DNA, child and elder abuse, and perimortem trauma

forensic odontology: issues of dentistry of interest in a court of law, including the use of antemortem and postmortem dental comparisons to establish identity and the analysis of bite marks

forensic psychiatry: the application of methods, theories, and practices of psychiatry and mental health to a court of law

forensic radiology: the use and interpretation of radiographic images of interest in a court of law

frequency: a characteristic of electromagnetic radiation waves, specifically the number of wave cycles per second, measured as hertz (Hz)

Frye test: outlined in *Frye v. United States* (293 F.2s 1013, 1923); two criteria used for governing admissibility of scientific evidence in court: (1) the expert qualifications of the presenting witness and (2) the general acceptance of the technique or theory within the relevant discipline

FX: accepted abbreviation of fracture

garroting: the use of a handheld weapon, typically a ligature such as a rope, scarf, or cord, to strangle someone

general intent: actions indicating recklessness or negligence

genocide: an international war crime involving the intent to destroy, in whole or in part, a national, ethnic, racial, or religious group

Global Positioning System (GPS): , a handheld tracking device that establishes its location relative to three satellites to provide the longitudinal and latitudinal coordinates of a location as well as its elevation

gloving: the imbibing of water into the skin following death, producing a wrinkled appearance of the skin; synonymous with maceration. Gloving of the hands and feet can result in skin slippage of the appendage as a whole.

governmental agencies: organizations with established personnel, protocols, and mandates, funded and overseen by an existing ruling body

grave profile: a map depicting a vertical representation of a body in a grave through a series of measurements from datum

greenstick fracture: an incomplete fracture, typically seen in juveniles

grid: a mapping technique in which stakes and string are used to superimpose a grid or series of equal measured squares over a scene

ground-penetrating radar (GPR): the use of electromagnetic wave propagation and scattering to image, locate, and quantify changes in the soil

gun: any device using pressurized gas or mechanical energy to force a projectile through a tube

haploid: the number of chromosomes in a sperm or ovum, half the number in somatic (diploid) cells

haplotype: a specific genetic sequence of mitochondrial DNA

Haversian canals: system of communication among osseous structures

Henssge's formula: a short-range method for using body temperature to ascertain time-since-death; temperature begins decreasing immediately after death at a rate of 1.5°C per hour until ambient temperature is reached

hesitation defect: a type of sharp force trauma involving partial depth incisions resulting from an incomplete action or attempt to wound

hidden exit: an atypical gunshot wound in which no plug or spall is produced; rather, the projectile exits through existing fractures or fractures caused by the entrance wound

histomorphometry: the assessment of morphological structures visualized in histological sections

historic remains: nonmedicolegal remains representing individuals whose deaths occurred during the historical period of a region

homicidal violence: cause of death in which modifications of the body are so severe that the victim could not have survived

homicide: death at the hands of another; implies responsibility

hospital autopsy: a postmortem examination outside the jurisdiction of the medicolegal authority after a natural death, conducted to confirm or refute a diagnosis

human biology: an academic field of biology that focuses on humans

humanitarian efforts: response to a mass or natural disaster that focuses primarily on the needs of the living, involving no medicolegal authority

HX: abbreviation meaning medical or other history

hyper-: prefix meaning too much

hypervariable region I (HVI): found on the D-loop of mitochondrial DNA, between nucleotide positions 16,024 and 16,365

hypervariable region II (HVII): found on the D-loop of mitochondrial DNA, between nucleotide positions 73 and 340

hypervariable region III (HVIII): found on the D-loop of mitochondrial DNA, between nucleotide positions 438 and 574

hypo-: prefix meaning too little

hypothetico-deductive method: an approach to making a scientific determination. A scientific inquiry proceeds by formulating a hypothesis that is intended to explain an observed phenomenon. From the hypothesis, explicit predictions of further phenomena are deduced that should be observable. Observations that run contrary to those predicted are taken as a conclusive falsification of the hypothesis; observations in agreement with those predicted are taken as corroborating the hypothesis. Given sufficient predictions and observations, it is then supposedly possible to compare the explanatory value of competing hypotheses by seeing how well they are sustained by their predictions (Encarta online encyclopedia, accessed August 4, 2006)

ICMP: International Commission on Missing Persons

ICTR: International Criminal Tribunal for Rwanda

ICTY: International Criminal Tribunal for the Former Yugoslavia

impacted fracture: a results of forcible intrusion of the diaphysis of a long bone into the metaphysis or epiphysis

implied intent: a decedent's state of mind as inferred from speech, conduct, or written word

incidence: the number of new cases of a specific disease occurring during a specific period of time in a specific population

incision: to divide tissue cleanly

inculpatory evidence: evidence supporting a defendant's involvement in a crime

indictment: a formal written accusation of a crime made by a prosecutor.

individual evidence: material traced to a single source

individualization: a principal component of personal identification, in which traits or features unique to the decedent are used to confirm identity

induction: reasoning that moves from specific to general

infraction: incomplete or partial thickness fracture

infrared photography: film developed to capture the infrared range of the electromagnetic spectrum

injury: a result from any trauma source other than weapons

inquest: inquiry convened by the medicolegal authority to establish facts relating to a death

in situ: contextual designation indicating the remains are exactly as deposited or that all alterations to a scene are the results of natural processes

Integrated Automated Fingerprint Identification System (IAFIS): computer-based search system to match suspect fingerprints against known prints, compiled and maintained by the FBI, and containing both civil and criminal prints

intensity: an aspect of electromagnetic radiation relating to the amount of energy relative to distance (see *inverse square law*)

intent: the state of mind accompanying an act; the mental resolve to act

intermediate range: the gunshot wound range of fire when the muzzle of the weapon was not in contact with the body at time of discharge but was close enough to result in tattooing or stippling

internal: originating within an organism

internal ballistics: the study of a projectile's travel before it leaves the barrel of a gun

internal validity: a model that is valid for a specific population but cannot be generalized to a wider population

interposed wound: an atypical gunshot wound in which the bullet strikes an intermediary target before striking the body

interval: units of measure that are fixed or standard, such as millimeters or degrees

intrinsic: belonging entirely to a part or organism

inverse square law: allows measurement of the intensity of electromagnetic radiation by stating that all radiation travels at the speed of light and diverges from its source. As the distance increases, the intensity decreases.

investigator bias: variability in search, collection, recovery, or curatorial methods among investigators that results in the incomplete recovery of remains or the failure to recognize or collect certain evidence

involuntary manslaughter: homicide with no demonstrated intent to kill or harm, committed with criminal negligence or during the commission of a crime

-itis: suffix indicating an inflammatory disorder

jail: a detention facility for those awaiting trial or convicted of misdemeanors

JPAC CILHI: Joint POW/MIA Accounting Command's Central Identification Laboratory in Hawaii

jurisdiction: legal authority of an agency or individual to perform an act within a defined geographic region

kerf: the groove created by a cutting tool

keyhole defect: an atypical gunshot wound occurring when a bullet strikes the skull and fragments, with only a fragment entering the skull while the rest deflects. The margins of the resulting defect are beveled both internally and externally.

Knight's formula: a short-range method of estimating time-since-death by means of body temperature; there is a 45-minute plateau after death in which body temperature remains constant, followed by a decrease of 1°C per hour until ambient temperature is reached.

kilovoltage (kVp): a measure of the energy of x-rays

lacerations: torn or split tissues, resulting when an applied force exceeds the mechanical strength of a tissue

lands: the elevated surfaces of the interior of a gun barrel, between the rifling grooves

latent prints: fingerprints left at a crime scene, often invisible to the naked eye, that require chemical or other processing to be rendered visible

lateral: side aspect or direction, or a body position in radiography—lying on one's left or right side

laws: aggregates of legislation, judicial precedent, and accepted legal principles in force in a prescribed geographic region

laws of stratigraphic succession: the deposition of strata represents the accumulation of materials over time in a relative temporal relationship, including geological (the natural deposition of strata within an area) and archaeological (the effects of human action within an area) stratification

law of superposition: strata or materials in their original depositional sequence will reflect the age of the strata relative to other strata; materials recovered lower in the sequence are said to be older than those above

Le Fort fracture: clinical name for three types of facial fractures involving the maxillae, zygomatics, and/or nasals

lesion: any pathological or traumatic discontinuity of tissue

ligature: a weapon, such as a rope or scarf, used to asphyxiate

likelihood ratio: a statistical test in which one expresses the ratio between the maximum of the likelihood function under the null hypothesis and the maximum with that constraint relaxed

limited autopsy: autopsy restricting invasive procedures to those necessary for evidence collection

linear fracture: a single simple fracture line

livor mortis: early postmortem change resulting in the purple discoloration of the skin of dependent parts of the body. The purple color results from unoxygenated hemoglobin in the blood. The pattern and distribution of livor mortis results from the settling of blood into the capillaries of the skin after circulation has ceased.

loading rate: the speed at which stress is applied during the development of a stress–strain curve

load type: how energy is transferred during blunt force trauma, including direct, indirect, and penetrating load types

Locard's principle: any two objects that come in contact with each other leave traces, one upon the other

long-range gunshot wounds: the determination of range of fire, long-range wounds are those in which the sole defects on the skin are those produced by the projectile

long-range TSD estimates: cover the period from 30 days postmortem, continuing indefinitely

luminol: a chemical compound used in conjunction with an alternative light source to test for the presence of blood or bodily fluid

maceration: the imbibing of water into the skin, or the rendering of cadavers into skeletal remains through mechanical means such as boiling

magnetic resonance imaging (MRI): technique that uses strong magnetic fields, pulsed radio waves, and the nuclei of hydrogen atoms in the subject's body to produce medical diagnostic images

magnification factor: a number representing the amount of distortion in a radiographic image. Depending on the information available, there are two ways to calculate the magnification factor (MF) of a radiographic image:

1. If the size of the specimen or object is known,

   $MF = I/O$

   where I = image size and O = object size

2. If the object size is unknown but a standard SID was used or the distances are known,

   $F = SID/SOD$

   where SID = distance from x-ray source to film and SOD = distance from x-ray source to object

magnitude: the amount of force applied to tissue

malice aforethought: the requisite mental state of the accused for the finding of first-degree murder

manifest intent: identified intent on the basis of indirect evidence or circumstance, such as repeated action

manner of death: the fashion in which the cause of death comes into being, namely, homicide, suicide, accident, natural, or undetermined

manslaughter: the unlawful killing of a human being without malice aforethought

marbling: the discoloration of subcutaneous blood vessels and skin as the result of putrefaction

mass disaster: a mass death event resulting from human action or inaction, such as a plane crash

mass graves: complex deposits in the archaeological record containing the remains of more than one individual

mass murder: the killing of at least three people at the same approximate time and location

material evidence: witness testimony or other forms of evidence based on personal knowledge or observation

material witness: witness testimony based on personal knowledge or observation

matter out of place: a theory of evidence recognition based on unusual nature or location of physical evidence

mechanisms of death: processes leading to death, including toxicological, anatomical/mechanical, and physiological

mechanisms (agents) of transport: processes or organisms capable of modifying scenes postmortem through the transport of bodies or skeletal elements to alternative locations; agents of transport can operate by means of water, faunal, and human action

medical examiner: an appointed official who is a licensed, board-certified forensic pathologist, performs autopsies, and rules on the cause and manner of death

medicolegal: describing the application of medical science to law

medicolegal remains: an individual for whom the cause and manner of death are determined and a death certificate is issued

medicolegal response: efforts in response to a mass or natural disaster that meet the standards of conventional medicolegal investigations, including death certification

medicolegal (ML) system: the formal mechanism for death investigation and certification within a specific geographic area, overseen by a coroner or a medical examiner

mens rea: "a guilty mind"; the accused shows an understanding of the unlawful nature of the crime through behavior; the mental element of a crime

midrange TSD estimates: cover the period from 3 days to 30 days postmortem

milliampere (mA): controls the quantity of x-rays by controlling the electrical current flowing through the x-ray tube during exposure

mimicry: postmortem artifact modification resulting from natural taphonomic processes that mimic or resemble specific by-products of human action

minimum number of individuals (MNI): an estimate derived from duplication of elements as well as differences in age, sex, or size of elements to determine the minimum number of individuals represented at a scene

misadventure: a homicide committed accidentally by a person performing a lawful act with no intent to harm or kill

misdemeanor: a crime punishable by fine, penalty, or confinement in a jail

modeling: the process of bone growth through the deposition of primary lamellar bone in parallel sheets to circumferential and endosteal surfaces and its subsequent subtraction from opposing regions

Mohan **criteria:** Canadian standard for the admissibility of scientific evidence in court, based on the 1994 ruling *Regina v. Mohan* (S.C.R. File No. 23063)

monotonic: injuries resulting from a single traumatic event

morbid: describing any deviance from the normative state, including a pathological condition or disease

morphological deaths: deaths resulting from a lethal physical change in the body; synonymous with anatomical deaths

morphological variation: differences in physical traits among individuals or groups

morphoscopic traits: morphological or nonmetric features recorded as categorical data, such as suture complexity or orbital shape

motive: personal justification for an action

MSCT: multislice computed tomography

murder: the unlawful killing of a human being with malice aforethought

musculus: Terminologia anatomica meaning muscle

myo-: root referring to muscle

myocardial infarction: the partial or complete blockage of the blood vessels of the heart, commonly known as a heart attack

NAGPRA: North American Graves Protection and Repatriation Act

NAME: National Association of Medical Examiners

National DNA Index System (NDIS): the highest level of the CODIS hierarchy, enabling police across the country to access DNA samples from crime scenes and reference samples from convicted offenders

national: describing group identity based on country of birth or residence

natural deaths: deaths resulting from a disease or pathological condition

natural disasters: events of mass death or destruction resulting from nonhuman causes such as weather or environmental change

near contact gunshot wound: wound that results when the muzzle of the gun was within 10 mm of the skin at time of discharge but not in physical contact

necrophagous: describing organisms that feed on dead biological material

negligence: any conduct that falls below the legal standard established to protect others against unreasonable risk of harm

nephro-: root relating to the kidney

neo-: prefix meaning new or recent

neonatal line: a prominent dark line in the enamel of deciduous dentition, observable under transmitted light or through SEM analysis

neoplasms: abnormal tissues that grow more rapidly than normal by cellular proliferation and continue to grow after the stimuli that initiated the new growth cease (Pugh, 2000)

neuro-: root referring to nerve, nerve tissue, or the nervous system

nervus: Terminologia anatomica meaning nerve

Nomina anatomica: former system of anatomical terminology, adopted in 1955 and used until replaced by Terminologia anatomica in 1985

nongovernmental organizations (NGOs): independent agencies created by individuals without direct governmental supervision

non-Haversian structures: remnants of osseous structures, lacking Haversian canals at their centers, used to evaluate age-at-death in histomorphometric analysis

nonmedicolegal remains: human remains that fall outside the jurisdiction of the medicolegal authority, including historic or prehistoric remains, anatomical specimens, trophy or souvenir skulls, shrunken heads, and improperly disposed of mortuary remains

norm or normal state: any tissue, organ, or organism that is functionally intact, free of defects, and capable of sustaining its functionality

norma: Terminologia anatomica meaning aspect or side

North American Graves Protection and Repatriation Act (NAGPRA): Public Law 101–601 (25 United States Code 3001), enacted in 1990, which protects prehistoric remains from excavation on federal or tribal lands

NTSB: National Transportation Safety Board

nuclear medicine: a form of medical imaging using radiation-emitting radionuclides and a radiation-detecting gamma camera

null hypothesis: the prediction that there are no statistically significant differences between two groups

objective: existing independent of a single mind; describing matter that is observable and verifiable

objectivity: lacking bias

oblique: describing an angular approach in radiographic imaging

obscuring processes: events causing postmortem modification of remains resulting in the progressive loss of information due to time passage and environmental factors

ODONTOSEARCH: computer database containing dental records and formulas

-oid: suffix meaning resembling

-oma: suffix designating a tumor or neoplasm

OMI: New Mexico Office of the Medical Investigator

on-scene consultations: the participation of an anthropologist at a death scene

open scenes: scenes in which the location or dispersal of remains is unknown

optical density: describes the degree of blackness of a radiographic image; darker images have higher optical densities

orderly placement: description of body position in a mass grave in which all individuals are afforded sufficient space, bodies are placed supine with purposeful arrangement of the limbs, and all heads are oriented in the same direction

ordinal: units of measure that can be ranked, such as small, medium, and large

ORISE: Oak Ridge Institute for Science and Education

os: Terminologia anatomica referring to bone

-osis: suffix indicating disease state or condition

ossuaries: formal secondary burial arrangements

osteo-: root meaning bone

osteobiography: estimated demographic data for an individual derived from examination of skeletal remains, including age, sex, stature, and ethnic affinity; synonymous with biological profile

osteons: the central canal containing blood vessels and the concentric osseous lamellae surrounding it, occurring in cortical bone; synonymous with *Haversian canals*

oviposition: the laying of eggs

paleodemography: deriving population demographics (such as age-at-death) from archaeological skeletal remains

parsimony: a style of argument that arrives at an understanding of events based on the minimum number of step needed to explain the available data

parturition scars or pits: defects or depressions evident on the innominate (often called "dorsal pits"), argued by some researchers to represent evidence of childbirth

paternity testing: DNA analysis to include or exclude a male as the father of a child

pathognomonic: a hallmark symptom or specific diagnostic

pathological fracture: trauma occurring in weakened or diseased bone

pathology: deviation from the norm that compromises or restricts functioning

penetrance: the proportion of individuals in a population with a specific genotype who express the characteristic in the phenotype

penetrating wound: a type of gunshot wound in which the bullet enters an object but does not exit

perfidy: a war crime involving a falsehood, such as feigning surrender under a flag of truce

perforating wound: a type of gunshot wound in which a bullet enters and exits an object

perimortem period: events surrounding death

period: the length of time needed to complete one wave cycle of electromagnetic radiation

petechiae: minute hemorrhagic spots in the skin

photographic superimposition: presumptive identification technique in which digital mixing is used to superimpose a photo of a decedent over an image of a skull

PHR: Physicians for Human Rights

physical anthropology: the study of mechanisms of biological evolution, genetic variation, human adaptability and variation, primatology, primate morphology, skeletal biology, and the fossil record of human evolution

physical evidence: evidence arrived at by scientific means

physiological deaths: deaths resulting from a disruption of function at the cellular or system level

pillage: a war crime involving the unlawful acquisition or destruction of material goods

pistol: self-loading handgun that shoots rounds of ammunition held in a magazine

plan view: the initial map generated at a scene, documenting the major structures and components of the scene

-plasia: suffix meaning formation

plexiform: describing the arrangement of horizontal, regular layers of cells in mammalian cortical bone

plug: a fragment of bone displaced by a penetrating projectile

point of impact: the origin of a blunt force injury, as indicated by primary fracture production

polymerase chain reaction (PCR): an enzymatic process used to replicate a specific region of DNA to allow for testing

polytonic: describing injuries resulting from multiple traumatic events

position: the physical orientation of the patient or specimen during radiography

positive identification: the highest standard of identification, in which individualization has been demonstrated through the comparison of antemortem and postmortem data

posterior probabilities: a calculation of the relationship between an unknown and a set of reference samples

postmortem interval: synonymous with time-since-death; the period between the death of an individual and the recovery of the remains by investigators

postmortem period: time frame from the end of the perimortem period to the discovery of the body

powder tattooing: the embedding into the skin of gunpowder grains, seen in intermediate-range gunshot wounds

power calculation: measures the ability of a test to avoid a type II error by detecting a null hypothesis that has at least an 80% chance of being false

precision: the degree to which a method is repeatable

prehistoric remains: nonmedicolegal remains of individuals whose deaths predate the historic period in a region

premeditation: conscious planning that precedes the commission of a crime

presumptive identification: the result that occurs when circumstances or general characteristic concordance provide investigators with a basis for identification that is not invalidated by the presence of exclusionary evidence

prevalence: the incidence or frequency of a pathological condition in a specific population

prima facie: judged on first disclosure; a fact presumed to be true unless disproved by evidence to the contrary

prison: a state or federal facility of confinement for convicted felons

projection: the direction of travel of the x-ray beam during radiography

prone: describes a body position, namely, lying on the abdomen, or face down

proteoglycans: glycosaminoglycans bound to protein chains that occur in the extracellular matrix of connective tissue

protocols: standard operating procedures or guidelines established by an authority or agency that dictate the steps investigators must follow in specific circumstances

provenience: a record of the source or origin of artifacts or materials such as skeletal remains

pseudo-: prefix indicating mimicking or the false appearance of a pathological condition

pseudoscent: a commercially available chemical signature that simulates the odor of decay, used in the training of scent detection dogs

psychological autopsy: the forensic psychiatric practice of using multiple sources of information such as medical records, interviews, and autopsy findings to reconstruct a decedent's state of mind

pulse shape: the area over which force was applied in blunt force trauma

putrefaction: the major component of decomposition, resulting in marbling, bloating, subcutaneous gas accumulation, and maceration; from endogenous and exogenous bacterial activity after death

racial: describing group identity derived from physical characteristics

radiating fractures: fractures evidenced by lines dispersing from the point of impact in blunt force trauma

radiation oncology: the use of radiant energy wavelengths and particles to treat malignant neoplastic disease

radiographic film: the film used in radiography is composed of a layer of emulsion applied to one or both sides of a transparent plastic base. The emulsion layer,which comprises silver halide crystals embedded in gelatin, is in turn covered by hard gelatin. Altering the size of the silver crystals changes the response characteristics of the film. Exposure to x-rays alters the silver halide crystals. Processing the film converts the exposed crystals into black metallic silver. Unexposed crystals are dissolved by the processing solution, resulting in clear areas on the film. Like photographic film, x-ray film comes in different speeds, contrasts, and latitudes.

radioimmunoassay: an immunochemical procedure that uses radiolabeled reactants to detect antigens or antibodies

radioisotopes: isotopes that change to more stable states by emitting radiation

radiolucent: describing substances or tissues with low attenuation values

radionuclides: radiation-emitting materials used in nuclear medicine to allow imaging; can be administered orally, intravenously, or through inhalation

radiopaque: describing substances or tissues that are highly attenuating

range: a measure of variability, namely, the difference between the highest and lowest scores in the distribution

range of fire: the distance between the muzzle of a gun and the object or target

rapid loading: high-magnitude stress over a short duration, resulting in comminuted or complex fractures as the bone responds like brittle material

rationality: a style of argument that requires events to follow a logical course

recidivism: the rate of re-offending by convicted assailants

reliability: the level of agreement between repeated measurements of the same variable; synonymous with repeatability and reproducibility

religious: describing group identity based on shared religious beliefs, rituals, or doctrines

remodeling: following skeletal maturity, bone growth follows an activation-resorption formation sequence that creates discrete bone structural units which alter bone morphology

repeated loading fracture: fracture that can result from repetitive motion injuries, including stress fractures

report writing: the formal documentation and presentation of the events surrounding the processing of a scene or the performance of a postmortem examination

requisitioning: the lawful taking of necessities such as food, fuel, and medicine from civilian populations for use by an army of occupation

res gestae: "things done"; the events relevant to a crime

resistivity: a form of geophysical testing that identifies disturbances in soil by examining soil variables such as water content and particle size

responsibilities: the duties and tasks assigned to an investigator as outlined in a job description or terms of employment

restriction fragment length polymorphism (RFLP): variations in the length of DNA fragments generated by the cleaving action of restriction enzymes

retinal folds: a condition considered pathognomic of shaken-baby syndrome, with defects resulting from the transmission of force through the attachments among the lens, vitreous, body, and retina of the eye, causing traction on the retina which splits

reverse parentage testing: DNA analysis to determine the identity of an unknown decedent by means of known samples from living next-of-kin

revolver: a handgun characterized by a revolving cylinder containing numerous firing chambers

rifle: a firearm with a rifled barrel that is fired from the shoulder

rifling: a series of lands and grooves inside the barrel of a gun, designed to impart spin to the bullet as it passes through the barrel

rigor mortis: the part of the triad of early postmortem changes in which chemical changes in the muscle fibers cause progressive hardening of the muscles. Lactic acid and other by-products of tissue metabolism initiate the process. The protoplasm gels as acid accumulates in the muscle, making the muscles rigid.

scale: an object introduced within the frame of a photograph against which viewers can judge the size and dimensions of the evidence in the shot

scanning electron microscopy (SEM): medical imaging technique in which a beam of electrons scans an object and produces an enlarged image on a cathode-ray tube

scat: animal feces

scent detection dogs: civilian or police canines that have been specially trained to detect and alert to the presence of human remains; synonymous with cadaver dogs

search strategy: the plan that is developed and executed to locate evidence or remains

second-degree murder: an act of homicide that was not premeditated

secular trends: patterns of change associated with time

sequelae: morbid conditions following a given condition or event

serial murder: the killing by a single individual of a minimum of three people at different times and locations

sharp force trauma: an injury or wound inflicted by a sharp object, includes chopping, stabbing, slashing, and incising wounds

shored exit: atypical gunshot wound in which the exit wound is abraded, mimicking an entrance wound; occurs when pressure on the skin from clothing or contact with solid objects causes the bullet to abrade the skin on exit

short-range TSD estimates: cover the period from death to 72 hours postmortem

short tandem repeat (STR): a form of DNA typing used in identification

shotgun: a firearm intended to be fired from the shoulder like a rifle but with a smooth bore designed to fire multiple pellets

shot shells: ammunition for shotguns; a plastic body with a crimped end and a metal head containing a primer

shrunken heads: nonmedicolegal remains, normally mummified skulls with retained scalps, often fabricated from nonhuman materials such as monkey or rodent skulls

simple fracture: fracture mainfesting no communication with the outside environment, also called closed fracture

sinciput: the top of the forehead at the midline; synonymous with vertex

slow loading: transfer of force that results in simple or linear fractures after passing through the elastic phase of Young's modulus of elasticity

smokeless powder: a type of firearm propellant involving a highly flammable solid formed through a chemical reaction

solid-phase, double-antibody radioimmunoassay: means of identifying an unknown antigen by allowing a protein isolated from bone to be bound in solution to polystyrene microtiter plates and then washed with a soy protein solution to prevent further protein binding. Antisera raised in animal species are introduced, and the antibodies bind to the antigens. The species of the unknown antigen can be determined through examining the binding of specific antigens to specific antisera.

source–image distance (SID): in radiography, the distance between the x-ray beam source and the film

source–object distance (SOD): in radiography, the distance between the x-ray beam source and the patient or specimen

souvenir skulls: nonmedicolegal remains, typically historic or archaeological skulls purchased oversees by tourists and imported illegally into the country

spall: the displaced fragments of bone resulting from a penetrating projectile; produces the beveling seen in gunshot wound entrance defects

stab wounds: sharp force traumas characterized by thrust or impaling force

standard deviation: a measure of variability; it is the square root of the variance

standard error (SE): typically references the standard error of the mean; it is a measure of the variation in means when a population is sampled repeatedly

statutes: acts established by the legislature, constituting the laws of a state that prohibit, declare, or command something

stenosis: the narrowing of a blood vessel

sterile soil: the undisturbed ground representing the lower limits of a grave as originally formed

stippling: the embedding of gunpowder grains into the skin

stipulation: a voluntary agreement between opposing parties regarding a relevant point or piece of evidence

striae: in sharp force trauma wounds, grooves or defects evident in the surface of the kerf

subjective: belonging only to reality as perceived, existing solely in the mind

subpoena: an order to produce documents, evidence, or testimony for the court's review

suicide: the intentional killing of oneself

superseding authority: hierarchy of judicial and law enforcement agencies that determines how jurisdiction is established and transferred, such that federal agencies supersede state agencies, which in turn supersede local authorities

supine: an individual lying on the back, face up

supporting evidence: testimony or evidence presented at trial (e.g., reports, résumés, autopsy photographs, maps, diagrams) that was not collected at the scene

syndrome: a concurrence of symptoms or a collection of signs associated with a morbid process

tampering: unauthorized or undocumented handling of evidence

taphonomy: the natural laws of burial, describing how an organism moves from the biosphere to the lithosphere

tattooing: the embedding of gunpowder grains into the skin; synonymous with stippling

tentative identification: the identification of the deceased based on circumstances or associated materials; synonymous with confirmatory or directed identification

tenure: the permanent appointment of a teacher or professor

terminal ballistics: the study of the travel of a projectile through a target

Terminologia anatomica (TA): adopted in 1998, TA is a system of anatomic nomenclature consisting of 1,700 terms describing anatomical form or function

terrorism: a war crime involving violence with a political or social purpose, designed to intimidate or incite fear. Terrorism requires three parties: the terrorist, the victim, and a third party to be intimidated by what happened to the victim.

testimony: evidence given by a witness, under oath, or affirmation, rather than evidence derived from another source such as documents

tetranucleotide: a compound of four nucleotides

thermal scanning: converts infrared radiation emitted from heat sources into electrical impulses that can be visualized on a monitor

third-degree murder: statutorily defined murder resulting from an act that does not constitute murder under common law; considered less heinous than first or second degree.

throttling: strangulation by hand (manual) or ligature

time-since-death (TSD): the length of time between the death of an individual and the recovery of the body; synonymous with elapsed time and postmortem interval (PMI)

tool marks: discrete patterns or impressions resulting from contact between a tool or weapon and an impression medium such as bone

toxicological deaths: deaths resulting from an overabundance of an exogenous agent

trajectory: the path of a bullet in flight

transfer evidence: physical evidence removed from one object and deposited on another; the basis of Locard's principle

transferred intent: the intent of the original act is deferred to the committed act

transparency: describing processes that are visible or accessible as a result of guidelines requiring public agencies to disclose information

treachery: a war crime involving an act of betrayal that causes harm, such as the killing of surrendered combatants

triad of early postmortem change: rigor mortis, livor mortis, and algor mortis (body temperature decrease)

triage: the medical screening of individuals to determine need, priority, and sequence of treatment

trophy skulls: nonmedicolegal remains, typically brought to the United States by returning soldiers as battle souvenirs

type I/alpha error: synonymous with "false negative"; this type of error rejects the null hypothesis when it is true

type II/beta error: synonymous with "false positive"; this type of error accepts the null hypothesis when it is false

typical autopsy: an external examination of the body, followed by radiographs and removal of the thoracic, abdominal, and cranial contents for examination

typicality probabilities: a statistical statement of the likelihood that an unknown belongs to one or none of a set of reference populations

ultrasound: a medical imaging technique using sound waves to produce images; synonymous with diagnostic sonography

undetermined: manner-of-death classification when the manner is equivocal or elusive

uniformitarianism: theory stating that the present is the key to the past; processes and phenomena observed today are believed to be identical to those seen in the past

vacated: describing a ruling or court order that has been canceled or rescinded

variable number of tandem repeats (VNTR): the core repeat unit, referred to as a minisatellite, ranging from 10 to 100 base pairs in length, used in DNA testing

variance: a measure of variability; it is the mean of the squares of all the deviation scores in a distribution

velocity: the rate of travel of electromagnetic radiation waves

vena: Terminologia anatomica meaning vein

vertex: anatomical landmark defined as the midline of the top of the forehead; synonymous with sinciput

Virchow technique: autopsy method involving the removal of individual organs

virtual autopsy: a noninvasive form of autopsy that uses advanced imaging techniques such as multislice computed tomography and magnetic resonance imaging to visualize internal structures, pathological conditions, and trauma

visual recognition: most common yet least reliable form of personal identification, involving pattern recognition based on memory or antemortem photographs

volition: the act of making a choice, an act of free will with an understanding or acceptance of the consequences of that act

voluntary manslaughter: an act of murder reduced to manslaughter because of extenuating circumstances or diminished capacity

wad: a layer of cotton or fiber separating the gunpowder from the projectiles in a shot shell

war crimes: violations of the laws of war or international humanitarian law that incur individual criminal responsibility

warrant: a court order authorizing law enforcement to act

wavelength: a characteristic of electromagnetic radiation, specifically, the distance between the peaks or valleys of waves

windowing: the manipulation of computed tomography images to reproduce the images in different planes

wipe: minute fragments of the projectile deposited in tissue as it passes through the body, visible grossly or on radiograph

work product: any tangible material or its intangible equivalent that is prepared by or for a lawyer or act of litigation

wound: a defect or lesion caused by a weapon; according to the legal definition, the defect must penetrate the skin

wound track: the path of a projectile through tissue

yaw: deviation of the long axis of a bullet from its line of flight

Young's modulus of elasticity: the stiffness of a material during the elastic portion of its stress–strain curve; the slope of the curve is the modulus of the material's elasticity

References and Related Bibliographic Items

Aaseng, N. 1995. *The O.J. Simpson Trial: What It Shows Us About Our Legal System.* New York: Walker.

Adams, B.J. 2003a. The diversity of adult dental patterns in the United States and the implications for personal identification. *Journal of Forensic Sciences* 48(3):497–503.

Adams, B.J. 2003b. Establishing personal identification based on specific patterns of missing, filled and unrestored teeth. *Journal of Forensic Sciences* 43(3):487–496.

Adams, B.J. and J.E. Byrd. 2006. Resolution of small-scale commingling: a case report from the Vietnam War. *Forensic Science International* 156(1):63–69.

Adams, B.J. and L.W. Konigsberg. 2005. Quantification of commingled human skeletal remains: determining the most likely number of individuals (MLNI). *Proceedings of the Annual Meeting of the American Academy of Forensic Sciences* 11:309–310.

Adams, B.J. and R.C. Maves. 2002. Radiographic identification using the clavicle of an individual missing from the Vietnam conflict. *Journal of Forensic Sciences* 47(2):369–373.

Adams, R.F. 1960. *A Fitting Death for Billy the Kid.* Norman: University of Oklahoma Press.

Adams, V.I. 2002. Medicolegal autopsies and autopsy toxicology. In *Handbook of Autopsy Practice*, 3rd edition. J. Ludwig, ed. Pp. 7–20. Totowa, NJ: Humana Press.

Adams, V.I. and J. Ludwig. 2002. Autopsy law. In *Handbook of Autopsy Practice*, 3rd edition. J. Ludwig, ed. Pp. 159–166. Totowa, NJ: Humana Press.

Aghayev, E., M.J. Thali, M. Sonnenschein, J. Hurlimann, C. Jackowski, T. Kilchoer and R. Dirnhofer. 2005. Fatal steamer accident: blunt force injuries and drowning in post-mortem MSCT and MRI. *Forensic Science International* 152(1):65–71.

Aguilar, J.C. 1983. Shored gunshot wound of exit. A phenomenon with identity crisis. *American Journal of Forensic Medicine and Pathology* 4(3):199–204.

Aiello, L.C. and B.A. Wood. 1994. Cranial variable as predictors of hominine body mass. *American Journal of Physical Anthropology* 95:409–426.

Airy, H. 1993. *Whatever Happened to Billy the Kid?* Santa Fe, NM: Sunshine Press.

Albanese, J. 2003. A metric method for sex determination using the hipbone and the femur. *Journal of Forensic Sciences* 48:1–11.

Albanese, J. 2006. The Grant Human Skeletal Collection and other contributions of J. C. B. Grant to anthropology. Electronic document, www.uwindsor.ca/users/a/albanese/Main.nsf/inToc/50D0257526098F2B85257020006C4FEA, accessed July 17, 2006.

Albanese, J. and S.R. Saunders. 2006. Is it possible to escape racial typology in forensic identification? In *Forensic Anthropology and Medicine: Complementary Sciences from Recovery to Cause of Death*. A. Schmitt, E. Cunha, J. Pinheiro, eds. Pp. 281–316. Totowa NJ: Humana Press.

Alonso, A., P. Martin, C. Albarran, P. Garcia, O. Garcia, L.F. de Simon, J. Garcia-Hirschfeld, M. Sancho, C. de La Rua and J. Fernandez-Piqueras. 2004. Real-time PCR designs to estimate nuclear and mitochondrial DNA copy number in forensic and ancient DNA studies. *Forensic Science International* 139(23):141–149.

Alonso, A., P. Martin, C. Albarran, P. Garcia, L. Fernandez de Simon, M. Jesus Iturralde, A. Fernandez-Rodriguez, I. Atienza, J. Capilla, J. Garcia-Hirschfeld, P. Martinez, G. Vallego, O. Garcia, E. Garcia, P. Real, D. Alvarez, A. Leon and M. Sancho. 2005. Challenges of DNA profiling in mass disaster investigations. *Croatian Medical Journal* 46(4):540–548.

Alter, R.E. 1975. *The Trail of Billy the Kid*. New York: Belmont Tower Books.

American Bar Association. 1990. *Guidelines for the Issuance of Search Warrants*. Chicago: ABA Section of Criminal Justice.

Anaya, A.P. and J.H. Earle. 1991. *I Buried Billy*. College Station, TX: Creative Publishing.

Andahl, R.O. 1978. The examination of saw marks. *Journal of the Forensic Science Society* 18(1–2):31–46.

Andelinovic, S., D. Sutlovic, I. Erceg Ivkosic, V. Skaro, A. Ivkosic, F. Paic, B. Rezic, M. Definis-Gojanovic and D. Primorac. 2005. Twelve year experience in identification of skeletal remains from mass graves. *Croatian Medical Journal* 46(4):530–539.

Andersen Torpet, L. 2005. DVI System International: software assisting in the Thai tsunami victim identification process. *Journal of Forensic Odontostomatology* 23(1):19–25.

Anderson, G.S. 1997. The use of insects to determine time of decapitation: a case-study from British Columbia. *Journal of Forensic Sciences* 42(5):947–950.

Anderson, G.S. 2004. Determining time of death using blow fly eggs in the early postmortem interval. *International Journal of Legal Medicine* 118(4):240–241.

Anderson, G.S. and S.L. Van Laerhoven. 1996. Initial studies on insect succession on carrion in southwestern British Columbia. *Journal of Forensic Sciences* 41(4): 617–625.

Andrews, L.B., N. Buenger, J. Bridge, L. Rosenow, D. Stoney, R.E. Gaensslen, T. Karamanski, R. Lewis, J. Paradis, A. Inlander and D. Gonen. 2004. Constructing ethical guidelines for biohistory. *Science* 304:215–216.

Angyal, M. and K. Derczy. 1998. Personal identification on the basis of antemortem and postmortem radiographs. *Journal of Forensic Sciences* 43(5):1089–1093.

Anslinger, K., G. Weichold, W. Keil, B. Bayer, and W. Eisenmenger. 2001. Identification of the skeletal remains of Martin Bormann by mtDNA analysis. *International Journal of Legal Medicine* 114(3):194–196.

Anton, S.C. 1988. Bony criteria for differentiation of metastatic carcinoma, multiple myeloma, major infectious diseases and hyperparathyroidism: a case study approach. In *Human Skeletal Biology: Contributions to the Understanding of California's Prehistoric Populations*. G.D. Richards, ed. Pp. 43–68. Archives of California Prehistory no. 24. Salinas, CA: Coyote Press.

Archer, M.S., R.B. Bassed, C.A. Briggs and M.J. Lynch. 2005. Social isolation and delayed discovery of bodies in houses: the value of forensic pathology, anthropology, odontology and entomology in the medico-legal investigation. *Forensic Science International* 151(2–3):259–265.

Argentine Forensic Anthropology Team. 2003. *Annual Report.* New York: Argentine Forensic Anthropology Team.

Argentine Forensic Anthropology Team. 2005. *Annual Report.* New York: Argentine Forensic Anthropology Team.

Arnaldos, M.I., M.D. Garcia, E. Romera, J.J. Presa and A. Luna. 2005. Estimation of postmortem interval in real cases based on experimentally obtained entomological evidence. *Forensic Science International* 149(1):57–65.

Ashizawa, K., T. Asami, M. Anzo, et al. 1996. Standard RUS skeletal maturation of Tokyo children. *Annals of Human Biology* 23(6):457–469.

Aturaliya, S. and A. Lukasewycz. 1999. Experimental forensic and bioanthropological aspects of soft tissue taphonomy: 1. Factors influencing postmortem tissue desiccation rate. *Journal of Forensic Sciences* 44(5):893–896.

Aufderheide, A.C. and C. Rodriguez-Martin. 1998. *The Cambridge Encyclopedia of Human Paleopathology.* Cambridge: Cambridge University Press.

Aulsebrook, W.A., M.Y. Iscan, J.H. Slabbert and P. Becker. 1995. Superimposition and reconstruction in forensic facial identification: a survey. *Forensic Science International* 75(2–3):101–120.

Austin, D. 1999. Video superimposition at the CA Pound Laboratory 1987 to 1992. *Journal of Forensic Sciences* 44(4): 695–699.

Auxier, J.A. and H.M. Prichard. 2001. The role of the expert witness: an update. *Health Physics* 81(3): 269–271.

Azizi, L., T.M. Garrick and C.G. Harper. 2006. Brain donation for research: strong support in Australia. *Journal of Clinical Neuroscience* 13(4):449–452.

Babapulle, C.J. and N.P. Jayasundera. 1993. Cellular changes and time since death. *Medicine, Science, and the Law* 33(3):213–222.

Baccino, E., D.H. Ubelaker, L.A C. Hayek, and A. Zerilli. 1999. Evaluation of seven methods of estimating age at death from mature human skeletal remains. *Journal of Forensic Sciences* 44: 931–936.

Bada, J.L. 1987. Paleoanthropological applications of amino acid racemization dating of fossil bones and teeth. *Anthropologischer Anzeiger* 45(1): 1–8.

Baik, S.O., J.M. Uku and M. Sikirica. 1991. A case of external beveling with an entrance gunshot wound to the skull made by a small caliber rifle bullet. *American Journal of Forensic Medicine and Pathology* 12(4):334–336.

Baker, P.T. and R.W. Newman. 1957. The use of bone weights for human identification. *American Journal of Physical Anthropology* 15:601–618.

Ballow, W. 1998. *Billy the Kid: A Graphic History.* Fort Worth, TX: Owlhoot Trail Publishing.

Banasr, A., G.L. de la Grandmaison and M. Durigon. 2003. Frequency of bone/cartilage lesions in stab and incised wounds fatalities. *Forensic Science International* 131(2–3):131–133.

Bang, G. 1989. Age changes in teeth: developmental and regressive. In *Age Markers in the Human Skeleton.* M.Y. Iscan, ed. Pp. 211–235. Springfield, IL: Charles C. Thomas.

Baraybar, J.P. 2004. When DNA is not available, can we still identify people? Recommendations for best practice. *Proceedings of the Annual Meeting of the American Academy of Forensic Sciences* 10:274–275.

Barker, A. 1993. *The Kid with Fast Hands: A Carefully Researched History of Billy the Kid with Fictional Dialogue and Incidental Action.* Pine Grove, CA: A. Barker.

Bartelink, E.J., J.M. Wiersema and R.S. Demaree. 2001. Quantitative analysis of sharp-force trauma: an application of scanning electron microscopy in forensic anthropology. *Journal of Forensic Sciences* 46(6):1288–1293.

Bass, W.M. 1968. Obituary of Charles Earnest Snow, 1910–1967. *American Journal of Physical Anthropology* 28:369–372.

Bass, W.M. 1969. Recent developments in the identification of human skeletal material. *American Journal of Physical Anthropology* 30:459–461.

Bass, W.M. 1979. Developments in the identification of human skeletal material. *American Journal of Physical Anthropology* 51:555–562.

Bass, W.M. 1983. The occurrence of Japanese trophy skulls in the United States. *Journal of Forensic Sciences* 28(3): 800–803.

Bass, W.M. and W.H. Birkby. 1978. Exhumation: the method could make the difference. *FBI Law Enforcement Bulletin* 47:6–11.

Bass, W.M. and J. Jefferson. 2003. *Death's Acre*. New York: Berkley Publishing Group.

Bassed, R. 2003. Identification of severely incinerated human remains: the need for a cooperative approach between forensic specialities. A case report. *Medicine, Science, and the Law* 43(4):356–361.

Bateman, C. 2001. Ongoing human rights abuses. *South African Medical Journal* 91(8):624–625.

Bell, B.B. 1996. *The Illustrated Life and Times of Billy the Kid*, 2nd edition. Phoenix, AZ: Boze Publications.

Bendheim, O.L. 1982. The psychiatric autopsy. *American Journal of Psychiatry* 139(10):1379.

Benecke, M. and R. Lessig. 2001. Child neglect and forensic entomology. *Forensic Science International* 120(1–2):155–159.

Benecke, M., E. Josephie and R. Zweihoff. 2004. Neglect of the elderly: forensic entomology cases and considerations. *Forensic Science International* 146(suppl): S195–S199.

Benedix, D.C. and W.R. Belcher. 2006. Research trends during the history of the Physical Anthropology section at the AAFS Annual Meetings. *Proceedings of the 58th Annual Meeting of the American Academy of Forensic Sciences* 12:287.

Bennett, K.A. 1993. Victim selection in the Jeffrey Dahmer slayings: an example of repetition in the paraphilias? *Journal of Forensic Sciences* 38:1227–1232.

Bernstein, M.L. 1999. Radiologic applications in forensic dentistry. In *Forensic Radiology*. B.G. Brogdon, ed. Pp. 97–140. Boca Raton, FL: CRC Press.

Berryman, H.E. and W.M. Gunther. 2000. Keyhole defect production in tubular bone. *Journal of Forensic Sciences* 45(2):483–487.

Berryman, H.E., W.M. Bass, S.A. Symes and O.C. Smith. 1997. Recognition of cemetery remains in the forensic setting. In *Forensic Taphonomy: The Postmortem Fate of Human Remains*. W.D. Haglund, M.H. Sorg, eds. Pp. 165–170. Boca Raton FL: CRC Press.

Bertin, J.E. and M.S. Henifin. 1994. Science, law, and the search for truth in the courtroom: lessons from *Daubert v. Merrell Dow*. *Journal of Laws, Medicine and Ethics* 22(1):6–20.

Betz, P. and W. Eisenmenger. 1995. Unusual suicides with electric saws. *Forensic Science International* 75:173–179.

Beunen, G., J. Lefevre, M. Ostyn, R. Renson, J. Simons and D. Van Gerven. 1990. Skeletal maturity in Belgian youths assessed by the Tanner-Whitehouse method (TW2). *Annals of Human Biology* 17(5):355–376.

Bevan, B.W. 1991. The search for graves. *Geophysics* 56(9):1310–1319.

Biasotti, A.A. 1964. The principles of evidence evaluation as applied to firearms and tool mark identification. *Journal of Forensic Sciences* 9(4):428–433.

Bidmos, M.A. and S.A. Asala. 2004. Sexual dimorphism of the calcaneus of South African blacks. *Journal of Forensic Sciences* 49(3):446–450.

Bidmos, M.A. and M.R. Dayal. 2004. Further evidence to show population specificity of discriminant function equations for sex determination using the talus of South African blacks. *Journal of Forensic Sciences* 49(6):1165–1170.

Bilge, Y., P.S. Kedici, Y.D. Alakoc, K.U. Ulkuer and Y.Y. Ilkyaz. 2003. The identification of a dismembered human body: a multidisciplinary approach. *Forensic Science International* 137(2–3):141–146.

Bini, C., S. Ceccard, D. Luiselli, G. Ferri, S. Pelotti, C. Colalongo, M. Falconi and G. Pappalardo. 2003. Different informativeness of the three hypervariable mitochondrial DNA regions in the population of Bologna (Italy). *Forensic Science International* 135(1):48–52.

Biwasaka, H., K. Saigusa and Y. Aoki. 2005. The applicability of holography in forensic identification: a fusion of the traditional optical technique and digital technique. *Journal of Forensic Sciences* 50(2):393–399.

Black, S.M. 2003a. Forensic anthropology—regulation in the United Kingdom. *Science & Justice* 43(4):187–192.

Black, S.M. 2003b. Supply and demand: the shifting expectations of forensic anthropology in the United Kingdom. *Science & Justice* 43(4):183–186.

Blackwell, R.J. and W.A. Crisci. 1975. Digital image processing technology and its application in forensic sciences. *Journal of Forensic Sciences* 20(2):288–304.

Blalock, H.M. 1972. *Social Statistics,* 2nd edition. London: McGraw-Hill Kogakusha Ltd.

Blazer, A.N., J.H. Blazer and P. Blazer. n.d. Papers of the Blazer family, 1857–1966 (bulk 1870–1955). Special Collections, University of Arizona Library.

Blitzer, H.L and J. Jacobia. 2002. *Forensic Digital Imaging and Photography.* San Diego: Academic Press.

Bloom, R.M. 2003. *Searches, Seizures, and Warrants: A Reference Guide to the United States Constitution.* Westport, CT: Praeger.

Bock, J.H. and D.O. Norris. 1997. Forensic botany: an under-utilized resource. *Journal of Forensic Sciences* 42(3):364–367.

Bocquet-Appel, J.J. and C. Masset. 1982. Farewell to paleodemography. *Journal of Human Evolution* 11:321–333.

Bodziak, W.J. 2000. *Footwear Impression Evidence: Detection, Recovery, and Examination,* 2nd edition. Boca Raton, FL: CRC Press.

Boemer, F., J.F. Vanbellinghen, V. Bours and R. Schoos 2006. Screening for sickle cell disease on dried blood: a new approach evaluated on 27,000 Belgian newborns. *Journal of Medical Screening* 13(3):132–136.

Bohan, T.L. and E.J. Heels. 1995. The case against *Daubert:* the new scientific evidence "standard" and the standards of several states. *Journal of Forensic Sciences* 40(6):1030–1044.

Bohnert, M., T. Rost, M. Faller-Marquardt, D. Ropohl and S. Pollak. 1997. Fractures of the base of the skull in charred bodies—post-mortem heat injuries or signs of mechanical traumatization? *Forensic Science International* 87(1):55–62.

Bohnert, M., U. Schmidt, M.G. Perdekamp and S. Pollak. 2002. Diagnosis of a captive-bolt injury in a skull extremely destroyed by fire. *Forensic Science International* 127(3):192–197.

Boldsen, J.L., G.R. Milner, L.W. Konigsberg and J.W. Wood. 2002. Transition analyses: a new method for estimating age from skeletons. In *Palaeodemography: I Age Distributions from Skeletal Samples,* R.D. Hoppa, J.W. Vaupel, eds. Pp. 73–106. Cambridge: Cambridge University Press.

Bolliger, S., M. Thali, C. Jackowski, E. Aghayev, R. Dirnhofer and M. Sonnenschein. 2005. Postmortem non-invasive virtual autopsy: death by hanging in a car. *Journal of Forensic Sciences* 50(2):455–460.

Bomberger, K.C., B.V. Kennedy, E.F. Huffine and T.D. Anderson. 2001. International Commission on Missing Persons—an international humanitarian mission resulting in a local, human identification effort. *Proceedings of the Annual Meeting of the American Academy of Forensic Sciences* 7:239–240.

Bonte, W. 1975. Tool marks in bone and cartilage. *Journal of Forensic Sciences* 20(2):315–325.

Bontrager, K.L. 2002. *Bontrager's Pocket Atlas—Handbook of Radiographic Positioning and Techniques*, 4th edition. Phoenix, AZ: Bontrager.

Boscolo, F.N., S.M. Almeida, F. Haiter Neto, A.E. Oliveira and F.M. Tuji. 2002. Fraudulent use of radiographic images. *Journal of Forensic Odontostomatology* 20(2):25–30.

Botega, N.J., K. Metze, E. Marques, A. Cruvinel, Z.V. Moraes, L. Augusto and L.A. Costa. 1997. Attitudes of medical students to necropsy. *Journal of Clinical Pathology* 50(1):64–66.

Bouhaidar, R. 2005. Forensic webwatch: forensic computing. *Journal of Clinical Forensic Medicine* 12(1):47–49.

Bourel, B., G. Tournel, V. Hedouin, M. Deveaux, M.L. Goff and D.Gosset. 2001. Morphine extraction in necrophagous insects remains for determining antemortem opiate intoxication. *Forensic Science International* 120(1–2):127–131.

Bouwman, A.S., E.R. Chilvers, K.A. Brown and T.A. Brown. 2006. Brief communication: identification of the authentic ancient DNA sequence in a human bone contaminated with modern DNA. *American Journal of Physical Anthropology* 131(3): 428–431.

Bowers, C.M. and R.J. Johansen. 2002. Digital imaging methods as an aid in dental identification of human remains. *Journal of Forensic Sciences* 47(2):354–359.

Bowling, A. 2002. *Research Methods in Health*, 2nd edition. Maidenhead (U.K.): Open University Press.

Boyd, R.M. 1979. Buried body cases. *FBI Law Enforcement Bulletin* 48:106.

Boylan, J. 1960. *The Old Lincoln County Courthouse, Lincoln, New Mexico*. Roswell, NM: Hall-Poorbaugh Press.

Boza-Arlotti, A., E.F. Huffine and R.J. Harrington. 2003. The influence of large-scale DNA testing on the traditional anthropological approach to human identification: the experience in Bosnia and Herzegovina. *Proceedings of the Annual Meeting of the American Academy of Forensic Sciences* 9:249.

Branch, L.L. and C.F. Rudulph. 1980. *Los Bilitos: The Story of Billy the Kid and His Gang*. New York: Carlton Press.

Brannon, R.B., W.M. Morlang and B.C. Smith. 2003. The Gander disaster: dental identification in a military tragedy. *Journal of Forensic Sciences* 48(6):1331–1335.

Brautbar, N. 1999. Science and the law: scientific evidence, causation, admissibility, reliability "Daubert" decision revisited. *Toxicology and Industrial Health* 15:532–551.

Breiham, C.W. and M. Ballert. 1970. *Billy the Kid; A Date with Destiny*. Seattle, WA: Hangman Press.

Breitmeier, D., U. Graefe-Kirci, K. Albrecht, M. Weber, H.D. Troger and W.J. Kleemann. 2004. Evaluation of the correlation between time corpses spent in in-ground graves and findings at exhumation. *Forensic Science International* 154(2–3):218–223.

Bremme, A.G.S. 2003. This grave speaks: forensic anthropology in Guatemala. *Proceedings of the Annual Meeting of the American Academy of Forensic Sciences* 9:277.

Brennan, T.A. and R. Kirschner.1992. Medical ethics and human rights violations: the Iraqi occupation of Kuwait and its aftermath. *Annals of Internal Medicine* 117(1):78–82.

Brenner, C.H. and B.S. Weir. 2003. Issues and strategies in the DNA identification of World Trade Center victims. *Theoretical Population Biology* 63(3):173–178.

Brickley, M.B. and R. Ferllini, eds. 2007a. *Forensic Anthropology: Case Studies from Europe.* Springfield, IL: Charles C. Thomas.

Brickley, M.B and R. Ferllini. 2007b. Forensic anthropology: developments in two continents. In *Forensic Anthropology: Case Studies from Europe.* M.B. Brickley, R. Ferllini, eds. Springfield, IL: Charles C. Thomas.

Bridges, P.S. 1989. Changes in activities with the shift to agriculture in the southeastern United States. *Current Anthropology* 30:385–394.

Brkic, H., D. Strinovic, M. Slaus, J. Skavic, D. Zecevic and M. Mileicevic. 1997. Dental identification of war victims from Petrinja in Croatia. *International Journal of Legal Medicine* 110(2):47–51.

Brkic, H., D. Strinovic, M. Kubat and V. Petrovecki. 2000. Odontological identification of human remains from mass graves in Croatia. *International Journal of Legal Medicine* 114(1–2):19–22.

Broeders, A.P. 2006. Of earprints, fingerprints, scent dogs, cot deaths and cognitive contamination—a brief look at the present state of play in the forensic arena. *Forensic Science International* 159(2–3):148–157.

Brogdon, B.G., ed. 1998a. *Forensic Radiology.* Boca Raton, FL: CRC Press.

Brogdon, B.G. 1998b. Radiological identification of individual remains. In *Forensic Radiology.* B.G. Brogdon, ed. Pp. 149–187. Boca Raton, FL: CRC Press.

Bromage, T.G. and A. Boyde. 1984. Microscopic criteria for the determination of directionality of cutmarks on bone. *American Journal of Physical Anthropology* 65(4):359–366.

Brooks, S.T. 1955. Skeletal age at death: the reliability of cranial and pubic age indicators. *American Journal of Physical Anthropology* 13:567–597.

Brooks, S.T. 1970. Obituary: Theodore Doney McCown. 1908–1969. A biographical sketch. *American Journal of Physical Anthropology* 32:165–168

Brooks, S.T. 1981. Teaching of forensic anthropology in the United States. *Journal of Forensic Sciences* 26(4):627–631.

Brooks, S.T. and R.H. Brooks.1984. Problems of burial exhumation, historical and forensic aspects. In *Human Identification. Case Studies in Forensic Anthropology.* T.A. Rathbun, J.E. Buikstra, eds. Pp. 64–86. Springfield, IL: Charles C. Thomas.

Brooks, S.T. and J.M. Suchey. 1990. Skeletal age determination based on the os pubis: a comparison of the Acsádi-Nemeskeri and Suchey-Brooks methods. *Human Evolution* 5:227–238.

Bruce, C.J. 2006. The role of expert evidence. Electronic document, www.transportationsafety.com, accessed May 5, 2006.

Buck, S.C. 2003. Searching for graves using geophysical technology: field tests with ground penetrating radar, magnetometry, and electrical resistivity. *Journal of Forensic Sciences* 48(1): 5–11.

Budimlija, Z.M., M.K. Prinz, A. Zelson-Mundorff, J. Wiersema, E. Bartelink, G. MacKinnon, B.L. Nazzaruolo, S.M. Estacio, M.J. Hennessey and R.C. Shaler.

2003. World Trade Center human identification project: experiences with individual body identification cases. *Croatian Medical Journal* 44(3):259–263.

Budowle, B., F.R. Bieber and A.J. Eisenberg. 2005. Forensic aspects of mass disasters: strategic considerations for DNA-based human identification. *Legal Medicine (Tokyo)* 7(4):230–243.

Buikstra, J.E. 2002. Forensic anthropology today. *General Anthropology Division Newsletter* 9(1):1–6. Washington, DC: American Anthropological Association.

Buikstra, J.E., J.L. King and K.C. Nystrom. 2003. Forensic anthropology and bioarchaeology in the *American Anthropologist:* rare but exquisite gems. *American Anthropologist* 105:38–52.

Buikstra, J.E. and M.K. Maples. 1999. The life and career of William R. Maples, Ph.D. *Journal of Forensic Sciences* 44(4):677–681.

Buikstra, J.E. and J.H. Mielke. 1985. Demography, diet, and health. In *The Analysis of Prehistoric Diets*. R.I. Gilbert, J.H. Mielke, eds. Pp. 359–422. Orland, FL: Academic Press.

Buikstra, J.E., and D.H. Ubelaker, eds. 1994. *Standards for Data Collection from Human Skeletal Remains*. Proceedings of a Seminar at the Field Museum of Natural History, Arkansas Archeological Survey Research Series no. 44. Fayetteville: Arkansas Archeological Survey.

Buikstra, J.E., G.R. Milner and J.L. Boldsen. 2005. Janaab' Pakal: La controversia de la edad cronológica re-visitada. In *Janaab' Pakal de Palenque. Vida y muerte de un gobernante maya*, V. Tiesler, A. Cucina, eds. Pp. 103–122. UNAM/UADY, México, D.F.

Bunce, C. 1997. Doctors involved in human rights' abuses in Kenya. *British Medical Journal* 314(7075):166.

Bunch, A.W. 2006. A preliminary investigation of decomposition in cold climate. *Proceedings of the American Academy of Forensic Sciences* 12:297–298.

Bunch, A.W. and C.G. Fielding. 2005. The use of World War II chest radiograph in identification of a missing-in-action U.S. marine. *Military Medicine* 170(3):239–242.

Burd, D.Q. and A.E. Gilmore. 1968. Individual and class characteristics of tools. *Journal of Forensic Sciences* 13(3):390–396.

Burd, D.Q. and R.S. Greene. 1957. Tool mark examination techniques. *Journal of Forensic Sciences* 2(3):297–310.

Burns, J.W. 1930. *Billy the Kid*. London: Geoffrey Bles.

Burns, W.N. 1926. *The Saga of Billy the Kid*. Garden City, NY: Doubleday, Page & Company.

Burrows, A.M., V.P. Zanella and T.M. Brown. 2003. Testing the validity of metacarpal use in sex assessment of human skeletal remains. *Journal of Forensic Sciences* 48:1–4.

Bushong, S.C. 1988. *Radiologic Science for Technologists*, 4th edition. St. Louis: Mosby.

Bushong, S.C. 2000. *Computed Tomography*. Boston: McGraw-Hill.

Butler, J.M. 2005. *Forensic DNA Typing: Biology, Technology and Genetics of STR Markers*, 2nd ed. Amsterdam: Elsevier Academic Press.

Buxton, R.B. 2002. *Introduction to Functional Magnetic Resonance Imaging: Principles and Techniques*. New York: Cambridge University Press.

Byrd, J.E. and B.J. Adams. 2003. Osteometric sorting of commingled human remains. *Journal of Forensic Sciences* 48(4):717–724.

Byrd, J.E., B.J. Adams, L.M. Leppo and R.J. Harrington. 2003. Resolution of large-scale commingling issues: lessons from CILHI and ICMP. *Proceedings from the Annual Meeting of the American Academy of Forensic Sciences* 9:248.

Byrd, J.H. and J.L. Castner. 2001. *Forensic Entomology: The Utility of Arthropods in Legal Investigations*. Boca Raton, FL: CRC Press.

Cadenas, A.M, M. Regueiro, T. Gayden, N. Singh, L.A. Zhivotovsky, P.A. Underhill and R.J. Herrera. 2007. Male amelogenin dropouts: phylogenetic context, origins and implications. *Forensic Science International* 166(2–3):155–163.

Callaway, W.J. 2002. *Mosby's Comprehensive Review of Radiography*, 3rd edition. St. Louis: Mosby-Year Book Publishing.

Cameriere, R., L. Ferrante, D. Mirtella, F.U. Rollo and M. Cingolani. 2005. Frontal sinuses for identification: quality of classifications, possible error and potential corrections. *Journal of Forensic Sciences* 50(4):770–773.

Campbell, C.S. 1998. Religion and the body in medical research. *Kennedy Institute for Ethics Journal* 8(3):275–305.

Campman, S.C., F.A. Springer and D.M. Henrikson. 2000. The chain saw: an uncommon means of committing suicide. *Journal of Forensic Sciences* 45(2):471–473.

Campobasso, C.P., J.G. Linville, J.D. Wells and F. Introna. 2005. Forensic genetic analysis of insect gut contents. *American Journal of Forensic Medicine and Pathology* 26(2):161–165.

Camps, F.E. 1959. Establishment of the time of death—a critical assessment. *Journal of Forensic Sciences* 4(1):73–82.

Canadian Psychiatric Association. 1986. *Guidelines on Ethics in Courtroom Testimony* (1978–5s). Electronic document, www.cpa-apcorg/publications/position_papers/ethics.asp, accessed June 17, 2006.

Cao, F., H.K. Huang, E. Pietka and V. Gilsanz. 1999. Digital hand atlas and Web-based bone age assessment: system design and implementation. *Computerized Medical Imaging and Graphics* 24: 297–307.

Capasso, L., K.A.R. Kennedy and C.A. Wilczak. 1999. *Atlas of Occupational Markers on Human Remains*. Teramo, Italy: Edigrafital S. p. A.

Caplan, R.M. 1990. How fingerprints came into use for personal identification. *Journal of the American Academy of Dermatology* 23(1):109–114.

Carlton, R.R. and A.M. Adler. 1992. *Principles of Radiographic Imaging*. Albany, NY: Delmar Publishers.

Carson, E.A., V.H. Stefan and J.F. Powell. 2000. Skeletal manifestations of bear scavenging. *Journal of Forensic Sciences* 45(3):515–526.

Castellano, M.A., E.C. Vaillanueva and R. von Frenckel. 1984. Estimating the date of bone remains: a multivariate study. *Journal of Forensic Sciences* 29:527–534.

Catts, E.P. and M.L. Goff. 1992. Forensic entomology in criminal investigations. *Annual Review of Entomology* 37:253–272.

Chaillet, N., M. Nyström and A. Demirjian. 2005. Comparison of dental maturity in children of different ethnic origins: international maturity curves for clinicians. *Journal of Forensic Sciences* 50:1164–1172.

Chamberlain, K. 1997. *Billy the Kid and the Lincoln County War: A Bibliography*. Albuquerque: Center for the American West, University of New Mexico.

Chan, A.H.W., C.M Crowder and T.L. Rogers. 2007. Variation in cortical bone histology within the human femur and its impact. *American Journal of Physical Anthropology* 132(1): 80–85.

Chandra Sekharan, P. 1985. Identification of skull from its suture pattern. *Forensic Science International* 27(3):205–214.

Chang, Y.M., L.A. Burgoyne and K. Both. 2003. Higher failure of amelogenin sex test in an Indian population group. *Journal of Forensic Sciences* 48:1–5.

Charles, D.K., K. Condon, J.M. Cheverud and J.E. Buikstra. 1989. Estimating age at death from growth layer groups in cementum. In *Age Markers in the Human Skeleton.* M. Y. Iscan, ed. Pp. 227–301. Springfield, IL: Charles C. Thomas.

Christensen, A.M. 2002. Experiments in the combustibility of the human body. *Journal of Forensic Sciences* 47(3):466–470.

Christensen, A.M. 2004a. The impact of Daubert: implications for testimony and research in forensic anthropology (and the use of frontal sinuses in personal identification). *Journal of Forensic Sciences* 49(3):427–430.

Christensen, A.M. 2004b. The influence of behavior on freefall injury patterns: possible implications for forensic anthropological investigations. *Journal of Forensic Sciences* 49(1):5–10.

Christensen, A.M. 2005. Testing the reliability of frontal sinuses in positive identification. *Journal of Forensic Science,* 50(1):18–22.

Cingolani, M., A. Osculati, A. Tombolini, A. Tagliabracci, G. Ghimenton and S.D. Ferrara. 1994. Morphology of sweat glands in determining time of death. *International Journal of Legal Medicine* 107(3):132–140.

Clark, D.H, ed. 1992. *Practical Forensic Odontology.* Boston: Oxford University Press.

Clark, M.A. and W. Micik. 1984. Confusing wounds of entrance and exit with an unusual weapon. *American Journal of Forensic Medicine and Pathology* 5(1):75–78.

Clark, S.P., B. Delahunt, K.J. Thomson and T.L. Fernando. 1989. Suicide by band saw. *American Journal of Forensic Medicine and Pathology* 10:332–334.

Clayton, T.M, J.P. Whitaker, R. Sparkes and P. Gill. 1998. Analysis and interpretation of mixed forensic stains using DNA STR profiling. *Forensic Science International* 91(1):55–70.

Cohen, J. 1960. A coefficient of agreement for nominal scales. *Educational and Psychological Measurement* 20(1):37–46.

Coleman, W.H. 1969. Sex differences in the growth of the human bony pelvis. *American Journal of Physical Anthropology* 31:25–152.

Cologlu, A.S., M.Y. Iscan, M.F. Yavuz and H. Sari. 1998. Sex determination from the ribs of contemporary Turks. *Journal of Forensic Sciences* 43:273–276.

Conaty, G.T. and R.R. Janes. 1997. Issues of repatriation: a Canadian view. *European Review of Native American Studies* 11(2):31–37.

Coppock, C.A. 2001. *Contrast: An Investigator's Basic Reference Guide to Fingerprint Identification Concepts.* Springfield IL: Charles C. Thomas.

Cordner, S. and R. Coupland. 2003. Missing people and mass graves in Iraq. *Lancet* 362(9392):1325–1326.

Coupland, R. and S. Cordner. 2003. People missing as a result of armed conflict. *British Medical Journal* 326 (7396):943–944.

Cox, M. 2000a. Aging adults from the skeleton. In *Human Osteology: In Archaeology and Forensic Science* M. Cox, S. Mays, eds. Pp. 62–81. London: Greenwich Medical Media.

Cox, M. 2000b. Assessment of parturition. In *Human Osteology: In Archaeology and Forensic Science* M. Cox, S. Mays, eds. Pp. 131–142. London: Greenwich Medical Media.

Cox, M. and S. Mays. 2000. *Human Osteology in Archaeology and Forensic Science.* London :Greenwich Medical Media.

Cox, M. and A. Scott. 1992. Evaluation of the obstetric significance of some pelvic characters in an 18th century British sample of known parity status. *American Journal of Physical Anthropology* 89:431–440.

Cox, R.J., S.L. Mitchell and E.O. Espinoza. 1994. CompuTOD, a computer program to estimate time of death of deer. *Journal of Forensic Sciences* 39(5):1287–1299.

Crowder, C. 2005. *Evaluating the Use of Quantitative Bone Histology to Estimate Adult Age at Death*. PhD Dissertation, University of Toronto.

Crowder, C. and D. Austin. 2005. Age ranges of epiphyseal fusion in the distal tibia and fibula of contemporary males and females. *Journal of Forensic Sciences* 50:1001–1007.

Cunha, E. and C. Cattaneo. 2006. Forensic anthropology and forensic pathology: the state of the art. In *Forensic Anthropology and Medicine: Complementary Sciences from Recovery to Cause of Death*. A. Schmitt, E. Cunha, J. Pinheiro, eds. Totowa, NJ: Humana Press.

Cunha, E., M.C. Mendonca and D.N. Vieira. 2005. Exhumation and identification of a particular individual in a mass grave. *Proceedings of the Annual Meeting of the American Academy of Forensic Sciences* 11:314.

Curran, J.M., C.M. Triggs, J. Buckleton and B.S. Weir. 1999. Interpreting DNA mixtures in structured populations. *Journal of Forensic Sciences* 44(5):987–995.

Curran, W.J. 1986. The forensic investigation of the death of Josef Mengele. *New England Journal of Medicine* 315(17):1071–1073.

Cwik, C.H. 1999. Guarding the gate: expert evidence admissibility. *Litigation* 25(4):6–12.

Dalstra, M., P.M. Cattaneo, and F. Beckmann. 2006. Synchrotron radiation-based microtomography of alveolar support tissues. *Orthodontic and Craniofacial Research* 9(4):199–205.

Damann, F.E., M.D. Leney and S.M. Edson. 2005. Separating commingled remains using ancient DNA analysis. *Proceedings of the Annual Meeting of the American Academy of Forensic Sciences* 11:314–315.

Davenport, G.C., J.W. Lindemann, T.I. Griffen and J.E. Borowski. 1988. Geotechnical applications: 3 Crime scene investigation techniques. *Geophysics: The Leading Edge of Exploration* 7(8):64–66.

Davis, G.J. and B.R. Peterson. 1996. Dilemmas and solutions for the pathologist and clinician encountering religious views of the autopsy. *Southern Medical Journal* 89(11):1041–1045.

Davis, J.R. 1984. *The Warrant Requirement in Crime Scene Searches*. Washington DC: FBI, U.S. Department of Justice Publication 717-C-5.

Dayal, M.R. and M.A. Bidmos. 2005. Discriminating sex in South African blacks using patella dimensions. *Journal of Forensic Sciences* 50:1–4.

Dean, D.E., N.E. Tatarek, J. Rich, B.G. Brogdon and R.H. Powers. 2005. Human identification from the ankle with pre- and postsurgical radiographs. *Journal of Clinical Forensic Medicine* 12(1):5–9.

Debenham, P.G. 1994. Genetics leaves no bones unturned. *Nature Genetics* 6:113–114.

De Greef, S. and G. Willems. 2005. Three-dimensional cranio-facial reconstruction in forensic identification: latest progress and new tendencies in the 21st century. *Journal of Forensic Sciences* 50(1):12–17.

de Gruchy, S. and T.L. Rogers. 2002. Identifying chop marks on cremated bone: a preliminary study. *Journal of Forensic Sciences* 47(5):933–936.

de la Grandmaison, G.L. and M. Durigon. 2001. Do medico-legal truths have more power than war lies? About the conflicts in the former Yugoslavia and in Kosovo. *Medicine, Science and the Law* 41(4):301–304.

De Letter, E.A. and M.H.A. Piette. 2001. An unusual case of suicide by means of a pneumatic hammer. *Journal of Forensic Sciences* 46(4):962–965.

Demirjian, A. 1978. Dentition. In *Human Growth,* Vol 2. F. Falkner, J.M. Tanner, eds. Pp. 413–444. New York: Plenum Press.

Denninghoff, K.R. 2000. Enrollment of sudden cardiac death victims into a limited cardiac autopsy study in the emergency department. *Journal of the National Medical Association* 92(1):36–38.

Department of Health and Human Services. 2003. *Medical Examiners' and Coroners' Handbook on Death Registration and Fetal Death Reporting.* Hyattsville, MDL: DHHS Publication no. (PHS) 2003–1110.

Detter, I., ed. 2000. *The Law of War,* 2nd edition. Cambridge: Cambridge University Press.

De Vito, C. and S.R. Saunders.1990. A discriminant function analysis of deciduous teeth to determine sex. *Journal of Forensic Sciences* 35:845–858.

Devos, D. 1995. *Basic Principles of Radiographic Exposure.* Philadelphia: Williams & Wilkins.

Di Lonardo, A.M., P. Darlu, M. Baur, C. Orrego and M.C. King.1984. Human genetics and human rights. Identifying the families of kidnapped children. *American Journal of Forensic Medicine and Pathology* 5(4):339–347.

DiMaio, V.J.M. 1999. *Gunshot Wounds: Practical Aspects of Firearms, Ballistics and Forensic Techniques,* 2nd edition. Boca Raton, FL: CRC Press.

DiMaio, V.J. and D. DiMaio. 2001. *Forensic Pathology,* 2nd edition. Boca Raton, FL: CRC Press.

Dirkmaat, D.C. and J.M. Adovasio. 1997. The role of archaeology in the recovery and interpretation of human remains from an outdoor forensic setting. In *Forensic Taphonomy: The Postmortem Fate of Human Remains.* W.D. Haglund, M.H. Sorg, eds. Pp. 39–64. Boca Raton, FL: CRC Press.

Dirkmaat, D.C., S.A. Symes, E. Vey and O.C. Smith. 2002. Recognizing child abuse in the thoracic region through a multidisciplinary approach. *Proceedings of the 54th Annual Meeting of the American Academy of Forensic Sciences* 8:248–249.

Djuric, M.P. 2004. Anthropological data in individualization of skeletal remains from a forensic context in Kosovo—a case history. *Journal of Forensic Sciences* 49(3):464–468.

Doretti, M. and C.C Snow. 2003. Forensic anthropology and human rights: the Argentine experience. In *Hard Evidence: Case Studies in Forensic Anthropology.* D.W. Steadman, ed. Pp. 290–310. Upper Saddle River, NJ: Prentice Hall.

Drawdy, S.M. 2003. Location, identification and repatriation of remains of victims of conflict: implications for forensic anthropology. *Proceedings from the Annual Meeting of the American Academy of Forensic Sciences* 9:276–277.

Dror, I.E., D. Charlton and A.E. Peron. 2006. Contextual information renders experts vulnerable to making erroneous identifications. *Forensic Science International* 156(1):74–78.

Duday, H. and M. Guillon. 2006. Understanding the circumstances of decomposition when the body is skeletonized. In *Forensic Anthropology and Medicine.* A. Schmitt, E. Cunha, J. Pinheiro, eds. Pp. 117–157. Totowa, NJ: Humana Press.

Duhig, C. 2003. Non-forensic remains: the use of forensic archaeology, anthropology and burial taphonomy. *Science and Justice* 43(4):211–214.

Dumancic, J., Z. Kaic, V. Njemirovskij, H. Brkic and D. Zecevic. 2001. Dental identification after two mass disasters in Croatia. *Croatian Medical Journal* 42(6):657–662.

Dupras, T.L., J.J. Schultz, S.M. Wheeler and L.J. Williams. 2006. *Forensic Recovery of Human Remains: Archaeological Approaches.* Boca Raton, FL: CRC Press.

Dykes, J. 1952. *Billy the Kid, the Bibliography of a Legend.* Albuquerque: University of New Mexico Press.

Dykstra, M.J. 1992. *Biological Electron Microscopy: Theory, Techniques and Troubleshooting.* New York: Plenum Press.

Earle, J.H. 1988. *The Capture of Billy the Kid.* College Station, TX: Creative Publishing.

Ebbesmeyer, C.C. and W.D. Haglund. 1994. Drift trajectories of a floating human body simulated in a hydraulic model of Puget Sound. *Journal of Forensic Sciences* 39(1):231–240.

Eckhert, W.G. and W.R. Teixeira. 1985. The identification of Josef Mengele. A triumph of international cooperation. *American Journal of Forensic Medicine and Pathology* 6(3):188–191.

Edelev, N.S. 1987. [Determination of the characteristics of the tracks by probe profilography in identifying the hacking weapon from traces on the bones: in Russian]. *Sudebno-Meditsinskaia Ekspertiza* 30(2):57–58.

Edge, W.D. 1984. Estimating the postmortem interval in big game animals. *Journal of Forensic Sciences* 29(4):1144–1149.

Edwards, H.L. 1995. *Goodbye to Billy the Kid.* College Station, TX: Creative Publishing.

Elkins, A. 1982. *Fellowship of Fear.* New York: Mysterious Press.

Emanovsky, P.D. 2004. Preliminary results on the use of cadaver dogs to locate Vietnam war-era human remains. *Proceedings of the Annual Meeting of the American Academy of Forensic Sciences* 10:319.

Ernest, R.N. 1991. Toolmarks in cartilage—revisited. *AFTE Journal* 23(4):958–959.

Etulain, R.W. 2002. *New Mexican Lives: Profiles and Historical Stories.* Albuquerque: UNM Center for the American West, University of New Mexico Press.

Evans, F.G. 1973. *Mechanical Properties of Bone.* Springfield IL: Charles C. Thomas.

Evans, W.E.D. 1963. *The Chemistry of Death.* Springfield IL: Charles C. Thomas.

Eyre-Walker, A. and P. Awadalla. 2001. Does human mtDNA recombine? *Journal of Molecular Evolution* 53(4–5):430–435.

Fable, E. 1980. *Billy the Kid, the New Mexican Outlaw.* College Station, TX: Creative Publishing.

Fackler, E. 1995. *Billy the Kid: The Legend of El Chivato.* New York: Forge.

Faerman, M., G. Kahila, P. Smith, C. Greenblatt, L. Stager, D. Filon and A. Oppenheim. 1997. DNA analysis reveals the sex of infanticide victims. *Nature* 385:212–213.

Falsetti, A. B. 1995. Sex assessment from metacarpals of the human hand. *Journal of Forensic Sciences* 40: 774–776.

Falsetti, A.B. 1999. A thousand tales of dead men: the forensic anthropology cases of William R. Maples, PhD. *Journal of Forensic Sciences* 44(4):682–686.

Falys, C.G., H. Schutkowski and D.A. Weston. 2005. The distal humerus—a blind test of Rogers' sexing technique using a documented skeletal collection. *Journal of Forensic Sciences* 50(6):1289–1293.

Fauber, T.L. 2004. *Radiographic Imaging and Exposure,* 2nd edition. St. Louis: Mosby.

Fazekas, I. G. and F. Kósa. 1978. *Forensic Fetal Osteology.* Budapest: Akadémiai Kiadó.

Federal Bureau of Investigation. 1985. *The Science of Fingerprints: Classification and Uses.* Washington, DC: Department of Justice, Federal Bureau of Investigation.

Fenton, T.W., J.L. deJong and R.C. Haut. 2003. Punched with a fist: the etiology of a fatal depressed cranial fracture. *Journal of Forensic Sciences* 48(2):277–281.

Fenton, T.W., V.H. Stefan, L.A. Wood and N.J. Sauer. 2005. Symmetrical fracturing of the skull from midline contact gunshot wounds: reconstruction of individual death histories from skeletonized human remains. *Journal of Forensic Sciences* 50(2):274–285.

Ferllini, R. 2003. The development of human rights investigations since 1945. *Science and Justice* 43(4):219–224.

Fiedler, S. and M. Graw. 2003. Decomposition of buried corpses, with special reference to the formation of adipocere. *Naturwissenschaften* 90(7):291–300.

Figgener, L. and C. Runte. 2003. Digital radiography and electronic data storage from the perspective of legal requirements for record keeping. *Journal of Forensic Odontostomatology* 21(2):40–44.

Findlay, A.B. 1977. Bone marrow changes in the post mortem interval. *Journal of the Forensic Science Society* 16:213–218.

Fisher, B.A.J., A. Svensson and O. Wendel. 1987. *Techniques of Crime Scene Investigation*, 4th edition. New York: Elsevier.

Fitzgerald, C.M. and J.C. Rose. 2000. Reading between the lines: dental development and subadult age assessment using the microstructural growth markers of teeth. In *Biological Anthropology of the Human Skeleton*. M.A. Katzenberg, S.R. Saunders, eds. Pp. 163–186. New York: Wiley-Liss.

Fitzpatrick, J.J., D.R. Shook, B.L. Kaufman, S.J. Wu, R.J. Kirschner, H. MacMahon, et al. 1996. Optical and digital techniques for enhancing radiographic anatomy for identification of human remains. *Journal of Forensic Sciences* 41(6):947–959.

Fleiss, J.L. 1971. Measuring nominal scale agreement among many rates. *Psychological Bulletin* 76(5):378–382.

Follet, H., K. Bruyere-Garnier, F. Peyrin, J.P. Roux, M.E. Arlot, B. Burt-Pichat, C. Rumelhart and P.J. Meunier. 2005. Relationship between compressive properties of human os calcis cancellous bone and microarchitecture assessed from 2D and 3D synchrotron microtomography. *Bone* 36(2):340–351.

Forrest, E.R. n.d. Papers of Earle Robert Forrest, 1895–1960, 1929–1969. Special Collections, University of Arizona Library.

Foster, K.R. and P.W. Huber. 1999. *Judging Science: Scientific Knowledge and the Federal Courts*. Boston: First MIT Press.

France, D.L. 1983. *Sexual Dimorphism in the Human Humerus*. PhD Dissertation, University of Colorado, Boulder.

France, D.L. 1998. Observational and metric analysis of sex in the skeleton. In *Forensic Osteology: Advances in the Identification of Human Remains* 2nd edition. K. J. Reichs, ed. Pp. 163–186. Springfield, IL: Charles C. Thomas.

France, D.L., T.J. Griffin, J.G. Swanburg, J.W. Lindemann, G.C. Davenport, V. Trammell, C.T. Armburst, B. Kondratieff, A. Nelson, K. Castellano and D. Hopkins. 1992. A multidisciplinary approach to the detection of clandestine graves. *Journal of Forensic Sciences* 37(6):1445–1458.

Franklin, D., C. Oxnard, P. O'Higgins, and I. Dadour. 2006. Sexual dimorphism in the subadult mandible; quantification using geometric morphometrics. *Proceedings of the 58th Annual Meeting of the American Academy of Forensic Sciences* 12:277.

Frayer D.W. and J.G. Bridgens. 1985. Stab wounds and personal identity determined from skeletal remains: a case from Kansas. *Journal of Forensic Sciences* 30(1):232–238.

French, J.L. 1931. *A Gallery of Old Rogues*. New York: A.H. King.

Frisbie, T. and R. Garrett. 2005. *Victims of Justice Revisited*. Chicago: Northwestern University Press.

Fulton, M.G. n.d. Papers of Maurice G. Fulton, 1829–1955 (bulk 1870–1954). Special Collections, University of Arizona Library, Tucson.

Fulton, M.G. and R.N. Mullin. 1968. *Maurice Garland Fulton's History of the Lincoln County War*. Tucson: University of Arizona Press.

Fulton, M.G. and R.N. Mullin. 1997. *History of the Lincoln County War*. Tucson: University of Arizona Press.

Gagliano-Candela, R. and L. Aventaggiato. 2001. The detection of toxic substances in entomological specimens. *International Journal of Legal Medicine* 114(4–5):197–203.

Gallois-Montbrun, F.G., D.R. Barres and M. Durigon. 1988. Postmortem interval estimation by biochemical determination in birds muscle. *Forensic Science International* 37:189–192.

Galloway, A. and T.L. Simmons. 1997. Education in forensic anthropology: appraisal and outlook. *Journal of Forensic Sciences* 42(5):796–801.

Galloway, A., W.H. Birkby, A.M. Jones, T.E. Henry and B.O. Parks. 1989. Decay rates of human remains in an arid environment. *Journal of Forensic Sciences* 34(3):607–616.

Garcelon, J.S. and J.P. O'Leary. 1994. "What will the poor boys at LSU do without me?" An account of the death of Senator Huey P. Long. *American Surgeon* 60(12):988–990.

Garner, B.A. 2001. *Black's Law Dictionary*, 2nd pocket edition. St. Paul, MN: West Group.

Garrett, P.F. and M.G. Fulton. 1927. *Pat F. Garrett's Authentic Life of Billy the Kid*. New York: Macmillan .

Gatowski, S.I., S.A. Dobbin, J.T. Richardson, G.P. Ginsburg, M.L. Merlino and V. Dahir. 2001. Asking the gatekeepers: a national survey of judges on judging expert evidence in a post-*Daubert* world. *Law and Human Behavior* 25(5):433–458.

Geiger, H.J. and R.M. Cook-Deegan. 1993. The role of physicians in conflicts and humanitarian crises. Case studies from the field missions of Physicians for Human Rights, 1988 to 1993. *Journal of the American Medical Association* 270(5):616–620.

Gerberth, V.J. 1983. *Practical Homicide Investigation: Tactics, Procedures and Forensic Techniques*. New York: Elsevier.

Ghosh, A.K. and P. Sinha. 2005. An unusual case of cranial image recognition. *Forensic Science International* 148(2–3):93–100.

Gilbert B.M. 1990. *Mammalian Osteology*. Columbia: Missouri Archaeological Society.

Gilbert, B.M. and T.W. McKern. 1973. A method of aging the female *os pubis*. *American Journal of Physical Anthropology* 38:31–38.

Giles, E. 1964. Sex determination by discriminant function analysis of the mandible. *American Journal of Physical Anthropology* 22:129–135.

Giles, E. and O. Elliot. 1962. Race identification from cranial measurements. *Journal of Forensic Sciences* 7(2): 147–157.

Giles, E. and O. Elliot. 1963. Sex determination by discriminant function analysis of crania. *American Journal of Physical Anthropology* 21: 53–68.

Giles, E. and P.H. Vallangdigham. 1991. Height estimation from foot and shoeprint length. *Journal of Forensic Sciences* 36:1134–1151.

Giles, J. 2004. Guatemalan forensic work brings award and death threats. *Nature* 427(6976):664.

Gill, P. 1997. The Romanovs. *Medicine Legal Journal* 65(3):122–132.

Gill, P. 2005. DNA as evidence—the technology of identification. *New England Journal of Medicine* 352(26):2669–2671.

Gill, P. and E. Hagelberg. 2004. Ongoing controversy over Romanov remains. *Science* 306(5695):407–410.

Gill, P., A.J. Jeffreys and D.J. Werrett. 1985. Forensic application of DNA 'fingerprints.' *Nature* 318(6046):577–579.

Gill, P., P.L. Ivanov, C. Kimpton, R. Piercy, N. Benson, G. Tully, I. Evett, E. Hagelberg and K. Sullivan. 1994. Identification of the remains of the Romanov family by DNA analysis. *Nature Genetics* 6:130–135.

Gill-King, H. 1997. Chemical and ultrastructural aspects of decomposition. In *Forensic Taphonomy: The Postmortem Fate of Human Remains*. W.D. Haglund, M.H. Sorg, eds. Pp. 93–108. Boca Raton, FL: CRC Press.

Glaister, J., and J.C. Brash.1937. *Medico-Legal Aspects of the Ruxton Case*. Edinburgh: Livingstone.

Glassman, D.M. and W.M. Bass. 1986. Bilateral asymmetry of long arm bones and jugular foramen: implications for handedness. *Journal of Forensic Sciences* 31:589–595.

Gold, J.A., M.J. Zaremski, E.R. Lev and D.H. Shefrin. 1993. *Daubert v Merrell Dow*: the Supreme Court tackles scientific evidence in the courtroom. *Journal of the American Medical Association* 270(24):2964–2967.

Goldthorpe, S.B. and P. McConnell. 2000. A new method of recording clinical forensic evidence. *Journal of Clinical Forensic Medicine* 7(3):127–129.

Gomber, D. 1998. *Lincoln County War: Heroes and Villains*. Lincoln, NM: Bandillo Publishing.

Gonzalez-Andrade, F. and D. Sanchez. 2005. DNA typing from skeletal remains following an explosion in a military fort—first experience in Ecuador (South America). *Legal Medicine (Tokyo)* 7(5):314–318.

Gordon, C.C., and J.E. Buikstra. 1992. Linear models for the prediction of stature from foot and boot dimensions. *Journal of Forensic Sciences* 37(3):771–782.

Graham, M.H. 2003. *Federal Rules of Evidence*, 6th edition. St. Paul, MN; Thomson West.

Grant, L. 2006. Gallery exhibit highlights the first international war crimes tribunal. *Harvard Law School Alumni Bulletin*. Electronic document, http://www.law.harvard.edu/alumni/bulletin/2006/spring/gallery.php, accessed July 7, 2006.

Graves, J.L. 2001. *The Emperor's New Clothes: Biological Theories of Race at the Millennium*. New Brunswick, NJ: Rutgers University Press.

Greulich, W.W. and S.I. Pyle. 1959. *Radiographic Atlas of Skeletal Development of the Hand and Wrist*. Stanford, CA: Stanford University Press.

Grignani, P., G. Peloso, A. Achilli, C. Turchi, A. Tagliabracci, M. Alu, G. Beduschi, U. Ricci, L. Giunti, C. Robino, S. Gino and C. Previdere. 2005. Subtyping mtDNA haplogroup H by SNaPshot minisequencing and its application in forensic individual identification. *International Journal of Legal Medicine* 7:1–6.

Grisbaum, G.A. and D. H. Ubelaker. 2001. *An Analysis of Forensic Anthropology Cases Submitted to the Smithsonian Institution by the Federal Bureau of Investigation from 1962 to 1994*. Smithsonian Contributions to Anthropology, no 45. Washington, DC; Smithsonian Institution Press.

Grivas, C. and D. Komar. 2007. Daubert and Kumho: implications for anthropologists in the courtroom. *Proceedings of the 59th Annual Meeting of the American Academy of Forensic Sciences*, 13:387.

Gunby, P. 1994. Medical team seeks to identify human remains from mass graves of war in former Yugoslavia. *Journal of the American Medical Association* 272(23): 1804, 1806.

Gustafson, G. 1950. Age determination on teeth. *Journal of the American Dental Association* 41:45–54.

Gustafson, G. 1966. *Forensic Anthropology*. London: Staples Press.

Gutman, R., ed. 1999. *Crimes of War*. New York: Norton.

Haacke, E.M., R.W. Brown, M.R. Thompson and R. Venkatesan. 1999. *Magnetic Resonance Imaging: Physical Principles and Sequence Design.* New York: Wiley.

Haglund, W.D. 1991. *Applications of Taphonomic Models to Forensic Investigations.* Ph.D. Dissertation, University of Washington.

Haglund, W.D. 1993. Disappearance of soft tissue and the disarticulation of human remains in aqueous environments. *Journal of Forensic Sciences* 38(4):806–815.

Haglund, W.D. 1997a. Dogs and coyotes: postmortem involvement with human remains. In *Forensic Taphonomy: The Postmortem Fate of Human Remains.* W.D. Haglund, M.H. Sorg, eds. Pp. 367–382. Boca Raton, FL: CRC Press.

Haglund, W.D. 1997b. Rodents and human remains. In *Forensic Taphonomy: The Postmortem Fate of Human Remains.* W.D. Haglund, M.H. Sorg, eds. Pp. 405–414. Boca Raton, FL: CRC Press.

Haglund, W.D. 1997c. Scattered skeletal human remains: search strategy considerations for locating missing teeth. In *Forensic Taphonomy: The Postmortem Fate of Human Remains.* W.D. Haglund, M.H. Sorg, eds. Pp. 383–394. Boca Raton, FL: CRC Press.

Haglund, W.D. and M.H. Sorg, eds. 2002. *Advances in Forensic Taphonomy: Method, Theory, and Archaeological Perspectives.* Boca Raton, FL: CRC Press.

Haglund, W.D. and M.H. Sorg, eds. 1997a. *Forensic Taphonomy: The Postmortem Fate of Human Remains.* Boca Raton, FL: CRC Press.

Haglund, W.D. and M.H. Sorg. 1997b. Introduction to Forensic Taphonomy. In *Forensic Taphonomy: The Postmortem Fate of Human Remains.* W.D. Haglund, M.H. Sorg, eds. Pp. 1–26. Boca Raton, FL: CRC Press.

Haglund, W.D. and D.T. Reay. 1993. Problems of recovering partial human remains at different times and locations: concerns for death investigators. *Journal of Forensic Sciences* 38(1):69–80.

Haglund, W.D., D.T. Reay and D.R. Swindler. 1988. Tooth mark artifacts and survival of bones in animal scavenged human skeletons. *Journal of Forensic Sciences* 33(4):985–997.

Haglund, W.D., D.T. Reay and D.R. Swindler. 1989. Canid scavenging/disarticulation sequence of human remains in the Pacific Northwest. *Journal of Forensic Sciences* 34(3):587–606.

Haglund, W.D., D.G. Reichert and D.T. Reay. 1990. Recovery of decomposed and skeletal human remains in the Green River murder investigation: implications for medical examiner/coroner and police. *American Journal of Forensic Medicine and Pathology* 11(1):35–43.

Hailman, J.P. and K.B. Strier. 1997. *Planning, Proposing, and Presenting Science Effectively.* Cambridge: Cambridge University Press.

Hall, D.W. 1997. Forensic Botany. In *Forensic Taphonomy: The Postmortem Fate of Human Remains.* W.D. Haglund, M.H. Sorg, eds. Pp. 353–366. Boca Raton, FL: CRC Press.

Hamlin, W.L. 1959. *The True Story of Billy the Kid; A Tale of the Lincoln County Wars.* Caldwell, ID: Caxton Printers.

Hampton, A. 2005. Testing determination of adult age at death using four criteria of the acetabulum. *Proceedings from the American Academy of Forensic Sciences Meeting* 11:319.

Hanihara, K. 1958. Sexual diagnosis of Japanese long bones by means of discriminant function. *Journal of the Anthropological Society Nippon* 66:187–196.

Hanihara, K. 1959. Sexual diagnosis of Japanese skulls and scapulae by means of discriminant function. *Journal of the Anthropological Society of Nippon* 67:20–29.

Hannibal, K. and R.S. Lawrence. 1996. The health professional as human rights promoter: ten years of Physicians for Human Rights (USA). *Health and Human Rights* 2(1):111–127.

Hanson, I. 2003. Advances in surveying and presenting evidence from mass graves, clandestine graves and surface scatters. *Proceedings from the Annual Meeting of the American Academy of Forensic Sciences* 9:269–270.

Hanzlick, R., J.C. Hunsake and G.J. Davis. 2002. *A Guide for Manner of Death Classification*, 1st edition. Atlanta : National Association of Medical Examiners.

Harkess, J.W., W.C. Ramsey and J.W. Harkess. 1991. Principles of fractures and dislocations. In *Rockwood and Green's Fractures in Adults*, 3rd edition. C.A. Rockwood, D.P.Green, R.W. Bucholz, eds. Pp. 1–80. Philadelphia: Lippincott.

Harruff, R.C. 1995. Comparison of contact shotgun wounds of the head produced by different gauge shotguns. *Journal of Forensic Sciences* 40(5):801–804.

Haskell, N.H., R.D. Hall, V.J. Cervenka and M.A. Clark. 1997. On the body: insect's life stage presence, their postmortem artifacts. In *Forensic Taphonomy: The Postmortem Fate of Human Remains*. W.D. Haglund, M.H. Sorg, eds. Pp. 415–448. Boca Raton, FL: CRC Press.

Hausmann, R. and P. Betz. 2002. Thermally induced entrance wound–like defect of the skull. *Forensic Science International* 128(3):159–161.

Hawkey, D. and C.F. Merbs. 1995. Activity-induced musculoskeletal stress markers (MSM) and subsistence strategy changes among Hudson Bay Eskimos. *International Journal of Osteoarchaeology* 5:324–338.

Hayes, S., R. Taylor and A. Paterson. 2005. Forensic facial approximation: an overview of current methods used at the Victorian Institute of Forensic Medicine/Victoria Police Criminal Identification Squad. *Journal of Forensic Odontostomatology* 23(2):45–50.

Hefner, J.T. and S.D. Ousley. 2006. Macroscopic traits and the statistical determination of ancestry II. *Proceedings of the 58th Annual Meeting of the American Academy of Forensic Sciences* 12:282–283.

Hefner, J.T., S.D. Ousley and M.D. Warren. in press. An historical perspective on nonmetric cranial variation: Hooton and the Harvard List. In *Skeletal Attribution of Race: Methods for Forensic Anthropology*, 2nd edition. G. Gill, S. Rhine, eds. Albuquerque, NM: Maxwell Museum of Anthropology Papers.

Heide, S., M. Kleiber, J. Frohlich, W. Burkert and K. Trubner. 1998. Unusual explanation for the death of a car passenger. *International Journal of Legal Medicine* 111(2):85–87.

Hein, P.M. and E. Schulz. 1990. Contrecoup fractures of the anterior cranial fossae as a consequence of blunt force caused by a fall. *Acta Neurochirurgica* 105(1–2):24–29.

Helfman, P.M. and J.L. Bada. 1976. Aspartic acid racemisation in dentine as a measure of ageing. *Nature* 262:279–281.

Helmer, R.P. 1987. Identification of the cadaver remains of Josef Mengele. *Journal of Forensic Sciences* 32(6):1622–1644.

Hendron, J.W. 1948. *The Story of Billy the Kid: New Mexico's Number One Desperado*. Santa Fe, NM: Rydal Press.

Henke J. and L. Henke. 2005. Which short tandem repeat polymorphisms are required for identification? Lessons from complicated kinship cases. *Croatian Medical Journal* 46(4):593–597.

Henschen, F. 1965. *The Human Skull, A Cultural History*. New York: Praeger.

Henssge, C. 1988. Death time estimation in case work: I. The rectal temperature time of death nomogram. *Forensic Science International* 38:209–236.

Henssge, C., B. Madea and E. Gallenkemper. 1988. Death time estimation in case work: II. Integration of different methods. *Forensic Science International* 39:77–87.

Henssge, C., L. Althaus, J. Bolt, A. Freislederer, H.T. Haffner, C.A. Henssge, B. Hoppe and V. Schneider. 2000. Experiences with a compound method for estimating the time since death: II. Integration of non-temperature-based methods. *International Journal of Legal Medicine* 113:320–331.

Herrmann, N.P. and J.L. Bennett. 1999. The differentiation of traumatic and heat-related fractures in burned bone. *Journal of Forensic Sciences* 44(3):461–469.

Hertzog, P. 1963. *Little Known Facts About Billy, the Kid*. Santa Fe, NM: Press of the Territorian.

Hesse, B. and P. Wapnish. 1985. *Animal Bone Archeology: from Objectives to Analysis*. Washington, DC: Taraxacum.

Hinkes, M.J. 1989. The role of forensic anthropology in mass disaster resolution. *Aviation Space Environmental Medicine* 60(7):A60–A63.

Hiss, J. and T. Kahana. 1995. The medicolegal implications of bilateral cranial fractures in infants. *Journal of Trauma* 38(1):32–34.

Hiss, J. and T. Kahana. 2000. Trauma and identification of victims of suicidal terrorism in Israel. *Military Medicine* 165(11):889–893.

Hoffman, J. M. 1979. Age estimation from diaphyseal lengths: two months to twelve years. *Journal of Forensic Sciences* 24:461–469.

Hogge, J.P., J.M. Messmer and Q.N. Doan. 1994. Radiographic identification of unknown human remains and interpreter experience level. *Journal of Forensic Sciences* 39(2):373–377.

Hogge, J.P., J.M. Messmer and M.F. Fierro. 1995. Positive identification by post-surgical defects from unilateral lambdoid synostectomy: a case report. *Journal of Forensic Sciences* 40(4):688–691.

Holck, P. 2005. What can a baby's skull withstand? Testing the skull's resistance on an anatomical preparation. *Forensic Science International* 151(2–3):187–191.

Holcomb, S.M.C. and L.W. Konigsberg. 1995. Statistical study of sexual dimorphism in the human fetal sciatic notch. *American Journal of Physical Anthropology* 97:113–125

Holland, M.M., C.A. Cave, C.A. Holland and T.W. Bille. 2003. Development of a quality, high throughput DNA analysis procedure for skeletal samples to assist with the identification of victims from the World Trade Center attacks. *Croatian Medical Journal* 44(3):264–272.

Holland, T.D. 1986. Sex determination of fragmentary crania by analysis of the cranial base. *American Journal of Physical Anthropology* 70:203–208.

Horner, K., D.S. Brettle and V.E. Rushton. 1996. The potential medico-legal implications of computed radiography. *British Dentistry Journal* 180(7):271–273.

Hoshower, L.M. 1998. Forensic archeology and the need for flexible excavation strategies: a case study. *Journal of Forensic Sciences* 43(1):53–56.

Hoshower, L.M. 1999. Dr. William R. Maples and the role of the consultants at the U.S. Army Central Identification Laboratory, Hawaii. *Journal of Forensic Sciences* 44(4):689–691.

Houck, M.M. 1998. Skeletal trauma and the individualization of knife marks in bones. In *Forensic Osteology: Advances in the Identification of Human Remains*, 2nd edition. K.J. Reichs, ed. Pp. 410–424. Spring field, IL: Charles C. Thomas.

Houck, M.M., D. Ubelaker, D. Owsley, E. Craig, W. Grant, R. Fram, T. Woltanski and K Sandness. 1996. The role of forensic anthropology in the recovery and analysis of Branch Davidian compound victims: assessing the accuracy of age estimations. *Journal of Forensic Sciences* 41:796–801.

Howells, W.W. 1973. Cranial variation in man. In *Papers of the Peabody Museum*, Vol. 67, Peabody Museum of Archeology and Ethnology, Harvard University, Cambridge, MA.

Howells, W.W. 1989. Skull shapes and the map. In *Papers of the Peabody Museum*, Vol. 78, Peabody Museum of Archeology and Ethnology, Harvard University, Cambridge, MA.

Howells, W.W. 1995. Who's who in skulls: ethnic identification of crania from measurements. In *Papers of the Peabody Museum*, Vol 82, Peabody Museum of Archeology and Ethnology, Harvard University, Cambridge, MA.

Hrdlicka, A. 1920. *Anthropometry*. Philadelphia: Wistar Institute of Anatomy and Biology.

Hrdlicka, A. 1939. *Practical Anthropometry*. Philadelphia: Wistar Institute of Anatomy and Biology.

Hsiao, T.H, H.P. Chang and K.M. Liu. 1996. Sex determination by discriminant function analysis of lateral radiographic cephalometry. *Journal of Forensic Sciences* 41:792–795.

Hsu, C.M., N.E. Huang, L.C. Tsai, L.G. Kao, C.H. Chao, A. Linacre and J.C. Lee. 1999. Identification of victims of the 1998 Taoyuan Airbus crash accident using DNA analysis. *International Journal of Legal Medicine* 113(1):43–46.

Hudson, B. and M.H. Brothers. 1946. *Billy the Kid; the Most Hated, the Most Loved Outlaw New Mexico Ever Produced*. Farmington, NM: Hustler Press.

Huffine, E., J. Crews, B. Kennedy, K. Bomberger and A. Zinbo. 2001. Mass identification of persons missing from the break-up of the former Yugoslavia: structure, function, and role of the International Commission on Missing Persons. *Croatian Medical Journal* 42(3):271–275.

Hughes, J.R. 2005. A reappraisal of the possible seizures of Vincent van Gogh. *Journal of Epilepsy Behavior* 6(4):504–510.

Hughes, S., R. Wright and M. Barry. 2005. Virtual reconstruction and morphological analysis of the cranium of an Egyptian mummy. *Australian Physical Engineering, Science and Medicine* 28(2):122–127.

Hulewicz, B. and G.W. Wilcher. 2003. The use of thoracolumbar and hip joint dysmorphism in identification. *Journal of Forensic Sciences* 48(4):842–847.

Hulley, S.B., S.R. Cummings, W.S. Browner, D. Grady, N. Hearst, and T.B. Newman. 2001. *Designing Clinical Research*, 2nd edition. Philadelphia: Lippincott Williams & Wilkins.

Humphrey, J.H. and D.L. Hutchinson. 2001. Macroscopic characteristics of hacking trauma. *Journal of Forensic Sciences* 46(2):228–233.

Hunt, D.R. and J. Albanese. 2005. History and demographic compositions of the Robert J. Terry Anatomical Collection. *American Journal of Physical Anthropology* 127:406–417.

Hunt, F. 1956. *The Tragic Days of Billy the Kid*. New York: Hastings House.

Hunter, J., C. Roberts and A. Martin. 1996. *Studies in Crime: An Introduction to Forensic Archaeology*. London: Batsford.

Hunter, J.R., M.B. Brickley, J. Bourgeois, W. Bouts, L. Bourguignon, F. Hubrecht, J. De Winne, H. Van Haaster, T. Hakbijl, H. De Jong, L. Smits, L.H. Van Wijngaarden and M. Luschen. 2001. Forensic archaeology, forensic anthropology and human rights in Europe. *Science and Justice* 41(3):173–178.

Huxley A.K. and M. Finnegan. 2004. Human remains sold to the highest bidder! A snapshot of the buying and selling of human skeletal remains on eBay, an Internet auction site. *Journal of Forensic Sciences* 49(1):17–20.

Imaizumi, M. 1974. Locating buried bodies. *FBI Law Enforcement Bulletin* 43(8):2–5.

Innan, H. and M. Nordborg. 2002. Recombination or mutational hot spots in human mtDNA? *Molecular Biology and Evolution* 19(7):1122–1127.

Inoue, M., A. Suyama, T. Matouka, T. Inoue, K. Okada and Y. Ineawa. 1994. Development of an instrument to measure postmortem lividity and its preliminary application to estimate the time since death. *Forensic Science International* 65(3):185–193.

Introna, F., G. Di Vella and C.P. Campobasso. 1998. Sex determination by discriminant analysis of patella measurements. *Forensic Science International* 95(1):39–45.

Introna, F., G. Di Vella and C.P. Compobasso. 1999. Determination of postmortem interval from old skeletal remains by image analysis of luminol test results. *Journal of Forensic Sciences* 44(3):535–538.

Introna, F., C.P. Campobasso and M.L. Goff. 2001. Entomotoxicology. *Forensic Science International* 120(1–2):42–47.

Iscan, M.Y. 1985. Osteometric analysis of sexual dimorphism in the sternal end of the rib. *Journal of Forensic Sciences* 30: 1090–1099.

Iscan, M.Y., ed. 1989. *Age Markers in the Human Skeleton*. Springfield, IL: Charles C. Thomas.

Iscan, M.Y. 2001. Global forensic anthropology in the 21st century. *Forensic Science International* 117(1–2):1–6.

Iscan, M.Y. and P. Miller-Shaivitz. 1984a. Determination of sex from the tibia. *American Journal of Physical Anthropology* 64:53–57.

Iscan, M.Y. and P. Miller-Shaivitz. 1984b. Discriminant function sexing of the tibia. *Journal of Forensic Sciences* 29:1087–1093.

Iscan, M.Y. and G. Quatrehomme. 1999. Medicolegal anthropology in France. *Forensic Science International* 100:17–35.

Iscan, M.Y., M. Yoshino and S. Kato. 1994. Sex determination from the tibia: standards for contemporary Japan. *Journal of Forensic Sciences* 39:785–792.

Isenberg, A.R. 2004. Forensic mitochondrial DNA analysis. In *Forensic Science Handbook*, Vol. II, 2nd edition. R. Saferstein, ed. Pp. 297–327. Upper Saddle River, NJ: Prentice-Hall.

Isometsa, E.T. 2001. Psychological autopsy studies, a review. *European Psychiatry* 16(7):379–385.

Ivanov, P.L., M.J. Wadhams, R.K. Roby, M.M. Holland, V.W. Weedn and T.J. Parsons. 1996. Mitochondrial DNA sequence heteroplasmy in the Grand Duke of Russia Georgij Romanov establishes the authenticity of the remains of Tsar Nicholas II. *Nature Genetics* 12(4):417–420.

Iwamura, E.S., C.R. Oliveira, J.A. Soares-Vieira, S.A. Nascimento and D.R. Munoz. 2005. A qualitative study of compact bone microstructure and nuclear short tandem repeat obtained from femur of human remains found on the ground and exhumed 3 years after death. *American Journal of Forensic Medicine and Pathology* 26(1):33–44.

Iwase, H., Y. Yamada, S. Ootani, Y. Sasaki, M. Nagao, K. Iwadate and T. Takatori. 1998. Evidence for an antemortem injury of a burned head dissected from a burned body. *Forensic Science International* 94(1–2):9–14.

Jackes, M. 2000. Building the bases for paleodemographic analysis: adult age estimation. In *Biological Anthropology of the Human Skeleton*, M.A. Katzenberg, S.R. Saunders, eds. Pp. 417–466. New York: Wiley-Liss.

Jacknis, I. 1996. Repatriation as social drama: the Kwakiutl Indians of British Columbia, 1922–1980. *American Indian Quarterly* 20(2):274–286.

Jaffe, H.L. 1958. *Tumors and Tumorous Conditions of the Bones and Joints*. Philadelphia: Lea & Febiger.

James, H. 2005a. Thai tsunami victim identification overview to date. *Journal of Forensic Odontostomatology* 23(1):1–18.

James, H. 2005b. Localized natural disasters and terrorist acts. *Journal of Forensic Odontostomatology* 23(1):1.

Jameson, J.H. 1997. *Presenting Archaeology to the Public: Digging for Truths*. London: AltaMira Press.

Jameson, W.C and F. Bean. 1998. *The Return of the Outlaw, Billy the Kid*. Plano: Republic of Texas Press.

Jani, C.B. and B.D. Gupta. 2004. An autopsy study on medico-legal evaluation of post-mortem scavenging. *Medicine, Science and the Law* 44(2):121–126.

Jantz, R.L. and L. Meadows Jantz. 2000. Secular change in craniofacial morphology. *American Journal of Human Biology* 12:327–338.

Jantz, R.L., D.R. Hunt, and L. Meadows. 1994. Maximum length of the tibia: how did Trotter measure it? *American Journal of Physical Anthropology* 93:525–528.

Jantz R.L. 2001. Cranial change in Americans: 1850–1975. *Journal of Forensic Sciences* 46:784–787.

Jayaprakash, P.T., G.J. Srinivasan and M.G. Amravaneswaran. 2001. Craniofacial morphanalysis: a new method for enhancing reliability while identifying skulls by photo superimposition. *Forensic Science International* 117(1–2): 121–143.

Jeffreys, A.J.,V. Wilson and S.L. Thein 1985. Individual-specific 'fingerprints' of human DNA nature 316(6023):76–79.

Jeffreys, A.J., M.J. Allen, E. Hagelberg and A. Sonnberg. 1992. Identification of the skeletal remains of Josef Mengele by DNA analysis. *Forensic Science International* 56:65–76.

Jehaes, E., R. Decorte, A. Peneau, J.H. Petrie, P.A. Boiry, A. Gilissen, J.P. Moisan, H. Van den Berghe, O. Pascal and J.J. Cassiman. 1998. Mitochondrial DNA analysis on remains of a putative son of Louis XVI, King of France and Marie-Antoinette. *European Journal of Human Genetics* 6(4):383–395.

Jenardo, D. 1970. *The True Story of Billy the Kid*. Columbus, MN: Saddlebag Press.

Jentzen, J., G. Palermo, L.T. Johnson, K.C. Ho, K.A. Stormo and J. Teggatz. 1994. Destructive hostility: the Jeffrey Dahmer case. A psychiatric and forensic study of a serial killer. *American Journal of Forensic Medicine and Pathology* 15(4): 283–294.

Joyce, C. and E. Stover. 1991. *Witnesses from the Grave: The Stories Bones Tell*. Boston: University of Massachusetts Press.

Jurmain, R.D. 1999. *Stories from the Skeleton: Behavioral Reconstruction in Human Osteology*. Vol. 1, *Interpreting the Remains of the Past*. Toronto and Amsterdam: Gordon & Breach.

Kadlec, R.F. 1987. *They "Knew" Billy the Kid: Interviews with Old-Time New Mexicans*. Santa Fe, NM: Ancient City Press.

Kahana, T., M. Freund and J. Hiss. 1997. Suicidal terrorist bombings in Israel—identification of human remains. *Journal of Forensic Sciences* 42(2):260–264.

Kahana, T., J. Almog, J. Levy, E. Shmeltzer, Y. Spier and J. Hiss. 1999. Marine taphonomy: adipocere formation in a series of bodies recovered from a single shipwreck. *Journal of Forensic Sciences* 44(5):897–901.

Kalmey, J. K. and T.A. Rathbun. 1996. Sex determination by discriminant function analysis of the petrous portion of the temporal bone. *Journal of Forensic Sciences* 41:865–867.

Karger, B. and B. Vennemann. 2001. Suicide by more than 90 stab wounds including perforation of the skull. *International Journal of Legal Medicine* 115(3):167–169.

Kaufman, H.H. 2001. The expert witness: neither *Frye* nor *Daubert* solved the problem: what can be done? *Science and Justice* 41(1):7–20.

Keierleber, J.A. and T.L. Bohan. 2005. Ten years after *Daubert*: the status of the states. *Journal of Forensic Sciences* 50(5):1154–1163.

Kemkes-Grottenthaler, A. 2001. The reliability of forensic osteology—a case in point. Case study. *Forensic Science International* 117(1–2):65–72.

Kemp, B.M. and D.G. Smith. 2005. Use of bleach to eliminate contaminating DNA from the surface of bones and teeth. *Forensic Science International* 154(1):53–61.

Kennedy, K.A.R. 1989. Skeletal markers of occupational stress. In *Reconstruction of Life from the Skeleton*. M.Y. Iscan, K.A.R. Kennedy, eds. Pp. 129–160. New York: Liss.

Kerley, E. 1965. The microscopic determination of age in human bones. *American Journal of Physical Anthropology* 23:149–165.

Kerley, E. R. 1978. Recent developments in forensic anthropology. *Yearbook of Physical Anthropology* 21:160–173.

Kerley, E. and D.H. Ubelaker. 1978. Revisions in the microscopic method of estimating age at death in human cortical bone. *American Journal of Physical Anthropology* 49:545–546.

Kiel, F.W. 1965. The psychiatric character of the assailant as determined by autopsy observations of the victim. *Journal of Forensic Sciences* 10(3):263–271.

Killam, E.W. 1990. *The Detection of Human Remains*. Springfield, IL: Charles C. Thomas.

King, C.A., M.Y. Iscan and S.R. Loth. 1998. Metric and comparative analysis of sexual dimorphism in the Thai femur. *Journal of Forensic Sciences* 43:954–958.

Kirschner, R.H. 1984. The use of drugs in torture and human rights abuses. *American Journal of Forensic Medicine and Pathology* 5(4):313–315.

Kirschner, R.H. 1989. The role of forensic scientists in the documentation of human rights abuses. *Legal Medicine*: 59–91.

Kirschner, R.H. and K.E. Hannibal. 1994. The application of the forensic sciences to human rights investigations. *Medicine and Law* 13(5–6):451–460.

Klepinger, L. 2006. *Fundamentals of Forensic Anthropology*. Hoboken, NJ: Wiley.

Klepinger, L. and E. Giles. 1998. Clarification or confusion: statistical interpretation in forensic anthropology. In *Forensic Osteology: Advances in the Identification of Human Remains* 2nd edition. K.J. Reichs, ed. Pp. 427–440. Springfield, IL: Charles C. Thomas.

Klonowski, E., P. Drukier and N. Sarajlic. 2004. Exhumation—and what after? ICMP model in Bosnia and Herzegovina. *Proceedings of the Annual Meeting of the American Academy of Forensic Sciences* 10:318.

Knight, A., L.A. Zhivotovsky, D.H. Kass, D.E. Litwin, L.D. Green and P.S. White. 2004. Ongoing controversy over Romanov remains. *Science* 306(5695):407–410.

Knight, B. 1986. The evolution of methods for estimating the time of death from body temperature. *Forensic Science International* 36(1,2):47–55.

Knight, B. and I. Lauder. 1969. Methods of dating skeletal remains. *Human Biology* 41(3):322–341.

Kobilinsky, L., T.F. Liotti and J. Oeser-Sweat. 2005. *DNA: Forensic and Legal Applications*. Hoboken, NJ: Wiley.

Komar, D. 1998. Decay rates in a cold climate region: a review of cases involving advanced decomposition from the Medical Examiner's office in Edmonton, Alberta. *Journal of Forensic Sciences* 43(1):57–61.

Komar, D. 1999a. *Forensic Taphonomy of a Cold Climate Region: A Field Study in Central Alberta and a Potential New Method of Determining Time Since Death*. Ph.D. Dissertation, Department of Anthropology, University of Alberta.

Komar, D. 1999b. The use of cadaver air scent detection dogs in locating scattered, scavenged human remains: preliminary field test results. *Journal of Forensic Sciences* 44(2):405–408.

Komar, D. 2001. Differential decay rates in single, multiple and mass graves in Bosnia. *Proceedings of the Annual Meeting of the American Academy of Forensic Sciences* 7:242–243.

Komar, D. 2002. The validity of using unique biological features as a method of identifying victims of war crimes in the former Yugoslavia. *Proceedings of the Annual Meeting of the American Academy of Forensic Sciences* 8:241–242.

Komar, D. 2003a. Lessons from Srebrenica: the contributions and limitations of physical anthropology in identifying victims of war crimes. *Journal of Forensic Sciences* 48(4):713–716.

Komar, D. 2003b. Twenty-seven years of forensic anthropology casework in New Mexico. *Journal of Forensic Sciences* 48(3):521–524.

Komar, D. 2004. Reassociating commingled remains separated by distance and time: the tale of Simon and Steven. *Proceedings from the Annual Meeting of the American Academy of Forensic Sciences* 10:315.

Komar, D. 2006. The (almost) exhumation of Billy the Kid: why we aren't digging him up (and why you shouldn't either). *Proceedings from the Annual Meeting of the American Academy of Forensic Sciences* 12:318–319.

Komar, D. and O. Beattie. 1998a. Effects of carcass size on decay rates of shade and sun exposed carrion. *Canadian Society of Forensic Science Journal* 31(1):35–43.

Komar, D. and O. Beattie. 1998b. Identifying bird scavenging in fleshed and dry remains. *Canadian Society of Forensic Science Journal* 31(3):177–188.

Komar, D. and O. Beattie. 1998c. Postmortem insect activity may mimic perimortem sexual assault clothing patterns. *Journal of Forensic Sciences* 43(4):792–796.

Komar, D. and S. Lathrop. 2006. Frequencies of morphological features in two contemporary forensic collections: implications for identification. *Journal of Forensic Sciences* 51(5):974–978.

Komar, D., S. Lathrop and W. Potter. 2007. Proposed classes of morphological autopsy findings for decomposed and skeletal remains in mass death investigations. *American Journal of Forensic Medicine and Pathology*, in press.

Komar, D. and W. Potter. 2007. Percentage of body recovered and its effect on identification rates and cause and manner of death determination. *Journal of Forensic Sciences*, 52(3):528–531.

Komar, D., O. Beattie, G. Dowling and B. Bannach. 1999. Hangings in Alberta, with special reference to outdoor hangings with decomposition. *Canadian Society of Forensic Science Journal* 32(2 &3):85–96.

Komlev, V.S., Peyrin, F., Mastrogiacomo, M., Cedola, A., Papadimitropoulos, A., Rustichelli, F. and R. Cancedda. 2006. Kinetics of in vivo bone deposition by bone marrow stromal cells into porous calcium phosphate scaffolds: an x-ray computed microtomography study. *Tissue Engineering* 12(12):3449–3458.

Konigsberg, L.W. and S.M. Hens. 1998. Use of ordinal categorical variables in skeletal assessment of sex from the cranium. *American Journal of Physical Anthropology* 107:97–112.

Konigsberg, L.W. and R.L. Jantz. 2002. Of posteriors, typicality, and individuality in forensic anthropology. *Proceedings of the 54th Annual Meeting of the American Academy of Forensic Sciences* 8:241.

Konigsberg, L.W., S.M. Hens, L.M. Jantz and W.L. Jungers. 1998. Stature estimation and calibration: Bayesian and maximum likelihood perspectives on physical anthropology. *Yearbook of Physical Anthropology* 41:65–92

Konigsberg, L.W., N.P. Herrmann and D.J. Wescott 2002. Commentary on: McBride OG, Dietz MJ, Vennemeyer MT, meadors SA, Benfer RA, Furbee NL. Bootstrap methods of sex determination from the os coxae using the ID3 algorithm. *Journal of Forensic Sciences* 47(2):424–427.

Koot, M.G., N.J. Sauer and T.W. Fenton. 2005. Radiographic human identification using bones of the hand: a validation study. *Journal of Forensic Sciences* 50(2):263–268.

Kosovar Research and Documentation Institute (KODI). 2004. *Missing Persons: The Right to Know*. Prishtina: KODI Report no. 7.

Kovarik, C., D. Stewart and C. Cockerell. 2005. Gross and histological postmortem changes of the skin. *American Journal of Forensic Medicine and Pathology* 26(4):305–308.

Krogman, W.M. 1939. A guide to the identification of human skeletal material. *FBI Law Enforcement Bulletin* 8: 3–31.

Krogman, W.M. 1962. *The Human Skeleton in Forensic Medicine*. Springfield, IL: Charles C. Thomas.

Krogman, W.M. and M.Y. Iscan. 1986. *The Human Skeleton in Forensic Medicine*, 2nd edition. Springfield, IL: Charles C. Thomas.

Krompecher, T., C. Bergerioux, C. Brandt-Casadevall and H.R. Gujer.1983. Experimental evaluation of rigor mortis: IV. Effect of various causes of death on the evolution of rigor mortis. *Forensic Science International* 22(1):1–9.

Kulshrestha, P. and D.K. Satpathy. 2001. Use of beetles in forensic entomology. *Forensic Science International* 120(1–2):15–17.

Kury, G., J. Weiner and J.V. Duval. 2000. Multiple self-inflicted gunshot wounds to the head: report of a case and review of the literature. *American Journal of Forensic Medicine and Pathology* 21(1):32–35.

Kustar, A. 2004. The facial reconstruction of Antal Simon, a Hungarian priest-teacher of the 19th century. *Homo* 55(1–2):77–90.

Kuwait National Committee for Missing and Prisoners of War Affairs. 2004. *Kuwait's Prisoners of War in Iraq: A Humanitarian Tragedy*. Kuwait: National Committee for M. & P.O.W. Affairs.

Lachat, C.K., J.H. Van Camp, P.S. Mamiro, F.O. Wayua, A.S. Opsomer, D.A. Roberfroid and P.W. Kolsteren. 2006. Processing of complementary food does not increase hair zinc levels and growth of infants in Kilosa district, rural Tanzania. *British Journal of Nutrition* 95(1):174–180.

Ladika, S. 2005. South Asia tsunami. DNA helps identify missing in the tsunami zone. *Science* 307(5709): 504.

Lamedin, H., E. Baccino, J.F. Humbert, J.C. Taverier, R.M. Nossintchouk and A Zerilli. 1992. A simple technique for age estimation in adult corpses: the two-criteria dental method. *Journal of Forensic Sciences* 37:1373–1379.

Landis, J.R. and G.G. Koch. 1977. The measurement of observer agreement for categorical data. *Biometrics* 33:159–174.

Lantz, P.E.1994. An atypical, indeterminate-range, cranial gunshot wound of entrance resembling an exit wound. *American Journal of Forensic Medicine and Pathology* 15(1):5–9.

Lantz, P.E., S.H. Sinal, C.A. Stanton and R.G. Weaver. 2004. Perimacular retinal folds from childhood head trauma. *British Medical Journal* 328:754–756.

Lassen, C., S. Hummel and B. Herrman. 1997. Molekulare Geschlectsbestimmung an skelettresten früh- und neugeborener Individualen des Gräberfeldes Aegerten, Schwiez. *Anthropologischer Anzeiger* 55:183–191.

Lasseter, A.E., K.P. Jacobi, R. Farley and L. Hensel. 2003. Cadaver dog and handler team capabilities in the recovery of human remains in the southeastern United States. *Journal of Forensic Sciences* 48(3):617–621.

Lau, G, W.F. Tan and P.H. Tan. 2005. After the Indian Ocean tsunami: Singapore's contribution to the international disaster victim identification effort in Thailand. *Annals of the Academy of Medicine in Singapore* 34(5):341–351.

Lavash, D.R. 1990. *Wilson & the Kid*. College Station, TX: Creative Publishing.

Levine, L.J. 1984. The role of the forensic odontologist in human rights investigations. *American Journal of Forensic Medicine and Pathology* 5(4):317–320.

Lewis, M.E. and G.N. Rutty. 2003. The endangered child: the personal identification of children in forensic anthropology. *Science and Justice* 43(4):201–209.

Limson, K.S. and R. Julian. 2004. Computerized recording of the palatal rugae pattern and an evaluation of its application in forensic identification. *Journal of Forensic Odontostomatology*, 22(1):1–4.

Lindholm, C. 2001. *Culture and Identity*. Boston: McGraw Hill.

Line, W.S., R. B. Stanley and J.H. Choi. 1985. Strangulation: a full spectrum of blunt neck trauma. *Annals of Otology, Rhinology, and Laryngology* 94(6): 542–546.

Linville, J.G., J. Hayes and J.D. Wells. 2004. Mitochondrial DNA and STR analyses of maggot crop contents: effect of specimen preservation technique. *Journal of Forensic Sciences* 49(2):341–344.

Listi, G.A. and H.E. Bassett. 2006. Test of an alternative method for determining sex from the os coxae: applications for modern Americans. *Journal of Forensic Sciences* 51:248–252.

Loth, S.R. and M.Y. Iscan. 1989. Morphological assessment of age in the adult: the thoracic region. In *Age Markers in the Human Skeleton*. M.Y. Iscan, ed. Pp. 105–135. Springfield, IL: Charles C. Thomas.

Loth, S.R. and M. Henneberg. 2001. Sexually dimorphic mandibular morphology in the first few years of life. *American Journal of Physical Anthropology* 115: 179–186.

Love, J.C. and S.A. Symes. 2004. Understanding rib fracture patterns: incomplete and buckle fractures. *Journal of Forensic Sciences* 49(6):1153–1158.

Lovejoy, C.O., R.S. Meindl, T.R. Pryzbeck and R.P. Mensforth. 1985b. Chronological metamorphosis of the auricular surface of the ilium: a new method for the determination of adult skeletal age at death. *American Journal of Physical Anthropology* 68:15–28.

Lovejoy, C.O., R.S. Meindl, R.P. Mensforth and T.J. Barton. 1985b. Multifactorial determination of skeletal age at death, a method and blind tests of its accuracy. *American Journal of Physical Anthropology* 68:1–14.

Lovell, N.C. 1989. Test of Phenice's technique for determining sex from the os pubis. *American Journal of Physical Anthropology* 68:15–28.

Lovis, W.A. 1992. Forensic archaeology as mortuary anthropology. *Social Science and Medicine* 34(2):113–117.

Lowenstein, J.M, J.D. Reuther, D.G. Hood, G. Scheuenstuhl, S.C. Gerlach and D.H. Ubelaker, 2006. Identification of animal species by protein radioimmunoassay of bone fragments and bloodstained stone tools. *Forensic Science International* 159(2–3): 182–188.

Ludwig, J. 2002a. *Handbook of Autopsy Practice*, 3rd edition. Totowa, NJ: Humana Press.

Ludwig, J. 2002b. Principles of autopsy techniques, immediate and restricted autopsies and other special procedures. In *Handbook of Autopsy Practice*, 3rd edition. J. Ludwig, ed. Pp. 3–6. Totowa, NJ: Humana Press.

Luke, J.L., D.T. Reay, J.W. Eisele and H.J. Bonnell. 1985. Correlation of circumstances with pathological findings in asphyxial deaths by hanging: a prospective study of 61 cases from Seattle, WA. *Journal of Forensic Sciences* 30(4):1140–1147.

Lyman, R.L. 1979. *Archaeological Faunal Analysis: A Bibliography*. Pocatello: Idaho State University.

Lynnerup, N.A. 1993. A computer program for the estimation of time of death. *Journal of Forensic Sciences* 38(4):816–820.

Lyon, P. 1969. *The Wild, Wild West; for the Discriminating Reader: A Chilling Illustrated History Presenting the Facts About a Passel of Low-Down Mischievous Personages Including Joaquin Murieta, Wild Bill Hickok, Jesse James, Bat Masterson, Wyatt Earp & Billy the Kid....* New York: Funk & Wagnalls.

Macilwain, C.1995. Forensic team digs up Haiti's deadly past. *Nature* 377(6547):278.

Madea, B. 1992. Estimating time of death from measurement of the electrical excitability of skeletal muscle. *Forensic Science Society* 32(2):117–129.

Maltoni, D., D. Maio, A.K. Jain and S. Prabhaker. 2003. *Handbook of Fingerprint Recognition*. New York: Springer.

Manheim, M. 1997. Decomposition rates of deliberate burials: a case study of preservation. In *Forensic Taphonomy: The Postmortem Fate of Human Remains*. W.D. Haglund, M.H. Sorg, eds. Pp. 469–482. Boca Raton, FC: CRC Press.

Mann, R.W. 1998. Use of bone trabeculae to establish positive identification. *Forensic Science International* 98(1–2):91–99.

Mann, R.W. and S.P. Murphy. 1990. *Regional Atlas of Bone Disease: A Guide to Pathological and Normal Variation in the Human Skeleton*. Springfield, IL: Charles C. Thomas.

Mann, R.W. and D.W. Owsley. 1992. Human osteology: key to the sequence of events in a postmortem shooting. *Journal of Forensic Sciences* 37(5):1386–1392.

Mann, R.W., R.I. Jantz., W.M. Bass and P. Willey. 1991. Maxillary suture obliteration: a visual method for estimating skeletal age. *Journal of Forensic Sciences* 36:781–791.

Maples, W.R. 1986. Trauma analysis by the forensic anthropologist. In *Forensic Osteology: Advances in the Identification of Human Remains*. K.J. Reichs, ed. Pp. 218–228. Springfield, IL: Charles C. Thomas.

Maples, W.R. 1989. The practical application of age estimation techniques. In *Age Markers in the Human Skeleton*. M.Y, Iscan, ed. Pp. 319–324. Springfield, IL: Charles C. Thomas.

Maples, W.R. and M. Browning. 1994. *Dead Men Do Tell Tales*. New York: Doubleday.

Maples, W.R. and P.M. Rice. 1979. Some difficulties in the Gustafson dental age estimations. *Journal of Forensic Sciences* 24:168–172.

Maples, W.R., B.P. Gatliff, H. Ludena, R. Benfer and W. Goza. 1989. The death and mortal remains of Francisco Pizarro. *Journal of Forensic Sciences* 34(4):1021–1036.

Marchetti, D., I. Boschi, M. Polacco and J. Rainio. 2005. The death of Adolf Hitler—forensic aspects. *Journal of Forensic Sciences* 50(5):1147–1153.

Marshall, T.K. and F.E. Hoare. 1962. Estimating the time of death: rectal cooling after death and its mathematical expression. *Journal of Forensic Sciences* 7(1):56–81.

Martin, R. 2000. *Beethoven's Hair*. New York: Broadway Books.

Mason, J.K. and B.N. Purdue. 2000. *The Pathology of Trauma*, 3rd edition. London: Arnold.

Masset, C. 1989. Age estimation on the basis of cranial sutures. In *Age Markers in the Human Skeleton*. M.Y. Iscan, ed. Pp. 71–103. Springfield, IL: Charles C. Thomas.

Masten, J. and J. Strzelczyk. 2001. Admissibility of scientific evidence post-*Daubert*. *Health Physics* 81(6):678–682.

Mayne Correia, P.M. 1997. Fire modification of bone: a review of the literature. In *Forensic Taphonomy: The Postmortem Fate of Human Remains*. W.D. Haglund, M.H. Sorg, eds. Pp. 275–293. Boca Raton, FL: CRC Press.

Maxeiner, H. 1998. "Hidden" laryngeal injuries in homicidal strangulation: how to detect and interpret these findings. *Journal of Forensic Sciences* 43(4):784–791.

McAleese, K. 1998 . The reinterment of Thule Inuit burials and associated artifacts: IdCr-14 Rose Island, Saglek Bay, Labrador. *Etudes Inuit Studies* 22(2):41–52.

McHenry, H. 1992. Body size and proportions in early hominids. *American Journal of Physical Anthropology* 87:407–431.

McKern, T.W. and T.D. Stewart. 1958. *Skeletal Age Changes in Young American Males*. Quartermaster Research and Development Center, Environment Protection Research Division, Technical Report EP-45. Natick, MA: Headquartes, Quartermaster Research and Development Center.

McLaughlin, S. and M.F. Bruce. 1990. The accuracy of sex identification in European skeletal remains using the Phenice criteria. *Journal of Forensic Sciences* 35:1384–1392.

Meadows, L. and R.L. Jantz. 1995. Allometric secular change in the long bones from the 1800s to the present. *Journal of Forensic Sciences* 40:762–767.

Meadows Jantz, L. and R.L. Jantz. 1999. Secular change in long bone length and proportion in the United States, 1800–1970. *American Journal of Physical Anthropology* 110:57–67.

Megyesi, M.S., S.P. Nawrocki and N.H. Haskell. 2005. Using accumulated degree-days to estimate the postmortem interval from decomposed human remains. *Journal of Forensic Sciences* 50(3):618–626.

Meindl, R.S. and C.O. Lovejoy. 1985. Ectocranial suture closure: a revised method for the determination of skeletal age at death based on the lateral–anterior sutures. *American Journal of Physical Anthropology* 68:57–66.

Meindl, R.S., C.O. Lovejoy, R.P. Mensforth and L. Don Carlos. 1985. Accuracy and direction of error in the sexing of the skeleton: implications for paleodemography. *American Journal of Physical Anthropology* 68: 79–85.

Melbye, J. and S.B. Jimenez. 1997. Chain of custody from the field to the courtroom. In *Forensic Taphonomy: The Postmortem Fate of Human Remains*. W.D. Haglund, M.H. Sorg, eds. Pp. 65–76. Boca Raton, FL: CRC Press.

Mellet, J.S., 1992. Location of human remains with ground-penetrating radar. Geological Society of Finland, Special Paper no.16. Pp. 359–365.

Mellor, C. 1980. The Canadian Medical Association Code of Ethics annotated for psychiatrists. The position of the Canadian Psychiatric Association. *Canadian Journal of Psychiatry* 25(5):432–438.

Mel'nikov, I.L. and K.N. Alybaeva. 1995. [The determination of time of death by the content of free amino acids in the liver and lungs of cadavers using high-performance liquid chromatograph: in Russian.] *Sudebnuo-Meditsinskaia Ekspertiza* 38(3):10–13.

Menez, L.L. 2005. The place of a forensic archaeologist at a crime scene involving a buried body. *Forensic Science International* 152(2–3):311–315.

Merbs, C.F. 1983. *Patterns of Activity-Induced Pathology in a Canadian Inuit Population, Archaeological Survey of Canada*, National Museum of Man, Mercury Series, no. 119. Ottawa: National Museums of Canada.

Micozzi, M.S. 1997. Frozen environments and soft tissue preservation. In *Forensic Taphonomy: The Postmortem Fate of Human Remains*. W.D. Haglund, M.H. Sorg, ed. Pp. 171–180. Boca Raton, FL: CRC Press.

Mikko, D. and B.J. Hornsby. 1995. On the cutting edge: II. An identification involving a knife. *AFTE Journal* 27(4):293.

Miller, M.L. 2003. Utilizing ground penetrating radar and three-dimensional imagery to enhance search strategies of buried human remains. *Proceedings from the Annual Meeting of the American Academy of Forensic Sciences* 9:276.

Miller, P.S. 1996. Disturbances in the soil: finding buried bodies and other evidence using ground penetrating radar. *Journal of Forensic Sciences* 41(4):648–652.

Milner, G.R. and C.S. Larsen. 1991. Teeth as artifacts of human behavior: intentional mutilation and accidental modification. In *Advances in Dental Anthropology*. M.A. Kelley, C.S. Larsen, eds. Pp. 357–378. New York: Wiley-Liss.

Mitchell, R.J., M. Kreskas, E. Baxter, L. Buffalino and R.A. Van Oorschot. 2006. An investigation of sequence deletions of amelogenin (AMELY), a Y-chromosome locus commonly used for gender determination. *Annals of Human Biology* 33(2):227–240.

Mittler, D. M. and S.G. Sheridan. 1992. Sex determination in subadults using auricular surface morphology: a forensic science perspective. *Journal of Forensic Sciences* 37:1068–1075.

Moessens, A.A., J.E. Starrs, C.E. Henderson and F.E. Inbau. 1995. *Scientific Evidence in Criminal and Civil Cases*, 4th edition. Westbury, NY: Foundation Press.

Montiel, R., A. Malgosa and P. Francalacci. 2001. Authenticating ancient human mitochondrial DNA. *Human Biology* 73(5):689–713.

Moore, G. 2006. *History of Fingerprints*. Electronic document, http://onin.com/fp/fphistory.html, accessed June 28, 2006.

Moorrees, C.F.A., E.A. Fanning and E.E. Hunt. 1963a. Age variation of formation stages for ten permanent teeth. *Journal of Dental Research* 42:1490–1501.

Moorrees, C.F.A., E.A. Fanning and E.E. Hunt. 1963b. Formation and resorption of three deciduous teeth in children. *American Journal of Physical Anthropology* 21:205–213.

Morrison, W.M. 1958. *Billy the Kid: Las Vegas Newspaper Accounts of His Career, 1880–1881*. Waco, TX: Morrison Books.

Morse, D. and R.C. Dailey. 1985. The degree of deterioration of associated death scene materials. *Journal of Forensic Sciences* 30(1):119–127.

Morse, D., D. Crusoe and H.G. Smith. 1976. Forensic archaeology. *Journal of Forensic Sciences* 21(2):323–332.

Motovilin, E.G. 1965. [Determination of some properties of the blade from stab injuries of bone]: in Russian. *Sudebno-Meditsinskaia Ekspertiza* 8(3):55–56.

Mulhern, D.M. and D.H. Ubelaker. 2001. Differences in osteon banding between human and nonhuman bone. *Journal of Forensic Sciences* 46(2):220–222.

Muller-Bolla, M., J. Laugier, L. Lupi-Pégurier, M. Bertrand, P. Staccini, M. Bolla, G. Quatrehomme and V. Alunni-Perret. 2005. Scanning electron microscopy analysis of experimental bone hacking trauma. *Journal of Forensic Sciences* 50(4): 796–801.

Mullin, R.N. 1967. *The Boyhood of Billy the Kid*. El Paso, TX: Texas Western Press.

Mundorff, A.Z. 2003. The role of anthropology during the identification of victims from the World Trade Center disaster. *Proceedings from the Annual Meeting of the American Academy of Forensic Sciences* 9: 277–278.

Mundorff, A.Z., R. Shaler, E.T. Bieschke and E. Mar. 2005. Marrying of anthropology and DNA: essential for solving complex commingling problems in cases of extreme fragmentation. *Proceedings from the Annual Meeting of the American Academy of Forensic Sciences* 11:315–316.

Murail, P., J. Bruzek, F. Houet and E. Cunha. 2005. DSP: a tool for probabilistic sex diagnosis using worldwide variability in hip-bone measurements. *Bulletins et Mémoires de la Société d' Anthropologie de Paris* 17:167–176.

Murphy, W.A. and G.S. Gantner. 1982. Radiologic examination of anatomic parts and skeletonized remains. *Journal of Forensic Sciences* 27(1):9–18.

Murphy, W.A., F.G. Spruill and G.S. Gantner. 1980. Radiologic identification of unknown human remains. *Journal of Forensic Sciences* 25(4):727–735.

Murray, R.C. and J.C. Tedrow. 1975. *Forensic Geology: Earth Sciences and Criminal Investigations*. New Brunswick, NJ: Rutgers University Press.

Muthusubramanian, M., K.S. Limson and R. Julian. 2005. Analysis of rugae in burn victims and cadavers to simulate rugae identification in cases of incineration and decomposition. *Journal of Forensic Odontostomatology* 23(1):26–29.

Myers, J.C., M.I. Okoye, D. Kiple, E.H. Kimmerle and K.J. Reinhard. 1999. Three-dimensional (3-D) imaging in post-mortem examinations: elucidation and identification of cranial and facial fractures in victims of homicide utilizing 3-D computerized imaging reconstruction techniques. *International Journal of Legal Medicine* 113(1):33–37.

Nawrocki, S.P. 1998. Regression formulae for estimating age at death from cranial suture closure. In *Forensic Osteology: Advances in the Identification of Human Remains* 2nd edition. K. J. Reichs, ed. Pp. 276–292. Springfield, IL: Charles C. Thomas.

Nawrocki, S.P., J.E. Pless, D.A. Hawley and S.A. Wagner. 1997. Fluvial transport of human crania. In *Forensic Taphonomy: The Postmortem Fate of Human Remains*. W.D. Haglund, M.H. Sorg, eds. Pp. 529–552. Boca Raton, FL: CRC Press.

Nightingale, E.O. 1990. The role of physicians in human rights. *Law, Medicine and Health Care* 18(1–2):132–139.

Nokes, L.D., T. Flint, S. Jaafar and B.H. Knight. 1992. The use of either the nose or outer ears as a means of determining the postmortem period of a human corpse. *Forensic Science International* 54(2):153–158.

Nolan, F. 1998. *The West of Billy the Kid*. Norman: University of Oklahoma Press.

O'Brien, T.G. 1997. Movement of bodies in Lake Ontario. In *Forensic Taphonomy: The Postmortem Fate of Human Remains*. W.D. Haglund, M.H. Sorg, eds. Pp. 559–566. Boca Raton, FL: CRC Press.

O'Connor, R. 1960. *Pat Garrett: A Biography of the Famous Marshal and the Killer of Billy the Kid*. Garden City, NY: Doubleday.

O'Connor, T.P. 2003. *The Analysis of Urban Animal Bone Assemblages: A Handbook for Archaeologists*. York: York Archaeological Trust.

Ogino, T., H. Ogino and B. Nagy. 1985. Application of aspartic acid racemization to forensic odontology: post mortem designation of age at death. *Forensic Science International* 29:259–267.

Ohtani, S., Y. Matsushima, Y. Kobayashi and T. Yamamoto. 2002. Age estimation by measuring the racemization of aspartic acid from total amino acid content of several types of bone and rib cartilage: a preliminary account. *Journal of Forensic Sciences* 47:32–36.

Ohtani, S., I. Abe, and T. Yamamoto. 2005. An application of D- and L-aspartic acid mixtures as standard specimens for the chronological age estimation. *Journal of Forensic Sciences* 50:1298–1302.

O'Rahilly, R. and F. Müller. 2001. *Human Embryology and Teratology*, 3rd edition. New York: Wiley-Liss.

Orlowski, J.P. and J.K. Vinicky. 1993. Conflicting cultural attitudes about autopsies. *Journal of Clinical Ethics* 4(2):195–197.

Ortner, D.J. 2003. *Identification of Pathological Conditions in Human Skeletal Remains*, 2nd edition. San Diego: Academic Press.

Osborne, D.L., T.L. Simmons and S.P. Nawrocki. 2004. Reconsidering the auricular surface as an indicator of age at death. *Journal of Forensic Sciences* 49:905–911.

OSCE. 1999a. *Kosovo/Kosova—As Seen, As Told*, Pt. I. Warsaw: OSCE Office for Democratic Institutions and Human Rights.

OSCE. 1999b. *Kosovo/Kosova—As Seen, As Told*, Pt. II. Warsaw: OSCE Office for Democratic Institutions and Human Rights.

Otero, M.A.1998. *The Real Billy the Kid: With New Light on the Lincoln County War*. Houston, TX: Arte Público Press.

Ousley, S. 1995. Should we estimate biological or forensic stature? *Journal of Forensic Sciences* 40:768–773.

Ousley, S. and J. Hefner. 2005. Morphoscopic traits and the statistical determination of ancestry. *Proceedings of the Annual Meeting of the American Academy of Forensic Sciences* 11:291–292.

Ousley, S.D. and R.L Jantz. 1996. *FORDISC 2.0: Personal Computer Forensic Discriminant Functions*. Knoxville: University of Tennessee.

Ousley, S.D. and R.L Jantz. 1998. The Forensic Data Bank: documenting skeletal trends in the United States. In *Forensic Osteology*, 2nd edition. K.J. Reichs, ed. Pp. 441–458. Springfield, IL: Charles C. Thomas.

Ousley, S. and R.L. Jantz. 2002. Social races and human populations: why forensic anthropologists are good at identifying races. *American Journal of Physical Anthropology*, annual meeting issue 34:121.

Ousley, S.D. and R.L. Jantz. 2005. FORDISC 3.0: *Personal Computer Forensic Discriminant Functions*. Knoxville: University of Tennessee.

Owsley, D.W. 1995. Techniques for locating burials, with emphasis on the probe. *Journal of Forensic Sciences* 40(5):735–740.

Owsley, D.W. and R.W. Mann. 1992. Positive personal identity of skeletonized remains using abdominal and pelvic radiographs. *Journal of Forensic Sciences* 37(1):332–336.

Owsley, D.W., D.H. Ubelaker, M.M. Houck, K.L. Sandness, W.E. Grant, E.A. Craig, T.J. Woltanski and N. Peerwani. 1995. The role of forensic anthropology in the recovery and analysis of Branch Davidian compound victims: techniques of analysis. *Journal of Forensic Sciences* 40(3):341–348.

Pagaduan-Lopez, J. 1991. Medical professionals and human rights in the Philippines. *Journal of Medical Ethics* 17(Suppl):42–50.

Paine, R.R. and B.P. Brenton. 2006. Dietary health does affect histological age assessment: an evaluation of the Stout and Paine (1992) age estimation equation using secondary osteons from the rib. *Journal of Forensic Sciences* 51:489–492.

PaleoDNA. 2006. Electronic document, www.paleodna.com/pria/index.htm, accessed May 1, 2006.

Payne, S. 2004. Handle with care: thoughts on the return of human bone collections. *Antiquity* 78(300):419–420.

Pear, D.L. 1974. Skeletal manifestation of the lymphomas and leukemias. *Seminars in Roentgenology* 9(3):229–240.

Perkins, H.S., K.J. Shepherd, J.D. Cortez and H.P. Hazuda. 2005. Exploring chronically ill seniors' attitudes about discussing death and postmortem medical procedures. *Journal of the American Geriatrics Society* 53(5):895–900.

Perry, W.L., W.M. Bass, W.S. Riggsby and K. Sirotkin. 1988. The autodegradation of deoxyribonucleic acid (DNA) in human rib bone and its relationship to the time interval since death. *Journal of Forensic Sciences* 33(1):144–153.

Petersen, H.D. and O.M. Vedel. 1994. Assessment of evidence of human rights violations in Kashmir. *Forensic Science International* 68(2):103–115.

Peterson, J., B.A. Shook, M.J. Wells and M. Rodriguez. 2006. Cupric keratosis: green seborrheic keratoses secondary to external copper exposure. *Cutis* 77(1):39–41.

Peyrin, F., C. Muller and Y. Carillon. 2001. Synchrotron radiation (mu)CT: a reference tool for the characterization of bone samples. In *Noninvasive Assessment of Trabecular Bone Architecture and the Competence of Bone*. S. Magundar, B.K. Bay, eds. Pp. 129–142. New York: Kluwer/Plenum.

Pfeiffer, S., R. Lazenby and J. Chiang. 1995. Brief communication. Cortical remodeling data are affected by sampling location. *American Journal of Physical Anthropology* 96:89–92.

Phenice, T. 1969. A newly developed visual method of sexing in the os pubis. *American Journal of Physical Anthropology* 30:297–301.

Phillips, V.M. and C.F. Scheepers. 1990. Comparison between fingerprint and dental concordant characteristics. *Journal of Forensic Odontostomatology* 8(1):17–19.

Phillips, V.M. and N.A. Smuts. 1996. Facial reconstruction: utilization of computerized tomography to measure facial tissue thickness in a mixed racial population. *Forensic Science International* 83(1):51–59.

Physicians for Human Rights. 1996. International Forensic Program, Rwanda 1995–1996, Project Findings. Electronic document, http://phrusa.org/research/forensics/rwanda/findings2.html, accessed June 28, 2006.

Pickering, R.B. and D.C. Bachman. 1997. *The Use of Forensic Anthropology*. Boca Raton, FL: CRC Press.

Pietrusewsky, M. 2000. Metric analysis of skeletal remains: methods and applications. In *Biological Anthropology of the Human Skeleton*. M.A. Katzenberg, S.R. Saunders, eds. Pp. 375–415. New York: Wiley-Liss.

Piette, M. 1989. The effect of the post-mortem interval on the level of creatine in vitreous humour. *Medicine, Science and the Law* 29(1):47–54.

Poe, J.W. 1933. *The Death of Billy the Kid*. Boston: Houghton-Mifflin.

Poe, J.W. and E.A. Brininstool. 1922. *The True Story of the Death of "Billy the Kid": Notorious New Mexico Outlaw*. Los Angeles: private printing by E.A. Brininstool.

Pollanen, M.S. and D.A. Chiasson. 1996. Fracture of the hyoid bone in strangulation: comparisons of fractured and unfractured hyoids from victims of strangulation. *Journal of Forensic Sciences* 41(1):110–113.

Pope, E.J. and O.C. Smith. 2004. Identification of traumatic injury in burned cranial bone: an experimental approach. *Journal of Forensic Sciences* 49(3):431–440.

Porter, A.M.W. 2002. Estimation of body size and physique from hominin skeletal remains. *Homo* 53:17–38.

Powell, M. L., D.C. Cook, G. Bogdan, J.E. Buikstra, M.M. Castro, P.D. Horne, D.R. Hunt, R.T. Koritzer, S.F.M. Souza, M.K. Sandford, L. Saunders, G.A.M. Sene, L. Sullivan and J.J. Swetnam. 2006. Invisible hands: women in bioarchaeology. In *Bioarchaeology: The Contextual Analysis of Human Remains*. J.E. Buikstra, L.A. Beck, eds. Burlington, MA: Elsevier-Academic Press.

Prahlow, J.A. and J.L. McClain. 1997. Lesions that simulate gunshot wounds. *Journal of Clinical Forensic Medicine* 4(3):121–125.

Pretty, I.A. and D.P. Hildebrand. 2005. The forensic and investigative significance of reverse paternity testing with absent maternal sample. *American Journal of Forensic Medicine and Pathology* 26(4):340–342.

Pretty, I.A., R.J. Pretty, B.R. Rothwell and D. Sweet. 2003. The reliability of digitized radiographs for dental identification: a web-based study. *Journal of Forensic Sciences* 48(6):1325–1330.

Price, G.G. 1940. *Death Comes to Billy the Kid*. Greensburg, KS: Signal Publishing.

Priestley, L. and M. Peterson. 1989. *Billy the Kid: The Good Side of a Bad Man*. Las Cruces, NM: Arroyo Press.

Primorac, D. 2004. The role of DNA technology in identification of skeletal remains discovered in mass graves. *Forensic Science International* 146(Suppl.):S63–S64.

Prince, D.A. and D.H. Ubelaker. 2002. Application of Lamendin's adult dental aging technique to a diverse skeletal sample. *Journal of Forensic Sciences* 47: 107–116.

Pugh, M.B., ed. 2000. *Stedman's Medical Dictionary*, 27th edition. Philadelphia: Lippincott Williams & Wilkins.

Quatrehomme, G. and M.Y. Iscan. 1997. Bevelling in exit gunshot wounds in bones. *Forensic Science International* 89(1–2):93–101.

Quatrehomme, G. and M.Y. Iscan. 1998. Gunshot wounds to the skull: comparison of entries and exits. *Forensic Science International* 94(1–2):141–146.

Quatrehomme, G. and M.Y. Iscan. 1999. Characteristics of gunshot wounds in the skull. *Journal of Forensic Sciences* 44(3):568–576.

Quatrehomme, G., A. Lacoste, P. Bailet, G. Grevin and A. Ollier. 1997. Contribution of microscopic plant anatomy to postmortem bone dating. *Journal of Forensic Sciences* 42(1):140–143.

Rainio, J., K. Lalu and A. Penttila. 2001a. Independent forensic autopsies in an armed conflict: investigation of the victims from Racak, Kosovo. *Forensic Science International* 116(2–3):171–185.

Rainio, J., K. Lalu, H. Ranta, K. Takamaa and A. Penttila. 2001b. Practical and legal aspects of forensic autopsy expert team operations. *Legal Medicine (Tokyo)* 3(4):220–232.

Rainov, N.G. and W.L. Burket. 1994. An unusual suicide attempt using a circular saw. *International Journal of Legal Medicine* 106:223–224.

Ransom, D. 2005. Human identification of victims of natural disasters. *Journal of Law and Medicine* 12(3):273–276.

Rao, V.J. and R. Hart. 1983. Tool mark determination in cartilage of stabbing victim. *Journal of Forensic Sciences* 28(3):794–799.

Rasch, P.J. and R.K. DeArment. 1995. *Trailing Billy the Kid*. Laramie, WY: National Association for Outlaw and Lawman History; Stillwater, OK: Western Publications.

Ratner, S.R. 2006. Categories of war crimes. Electronic document, www.crimesofwar. org/thebook/war- crimes-categories.html, accessed June 10, 2006.

Raul, J.S., L. Berthelon, A. Tracqui and B. Ludes. 2002. Penetration of a piece of World War II rifle grenade initially suspected as a stab wound. *American Journal of Forensic Medicine and Pathology* 23(3):277–280.

Rees, P.O. and K.R. Cundy. 1969. A method for the comparisons of tool marks and other surface irregularities. *Journal of the Forensic Society* 9:153–156.

Reichs, K.J. 1992. Forensic anthropology in the 1990s. *American Journal of Forensic Medicine and Pathology* 13(2):146–153.

Reichs, K.J. 1995. A professional profile of diplomates of the American Board of Forensic Anthropology: 1984–1992. *Journal of Forensic Sciences* 40(2):176–182.

Reichs, K.J. 1997. *Déja Dead*. New York: Scribner.

Reichs, K.J. 1998a. Forensic anthropology: a decade of progress. In *Forensic Osteology: Advances in the Identification of Human Remains* 2nd edition. K. J. Reichs, ed. Pp. 13–33. Springfield IL: Charles C. Thomas.

Reichs, K.J., ed. 1998b. *Forensic Osteology: Advances in the Identification of Human Remains*, 2nd edition. Springfield: Charles C. Thomas.

Reichs, K.J. 1998c. Postmortem dismemberment: recovery, analysis and interpretation. In *Forensic Osteology: Advances in the Identification of Human Remains*, 2nd edition. K.J. Reichs, ed. Pp. 353–388. Springfield IL: Charles C. Thomas.

Reis, C., A.T. Ahmed, L.L. Amowitz, A.L. Kushner, M. Elahi and V. Iacopino. 2004. Physician participation in human rights abuses in southern Iraq. *Journal of the American Medical Association* 291(12):1480–1486.

Rennick, S.L., T.W. Fenton and D.R. Foran. 2005. The effects of skeletal preparation techniques on DNA from human and non-human bone. *Journal of Forensic Sciences* 50(5):1016–1019.

Renz, H. and R.J. Radlanski. 2006. Incremental lines in root cementum of human teeth—a reliable age marker? *Homo* 57:29–50.

Reuhl, J. and H. Bratzke. 1999. Death caused by a chain saw—homicide, suicide or accident? *Forensic Science International* 105:45–59.

Reynolds, B. 1980. Repatriation of Canadian Indian and Eskimo collections. In *Preserving Indigenous Cultures: A New Role for Museums*. R. Edwards, J. Stewart, eds. Pp. 161–167. Canberra: Australian Government Publishing Service.

Rhine, J.S. 1990a. Coming to terms with facial reproduction. *Journal of Forensic Sciences* 35(4):960–963.

Rhine, Stanley. 1990b. Non-metric skull racing. In *Skeletal Attribution of Race*. G.W. Gill, S. Rhine, eds. Pp. 9–20. Albuquerque, NM: Maxwell Museum of Anthropology Papers, no. 4.

Rickards, O., C. Martinez-Labarga, M. Favaro, D. Frezza and F. Mallegni. 2001. DNA analyses of the remains of the Prince Branciforte Barresi family. *International Journal of Legal Medicine* 114(3):141–146.

Riddick, L. 1998. Identification of the dead. In *Forensic Radiology*. B.G. Brogdon, ed. Pp. 55–61. Boca Raton, FL: CRC Press.

Ríos Frutos, L. 2002. Determination of sex from the clavicle and scapula in a Guatemalan contemporary rural indigenous population. *Journal of Forensic Medicine and Pathology* 23:284–288.

Rissech, C., G.F. Estabrook, E. Cunha and A. Malgosa. 2006. Using the acetabulum to estimate age at death of adult males. *Journal of Forensic Sciences* 51(2):213–229.

Robbins, L.M. 1978. The individuality of human footprints. *Journal of Forensic Sciences* 23(4): 778–785.

Robbins, L.M. 1984. Making tracks. *Law Enforcement Communications* 12(14–15).

Robbins, L.M. 1985. *Footprints, Collections, Analysis and Interpretation*. Springfield, IL: Charles C. Thomas.

Robbins, L.M. 1986. Estimating height and weight from size of footprints. *Journal of Forensic Sciences* 31(1):143–152.

Roberts, A. and R. Guieff. 2000. *Documents on the Laws of War*. Oxford: Oxford University Press.

Robling, A.G. and S.D. Stout. 2000. Histomorphometry of human cortical bones: applications to age estimation. . In *Biological Anthropology of the Human Skeleton*. M.A. Katzenberg, S.R. Saunders, eds. Pp. 187–213. New York City: Wiley-Liss.

Rodriguez, W.C. 2005. Methods and techniques for sorting commingled remains: anthropological and physical attributes. *Proceedings of the Annual Meeting of the American Academy of Forensic Sciences* 11:312.

Rodriguez, W.C. and W.M. Bass. 1983. Insect activity and its relationship to decay rates of human cadavers in East Tennessee. *Journal of Forensic Sciences* 28(2):423–432.

Rodriguez, W.C. and W.M. Bass. 1985. Decomposition of buried bodies and methods that may aid in their location. *Journal of Forensic Sciences* 30(3):836–852.

Rogers, N.L., K. Field, R.C. Froede and B. Towne. 2005. The belated autopsy and identification of an eighteenth-century naval hero—the saga of John Paul Jones. *Journal of Forensic Sciences* 50(2):487.

Rogers, T. 1999. A visual method of determining the sex of skeletal remains using the distal humerus. *Journal of Forensic Sciences* 44:57–60.

Rogers T.L. 2004. Crime scene ethics: souvenirs, teaching materials, and artifacts. *Journal of Forensic Sciences* 49(2):307–311.

Rogers, T.L. 2005a. Determining the sex of human remains through cranial morphology. *Journal of Forensic Sciences* 50(3):493–500.

Rogers, T.L. 2005b. Recognition of cemetery remains in a forensic context. *Journal of Forensic Sciences* 50(1):5–11.

Rogers, T.L. and T.T. Allard. 2004. Expert testimony and positive identification of human remains through cranial suture patterns. *Journal of Forensic Sciences* 49(2):203–207.

Rogers, T. and S. Saunders. 1994. Accuracy of sex determination using morphological traits of the human pelvis. *Journal of Forensic Sciences* 39:1047–1056.

Rogev, M.E. 1993. The medicolegal identification of Josef Mengele. *Legal Medicine*: 115–150.

Rollo, F., M. Mascetti and R. Cameriere. 2005. Titian's secret: comparison of Eleonara Gonzaga della Rovere's skull with the Uffizi portrait. *Journal of Forensic Sciences* 50(3):602–607.

Rosenberg, M.L., L.E. Davidson, J.C. Smith, A.L. Berman, H. Buzbee and G. Gantner. 1988. Operational criteria for the determination of suicide. *Journal of Forensic Sciences* 33:1445–1456.

Ross, A.H. 1996. Caliber estimation from cranial entrance defect measurements. *Journal of Forensic Sciences* 41(4):629–633.

Ross, A. H. 2002. Population-specific identification criteria for Cuban Americans in south Florida. *Proceedings of the Annual Meeting of the American Academy of Forensic Sciences* 8:233.

Rossen, J.M.T. 1985. *Billy the Kid: The Untold Story*. Newport Beach, CA: Teel Rossen Publishing.

Rothschild, B.M. and C. Rothschild. 1995. Comparison of radiologic and gross examination for detection of cancer in defleshed skeletons. *American Journal of Physical Anthropology* 96(4):357–363.

Rothschild, M.A. and V. Schneider. 1997. On the temporal onset of postmortem animal scavenging. "Motivation" of the animal. *Forensic Science International* 89(1–2):57–64.

Rouge-Maillart, C., N. Telmon, C. Rissech, A. Malgosa and D. Rouge. 2004. The determination of male adult age at death by central and posterior coxal analysis—a preliminary study. *Journal of Forensic Sciences* 49(2):203–209.

Ruff, C.B. 2000. Body mass prediction from skeletal frame size in elite athletes. *American Journal of Physical Anthropology* 113: 507–517.

Ruff, C.B. 2005. Body mass prediction from stature and bi-iliac breadth in two high latitude populations, with application to earlier higher latitude humans. *Homo* 48: 381–392.

Sagan, C. 1993. Science and pseudo-science. In *Headline News, Science Views II*. D. Jarmul, ed. Pp. 3–5. Washington, DC: National Academy Press.

Saks, M.J. and J.J. Koehler. 2005. The coming paradigm shift in forensic identification science. *Science* 309(5736):892–895.

Sanderson, P.L., I.C. Cameron, G.R. Holt and D. Stanley. 1997. Ulnar variance and age. *Journal of Hand Surgery* 22(1):21–24.

Sanner, M.A. 1994. In perspective of the declining autopsy rate. Attitudes of the public. *Archive of Pathology and Laboratory Medicine* 118(9):878–883.

Sanner, M.A. 1997. Registered bone marrow donors' views on bodily donations. *Bone Marrow Transplantation* 19(1):67–76.

Sansare, K. 1995. Forensic odontology, historical perspective. *Indian Journal of Dental Research* 6(2):55–57.

Sauer, N.J. 1998. The timing of injuries and manner of death: distinguishing among antemortem, perimortem, and postmortem trauma. In *Forensic Osteology: Advances in the Identification of Human Remains*, 2nd edition. K.J. Reichs, ed. Pp. 321–332. Springfield IL: Charles C. Thomas.

Sauer, N.J., S.S. Dunlap and L.R. Simson. 1988. Medicolegal investigation of an eighteenth-century homicide. *American Journal of Forensic Medicine and Pathology* 9(1):66–73.

Saul, Frank P. 1972. The human skeletal remains of Altar de Sacrificios: an osteobiographic analysis. Vol. 63(2), *Memoirs of the Peabody Museum of Archaeology and Ethnology*. Peabody Museum of Archaeology and Ethnology, Harvard University, Cambridge, MA.

Saunders, S.R. 2000. Subadult skeletons and growth-related studies. In *Biological Anthropology of the Human Skeleton*. M.A. Katzenberg, S.R. Saunders, eds. Pp. 135–161. New York: Wiley-Liss.

Saunders, S.R., C. Fitzgerald, T. Rogers, C. Dudar and H. McKillop. 1992. A test of several methods of skeletal age estimation using a documented archaeological sample. *Canadian Society of Forensic Science Journal* 25:97–118.

Scanland, J.M. 1952. *Life of Pat F. Garrett and the Taming of the Border Outlaw.* Colorado Springs, CO: J.J. Lipsey.

Schaefer, M.C. and S.M. Black. 2005. Comparison of ages of epiphyseal union in North American and Bosnian skeletal material. *Journal of Forensic Sciences* 50:777–784.

Scheuer, L. 2002. Brief communication: a blind test of mandibular morphology for sexing mandibles in the first few years of life. *American Journal of Physical Anthropology* 119:189–191.

Scheuer, L. and S. Black. 2000. *Developmental Juvenile Osteology.* London: Academic Press Ltd.

Scheuer, L. and S. Black. 2004. *The Juvenile Skeleton.* London: Elsevier Academic Press.

Scheuer, L. and N. Elkington. 1993. Sex determination from metacarpals and the first proximal phalanx. *Journal of Forensic Sciences* 38:769–778.

Schick, Elizabeth A., ed. 1997. Clyde Collins Snow. *Current Biography Yearbook: 1997.* Pp. 526–529. New York: H.W. Wilson.

Schmitt, A. 2004. Age-at-death assessment using the os pubis and auricular surface of the ilium: a test on an identified Asian sample. *International Journal of Osteoarchaeology* 14:1–6.

Schmitt, A., P. Murail, E. Cunha and D. Rougé. 2002. Variability of the pattern of aging on the human skeleton: evidence from bone indicators and implication on age at death estimation. *Journal of Forensic Sciences* 47(6):1203–1209.

Schmitt, A, E. Cunha, and J. Pinheiro, eds. 2006. *Forensic Anthropology and Medicine: Complementary Sciences from Recovery to Cause of Death.* Totowa NJ: Humana Press.

Schoenly, K., K. Griest and S. Rhine. 1991. An experimental field protocol for investigating the postmortem interval using multidisciplinary indicators. *Journal of Forensic Sciences* 36:1395–1415.

Schuck, M., G. Beier, E. Liebhardt and W. Spann. 1979. On the estimation of lay-time by measurements of rigor mortis. *Forensic Science International* 149(3):171–176.

Schuetz, J.E. and L.S. Lilley, eds. 1999. *The O.J. Simpson Trials: Rhetoric, Media and the Law.* Carbondale: Southern Illinois University Press.

Schulter-Ellis, F.P. 1980. Evidence of handedness on documented skeletons. *Journal of Forensic Sciences* 25:624–630.

Schultz, J.J. 2006. Forensic GPR: using ground-penetrating radar to search for buried bodies. *Proceedings of the Annual Meeting of the American Academy of Forensic Sciences* 12:278.

Schwidetzky, I. 1954. Forensic anthropology in Germany. *Human Biology* 26:1–20.

Seeram, E. 2001. *Computed Tomography: Physical Principles, Clinical Applications and Quality Control,* 2 nd edition. Philadelphia: Saunders.

Segerberg-Konttinen, M. 1984. Suicide by the use of a chain saw. *Journal of Forensic Sciences* 29:1249–1252.

Shahrom, A.W., P. Vanezis, R.C. Chapman, A. Gonzales, C. Blenkinsop and M.L. Rossi. 1996. Techniques in facial identification: computer-aided facial reconstruction using a laser scanner and video superimposition. *International Journal of Legal Medicine* 108(4):194–200.

Shen, M., X.Q. Liu, W. Liu, P. Xiang and B. Shen. 2006. [Study on appraisement and determination of GHB levels in hair: in Chinese]. *Fa Yi Xue Za Zhi* 22(1):48–51.

Shumard, G. 1969. *Billy the Kid: The Robin Hood of Lincoln County?* Deming, NM: Cambray Enterprises.

Siegel, S. 1956. *Social Statistics for the Behavioral Sciences*. London: McGraw-Hill.

Silvaram, S., V.N. Sehgal and R.P. Singh. 1977. Unusual instrument marks on bones. *Forensic Science* 9:109–110.

Simon, R.I. 1998. Murder masquerading as suicide: postmortem assessment of suicide risk factors at the time of death. *Journal of Forensic Sciences* 43(6):1119–1123.

Simonsen, J. 1988. Patho-anatomic findings in neck structures in asphyxiation due to hanging: a survey of 80 cases. *Forensic Science International* 38(1–2):83–91.

Siringo, C.A. 1920. *History of "Billy the Kid."* Santa Fe, NM.

Skinner, M.1987. Planning the archaeological recovery of evidence from recent mass graves. *Forensic Science International* 34(4):267–287.

Skinner, M. 1988. Method and theory in deciding identity of skeletonized human remains. *Canadian Society of Forensic Science Journal* 21(3):114–134.

Skinner, M. and R.A. Lazenby. 1983. *Found! Human Remains: A Field Manual for the Recovery of the Recent Human Skeleton*. Burnaby, BC, Canada: Archaeology Press, Simon Fraser University.

Skinner, M. and T. Dupras. 1993. Variation in birth timing and location of the neonatal line in human enamel. *Journal of Forensic Sciences* 38:1383–1390.

Skinner, M., D. Alempijevic and M. Djuric-Srejic. 2003. Guidelines for international forensic bio-archaeology monitors of mass grave exhumations. *Forensic Science International* 134(2–3):81–92.

Skolnick, A.A.1992. Game's afoot in many lands for forensic scientists investigating most-extreme human rights abuses. *Journal of the American Medical Association* 268(5):579–80; 583.

Slaus, M., D. Strinovic, J. Skavic and V. Petrovecki. 2003. Discriminant function sexing of fragmentary and complete femora: standards for contemporary Croatia. *Journal of Forensic Sciences* 48:1–4.

Sledzik, P.S. and M.S. Micozzi. 1997. Autopsied, embalmed and preserved human remains: distinguishing features in forensic and historic contexts. In *Forensic Taphonomy: The Postmortem Fate of Human Remains*. W.D. Haglund, M.H. Sorg, eds. Pp. 483–496. Boca Raton, FL: CRC Press.

Sledzik, P.S. and S. Ousley. 1991. Analysis of six Vietnamese trophy skulls. *Journal of Forensic Sciences* 36(2):520–530.

Smith, B.H. 1991. Standards of human tooth formation and dental age assessment. In *Advances in Dental Anthropology*. M. Kelley, C.S. Larsen, eds. Pp. 143–168. New York: Wiley-Liss.

Smith, D.R., K.G. Limbird and J.M. Hoffman. 2002. Identification of human remains by comparison of bony details of the cranium using computerized tomographic (CT) scans. *Journal of Forensic Sciences* 47(5):937–939.

Smith, G.S. 1979. *Mammalian Zooarchaeology, Alaska: A Manual for Identifying and Analyzing Mammal Bones from Archaeological Sites in Alaska*. Fairbanks: University of Alaska Cooperative Park Studies Unit.

Smith, L. 2004. The repatriation of human remains: problem or opportunity? *Antiquity* 78(300):404–413.

Smith, P. and G. Avishai. 2005. The use of dental criteria for estimating postnatal survival in skeletal remains of infants. *Journal of Archaeological Science* 32:83–89.

Smith, S.A. 1939. Studies in identification, no. 3. *Police Journal of London* 12:274–285.

Smith, S.A. 1959. *Mostly Murder*. New York: McKay.

Snodgrass, J.J. and A.Galloway. 2003. Utility of dorsal pits and pubic tubercle height in parity assessment. *Journal of Forensic Sciences* 48:1226–1230.

Snow, C.C. 1982. Forensic anthropology. *Annual Review of Anthropology* 11:97–131.

Snow, C.C., B.P. Gatliff and K.R. McWilliams. 1970. Reconstruction of facial features from the skull: an evaluation of its usefulness in forensic anthropology. *American Journal of Physical Anthropology* 33(2):221–228.

Snow, C.C., L. Levine, L. Lukash, L.G. Tedeschi, C. Orrego and E. Stover. 1984. The investigation of the human remains of the "disappeared" in Argentina. *American Journal of Forensic Medicine and Pathology* 5(4):297–299.

Snyder, L.1977. *Homicide Investigation: Practical Information for Coroners, Police Officers and Other Investigators*, 3rd edition. Springfield, IL: Charles C. Thomas.

Sonnichsen, C.L and W.V. Morrison. 1955. *Alias Billy the Kid "...I Want to Die a Free Man...."* Albuquerque: University of New Mexico Press.

Soomer, H., H. Ranta, M.J. Lincoln, A. Tenttila and E. Leibur. 2003. Reliability and validity of eight dental age estimation methods for adults. *Journal of Forensic Sciences* 48(1):149–152.

Sorg, M.H., J.H. Dearborn, E.I. Monahan, H.F. Ryan, K.G. Sweeney and E. David. 1997. Forensic taphonomy in marine contexts. In *Forensic Taphonomy: The Postmortem Fate of Human Remains*. W.D. Haglund, M.H. Sorg, eds. Pp. 567–604. Boca Raton, FL: CRC Press.

Sparks, D.L., P.R. Oeltgen, R.J. Kryscio and J.C. Hunsaker. 1989. Comparison of chemical methods for determining postmortem interval. *Journal of Forensic Sciences* 34(1):197–206.

Spence, M.W., M.J. Shkrum, A. Ariss and J. Regan.1999. Craniocervical injuries in judicial hangings: an anthropologic analysis of six cases. *American Journal of Forensic Medicine and Pathology* 20(4):309–322.

Spennemann, D.H.R. and B. Franke. 1995a. Archaeological techniques for exhumations: a unique data source for crime scene investigations. *Forensic Science International* 74:5–15.

Spennemann, D.H.R. and B. Franke. 1995b. Decomposition of buried human bodies and associated death scene materials on coral atolls in the tropical Pacific. *Journal of Forensic Sciences* 40(3):356–367.

Spicer, J. 1975. *Billy the Kid*. San Francisco: Oyster Press.

Spitz, W.E. and R.S. Fisher. 1980. *Medicolegal Investigation of Death*, 2nd edition. Springfield, IL: Charles C. Thomas.

Springer, E. 1995. Toolmark examinations—a review of its development in the literature. *Journal of Forensic Sciences* 40(6):964–968.

Sprogoe-Jakobsen, S., A. Eriksson, H.P. Hougen, P.J. Knudsen, P. Leth and N. Lynnerup. 2001. Mobile autopsy teams in the investigation of war crimes in Kosovo 1999. *Journal of Forensic Sciences* 46(6):1392–1396.

Staiti, N., D. Di Martino and L. Saravo. 2004. A novel approach in personal identification from tissue samples undergone different processes through STR typing. *Forensic Science International* 146 (Suppl.):S171–S173.

Steadman, D.W., ed. 2003. *Hard Evidence: Case Studies in Forensic Anthropology*. Upper Saddle River, NJ: Prentice Hall.

Steadman, D.W. and W.D. Haglund. 2001. The scope of anthropological contributions to human rights investigations. *Proceedings of the Annual Meeting of the American Academy of Forensic Sciences* 7:237–238.

Steadman, D.W. and W.D. Haglund. 2005. The scope of anthropological contributions to human rights investigations. *Journal of Forensic Sciences* 50(1):23–30.

Steadman, D.W., B.J. Adams and L. Konigsberg. 2002. The statistical basis for positive identifications in forensic anthropology. *American Journal of Physical Anthropology, Annual Meeting Issue* 34:146.

Steadman, D.W., B.J. Adams and L.W. Konigsberg. 2006a. Statistical basis for positive identification in forensic anthropology. *American Journal of Physical Anthropology*, 131(1):15–26.

Steadman, D.W., L.L. DiAntonio, J.J. Wilson, K.E. Sheridan and S.P. Tammariello. 2006b. The effects of chemical and heat maceration techniques on the recovery of nuclear and mitochondrial DNA from bone. *Journal of Forensic Sciences* 51(1):11–17.

Steele, D.G. 1976. The estimation of sex on the basis of the talus and calcaneus. *American Journal of Physical Anthropology* 45:581–588.

Steele, J. 2000. Skeletal indicators of handedness. In *Human Osteology: In Archaeology and Forensic Science*. M. Cox, S. Mays, eds. Pp. 307–323. London:Greenwich Medical Media .

Steinbock, R.T. 1976. *Paleopathological Diagnosis and Interpretation: Bone Diseases in Ancient Human Populations*. Springfield, IL: Charles C. Thomas.

Stephan, C.N. 2003. Anthropological facial reconstruction—recognizing the fallacies, "unembracing" the errors and realizing method limits. *Science and Justice* 43(4):193–200.

Stephan, C.N. and M. Henneberg. 2006. Recognition of forensic facial approximation: case-specific examples and empirical tests. *Forensic Science International* 156(2–3):182–191.

Stewart, T. D. 1976. Evidence of handedness in the bony shoulder joint. *Paper presented at the 28th Annual Meeting of the American Academy of Forensic Sciences*, Chicago IL.

Stewart, T. D. 1979a. *Essentials of Forensic Anthropology, Especially as Developed in the United States*. Springfield, IL: Charles C. Thomas.

Stewart, T.D. 1979b. Forensic anthropology. In *The Use of Anthropology*. W. Goldschmidt, ed. Pp. 169–183. Special Publication no. 11 of the American Anthropological Association. Washington, DC: American Anthropological Association.

Stewart, T.D. 1984. Perspective on the reporting of forensic cases. In *Human Identification: Case Studies in Forensic Anthropology*. T.A. Rathbun, J.E. Buikstra, eds. Pp. 15–18. Springfield, IL: Charles C. Thomas.

Stojanowski, C.M. 1999. Sexing potential of fragmentary and pathological metacarpals. *American Journal of Physical Anthropology* 109: 245–52.

Stone, A.C. 2000. Ancient DNA from skeletal remains. In *Biological Anthropology of the Human Skeleton*. M.A. Katzenberg, S.R. Saunders, eds. Pp. 351–371. New York: Wiley-Liss.

Stone, A.C., J.E. Starrs and M. Stoneking. 2001. Mitochondrial DNA analysis of the presumptive remains of Jesse James. *Journal of Forensic Sciences* 46(1):173–176.

Stone, A.C., G. Milner, S. Pääbo, and M. Stoneking. 1996. Sex determination of ancient human skeletons using DNA. *American Journal of Physical Anthropology* 99:231–238.

Stone, R. 2004. Buried, recovered, lost again? The Romanovs may never rest. *Science* 303(5659):753–757.

Stover, E., W.D. Haglund and M. Samuels. 2003. Exhumation of mass graves in Iraq. *Journal of the American Medical Association* 290(5):663–666.

Stubblefield, P.R. 1999. Homicide or accident off the coast of Florida: trauma analysis of mutilated human remains. *Journal of Forensic Sciences* 44(4):716–719.

Stubblefield, P.R. 2003. Body weight estimation in forensic anthropology. *Proceedings from the Annual Meeting of the American Academy of Forensic Sciences* 9:262–263.

Sturner, W.Q., M.A. Herrmann, C. Boden, T.P. Scarritt, R.E. Sherman, T.S. Harmon and K.B. Wood. 2000. The *Frye* hearing in Florida: an attempt to exclude scientific evidence. *Journal of Forensic Sciences* 45(4):908–910.

Suchey, J.M. and D. Katz. 1998. Applications of pubic age determination. In *Forensic Osteology: Advances in the Identification of Human Remains* 2nd edition. K. J. Reichs, ed. Pp. 204–236. Springfield, IL: Charles C. Thomas.

Suchey, J.M., D.V. Wiseley, R.F. Green, and T.T. Noguchi. 1979. Analysis of dorsal pitting in the os pubis in an extensive sample of modern American females. *American Journal of Physical Anthropology* 51:517–540.

Sullivan, T. 2005. Letter to the editor. *Ruidoso News*, Ruidoso, NM, October 21, 2005.

Suskewicz, J.A. 2004. Estimation of living body weight using measurements of anterior iliac spine breadth and stature. *Proceedings of the Annual Meeting of the American Academy of Forensic Sciences* 10:322–323.

Suter, K. 2002. Progress in the international protection of human rights. *Medicine, Conflict and Survival* 18(3):283–98.

Sutherland, L.D. and J.M. Suchey. 1991. Use of the ventral arc in pubic sex determination. *Journal of Forensic Sciences* 36:501–511.

Svanholm, H., H. Starklint, H.J.G. Gundersen, J. Fabricius, H. Barlebo and S. Olsen. 1989. Reproducibility of histomorphological diagnosis with special reference to the kappa statistic. *Acta Pathologica, Microbiologica, et Immunologica* 97:689–698.

Swan, L.K. and C.N. Stephan. 2005. Estimating eyeball protrusion from body height, interpupillary distance, and inter-orbital distance in adults. *Journal of Forensic Sciences* 50(4):774–776.

Swartley, R.1999. *The Billy the Kid Travel Guide*. Las Cruces, NM: Frontier Image Press.

SWGDAM. 2003. Guidelines for mitochondrial DNA (mtDNA) nucleotide sequence interpretations. *Forensic Science Communications;* 5(2): online. Electronic document, *http://www.fbi.gov/hq/lab/fsc/backissu/april2003/swgdammitodna.htm*, accessed May 29, 2006.

Swift, B. 1998. Dating human skeletal remains: investigating the viability of measuring the equilibrium between ^{210}Po and ^{210}Pb as a means of estimating the postmortem interval. *Forensic Science International* 98(1–2):119–26.

Symes, S.A. 1992. *Morphology of Saw Marks in Human Bone: Identification of Class Characteristics*. PhD Dissertation. Knoxville: Department of Anthropology, University of Tennessee.

Symes, S.A., O.C. Smith, H.E. Berryman, C.E. Peters, L.A. Rockhold, S.J. Haun, J.T. Francisco and T.P. Sutton.1996. *Bones: Bullets, Burns, Bludgeons, Blunders, and Why*. Workshop presented at the 48th Annual Meeting of the American Academy of Forensic Sciences, Nashville, TN.

Symes, S.A., H.E. Berryman and O.C. Smith. 1998. Saw marks in bone: introduction and examination of residual kerf contour. In *Forensic Osteology*, 2nd edition. K.J. Reichs, ed. Pp. 389–409. Springfield, IL: Charles C. Thomas.

Synstelien, J.A. and M.D. Hamilton. 2003. Expressions of handedness in the vertebral column. *Proceedings from the Annual Meeting of the American Academy of Forensic Sciences* 9:299.

Takatsu, A., N. Suzuki, A. Hattori and A. Shigeta. 1999. The concept of the digital morgue as a 3D database. *Legal Medicine (Tokyo)* 1(1):29–33.

Tanikawa, S. and H. Wright. 1980. *Billy the Kid*. Madison, WI: Bieler Press.

Tanner, J.M., R.H. Whitehouse, W.A. Marshall, M.J.R. Healy and H. Goldstein. 1975. *Assessment of Skeletal Maturity and Prediction of Adult Height (TW-2Method)*. New York: Academic Press.

Tanner, J.M., M.J.R. Healy, H. Goldstein and N. Cameron. 2001. *Assessment of Skeletal Maturity and Prediction of Adult Height (TW-3Method)*. London: W. B. Saunders.

Tantawi, T.I. and B. Greenburg. 1993. The effect of killing and preservative solutions on estimates of maggot age in forensic cases. *Journal of Forensic Sciences* 38(3):702–707.

Tatum, S. 1997. *Inventing Billy the Kid: Visions of the Outlaw in America, 1881–1981*. Tucson: University of Arizona Press.

Taylor, J.V., L. Roh and A.D. Goldman. 1984. Metropolitan Forensic Anthropology Team (MFAT) case studies in identification: 2. Identification of a Vietnamese trophy skull. *Journal of Forensic Sciences* 29(4):1253–1259.

Taylor, M.S., A. Challed-Spong and E.A. Johnson. 1997. Co-amplification of the amelogenin and HLA DQα?genes: optimization and validation. *Journal of Forensic Sciences* 42(1):130–136.

Taylor, R.E., J.M. Suchey, L.A. Payen and P.J. Slota. 1989. The use of radiocarbon (^{14}C) to identify human skeletal material of forensic science interest. *Journal of Forensic Sciences* 34(5):196–205.

Tedeschi, L.G. 1984. Human rights and the forensic scientist. *American Journal of Forensic Medicine and Pathology* 5(4):295–296.

Thali, M.J., R. Dirnhofer, R. Becker, W. Oliver and K. Potter. 2004. Is "virtual histology" the next step after the "virtual autopsy"? Magnetic resonance microscopy in forensic medicine. *Magnetic Resonance Imaging* 22(8):1131–1138.

Thali, M.J., K. Yen, T. Plattner, W. Schweitzer, P. Vock, C. Ozdoba and R. Dirnhofer. 2002a. Charred body: virtual autopsy with multi-slice computed tomography and magnetic resonance imaging. *Journal of Forensic Sciences* 47(6):1326–1331.

Thali, M.J., B.P. Kneubuehl, P. Vock, G. Allmen and R. Dirnhofer. 2002b. High-speed documented experimental gunshot to a skull-brain model and radiologic virtual autopsy. *American Journal of Forensic Medicine and Pathology* 23(3):223–228.

Thali, M.J., K. Yen, P. Vock, C. Ozdoba, B.P. Kneubuehl, M. Sonnenschein and R. Dirnhofer. 2003a. Image-guided virtual autopsy findings of gunshot victims performed with multi-slice computed tomography and magnetic resonance imaging and subsequent correlation between radiology and autopsy findings. *Forensic Science International* 138(1–3):8–16.

Thali, M.J., K. Yen, W. Schweitzer, P. Vock, C. Boesch, C. Ozdoba, G. Schroth, M. Ith, M. Sonnenschein, T. Doernhoefer, E. Scheurer, T. Plattner and R. Dirnhofer. 2003b. Virtopsy, a new imaging horizon in forensic pathology: virtual autopsy by postmortem multislice computed tomography (MSCT) and magnetic resonance imaging (MRI)—a feasibility study. *Journal of Forensic Sciences* 48(2):386–403.

Thompson, D.D. 1978. *Age Related Changes in Osteon Remodeling and Bone Mineralization*. PhD Dissertation, University of Connecticut, Storrs.

Thompson, D.D. 1979. The core technique in the determination of age at death in skeletons. *Journal of Forensic Sciences* 24:902–915.

Thompson, D.D. 1982. Forensic anthropology. In *A History of American Physical Anthropology 1930–1980*. F. Spencer, ed. Pp. 357–369. New York: Academic Press.

Thomsen, J.L. 2000. The role of the pathologist in human rights abuses. *Journal of Clinical Pathology* 53(8):569–572.

Thomsen, J.L. and J. Voigt. 1988. Forensic medicine and human rights. *Forensic Science International* 36(1–2):147–151.

Thomsen, J.L., J. Gruschow and E. Stover. 1989. Medicolegal investigation of political killings in El Salvador. *Lancet* 1(8651):1377–1379.

Tiesler, V., A. Cucina and A.R. Pacheco. 2004. Who was the Red Queen? Identity of the female Maya dignitary from the sarcophagus tomb of Temple XIII, Palenque, Mexico. *Homo* 55(1–2):65–76.

Todd, T.W. 1920. Age changes in the pubic bone: I, The male white pubis. *American Journal of Physical Anthropology* 3:286–334.

Todd, T.W. and D.W. Lyon. 1924. Endocranial suture closure, its progress and age relationship: I. adult males of white stock. *American Journal of Physical Anthropology* 7:325–384.

Tomczak, P.D. and J.E. Buikstra. 1999. Analysis of blunt trauma injuries: vertical deceleration versus horizontal deceleration injuries. *Journal of Forensic Sciences* 44(2):253–262.

Trancho, G.J., B. Robledo, I. López-Bueis and J. Sanchez. 1997. Sexual determination of the femur using discriminant functions. Analysis of a Spanish population of known sex and age. *Journal of Forensic Sciences* 42:181–185.

Trotter, M. 1970. Estimation of stature from intact long bones. In *Personal Identification in Mass Disasters*. T.D. Stewart, ed. Pp. 71–83. Washington, DC: Smithsonian Institution.

Trotter, M and G. C. Gleser. 1958. A re-evaluation of estimation of stature based on measurements of stature taken during life and of long bones after death. *American Journal of Physical Anthropology* 16:79–123.

Tucker, B.K., D.L. Hutchinson, M.F. Gilliland, T.M. Charles, H.J. Daniel and L.D. Wolfe. 2001. Microscopic characteristics of hacking trauma. *Journal of Forensic Sciences* 46(2):234–240.

Tuller, H. and J. Sterenberg. 2005. Not for the passive: the active application of electronic resistivity in the excavation of a mass grave. *Proceedings of the Annual Meeting of the American Academy of Forensic Sciences* 11:296.

Tuller, H., U. Hofmeister and S. Daley. 2005. The importance of body deposition recording in event reconstruction and the re-association and identification of commingled remains. *Proceedings of the Annual Meeting of the American Academy of Forensic Sciences* 11:313.

Tunstill, W.A. 1988. *Billy the Kid and Me Were the Same*. Roswell, NM: Western History Research Center.

Turner, G.E. 1974. *Secrets of Billy the Kid*. Amarillo, TX: Baxter Lane.

Turner, W., P. Tu, T. Kelliher and R. Brown. 2005. Computer-aided forensics: facial reconstruction. *Studies in Health Technology and Informatics* 119:550–555.

Tuska, J. 1994. *Billy the Kid, His Life and Legend*. Westport, CT: Greenwood Press.

Tyrrell, A.J., M.P. Evison, A.T. Chamberlain and M.A. Green. 1997. Forensic three-dimensional facial reconstruction: historical review and contemporary developments. *Journal of Forensic Sciences* 42(4):653–661.

Ubelaker, D.H. 1989. *Human Skeletal Remains: Excavation, Analysis, Interpretation*, 2nd edition. Washington, DC: Taraxacum.

Ubelaker, D.H. 1990. Positive identification of American Indian skeletal remains from radiographic comparison. *Journal of Forensic Sciences* 35(2):466–472.

Ubelaker, D.H. 1992. Hyoid fracture and strangulation. *Journal of Forensic Sciences* 37(5): 1216–1222.

Ubelaker, D.H. 1996a. The remains of Dr. Carl Austin Weiss: anthropological analysis. *Journal of Forensic Sciences* 41(1):60–79.

Ubelaker, D.H. 1996b. Skeletons testify: anthropology in forensic science. AAPA luncheon address: April 12, 1996. *Yearbook of Physical Anthropology* 39:229–244.

Ubelaker, D.H. 1997. Forensic anthropology. In *History of Physical Anthropology*, Vol. 1, *A-L*. F. Spencer, ed. Pp. 392–395. New York: Garland Publishing .

Ubelaker, D.H. 2000a. The forensic anthropology legacy of T. Dale Stewart (1901–1997). *Journal of Forensic Sciences* 45(2):245–252.

Ubelaker, D.H. 2000b. Methodological considerations in the forensic applications of human skeletal biology. In *Biological Anthropology of the Human Skeleton*. M.A. Katzenberg, S.R. Saunders, eds. Pp. 41–67. New York: Wiley-Liss.

Ubelaker, D.H. 2000c. T. Dale Stewart's perspective on his career as a forensic anthropologist at the Smithsonian. *Journal of Forensic Sciences* 45(2):269–278.

Ubelaker, D.H. 2001. Contributions of Ellis R. Kerley to forensic anthropology. *Journal of Forensic Sciences* 46(4):773–776.

Ubelaker, D.H. 2005. Advances in the assessment of commingling within samples of human remains. *Proceedings of the Annual Meeting of the American Academy of Forensic Sciences* 11:317.

Ubelaker, D.H. and B.J. Adams. 1995. Differentiation of perimortem and postmortem trauma using taphonomic indicators. *Journal of Forensic Sciences* 40(3): 509–512.

Ubelaker, D.H. and D.R. Hunt. 1995. The influence of William M. Bass III on the development of American forensic anthropology. *Journal of Forensic Sciences* 40(5):729–734.

Ubelaker, D.H., D.W. Owsley, M.M. Houck, E. Craig, W. Grant, T. Woltanksi, R. Fram, K. Sandness and N. Peerwani. 1995. The role of forensic anthropology in the recovery and analysis of Branch Davidian Compound victims: recovery procedures and characteristics of the victims. *Journal of Forensic Sciences* 40(3):335–340.

Ubelaker, D.H., A. H. Ross and S. M. Graver. 2002a. Application of forensic discriminant functions to a Spanish cranial sample. *Forensic Science Communications* 4(3):1–6.

Ubelaker, D.H., D.C. Ward, V.S. Braz and J. Stewart. 2002b. The use of SEM/EDS analysis to distinguish dental and osseous tissue from other materials. *Journal of Forensic Sciences* 47(5):940–943.

Ubelaker, D.H., J.M. Lowenstein and D.G. Hood. 2004. Use of solid-phase double-antibody radioimmunoassay to identify species from small skeletal fragments. *Journal of Forensic Sciences* 49(5):924–929.

Ubelaker, D.H., B.A. Buchholz and J.E. Stewart. 2006. Analysis of artificial radiocarbon in different skeletal and dental tissue types to evaluate date of death. *Journal of Forensic Sciences* 51(3):484–488.

United Nations. 1948. Article II and III, Convention on the Prevention and Punishment of Genocide. Resolution 260(III) A. United Nations General Assembly, December 9, 1948.

United Nations. 1988. International Covenant on Civil and Political Rights. *Annual Review of Population Law* 15:148.

United Nations. 1995. *Guidelines for the Conduct of United Nations Inquiries into Allegations of Massacres*. New York: United Nations Office of Legal Affairs.

University of Arizona. 1991. The Maurice Garland Fulton collection of New Mexicana. Tucson: Special Collections, University Library, University of Arizona.

Usher, B. M. 2002. Reference samples: the first step in linking biology and age in the human skeleton. In *Palaeodemography: I. Age Distributions from Skeletal Samples*, R.D. Hoppa, J.W. Vaupel, eds. Pp 29–47. Cambridge: Cambridge University Press.

Utley, R.M. 1986. *Four Fighters of Lincoln County*. Albuquerque: University of New Mexico Press.

Utley, R.M. 1989. *Billy the Kid*. Lincoln: University of Nebraska Press.

Valdez, J.P. and B.E. Hefner.1995. *Billy the Kid: "Killed" in New Mexico—Died in Texas*. Dallas: Outlaw Publications.

Van Den Oever, R. 1976. A review of the literature as to the present possibilities and limitations in estimating the time of death. *Medicine, Science and the Law* 16(4):269–276.

Vanezis, P. 1999. Investigation of clandestine graves resulting from human rights abuses. *Journal of Clinical Forensic Medicine* 6(4):238–242.

Van Oorschot, R.A., S. Treadwell, J. Beaurepaire, N.L. Holding and R.J. Mitchell. 2005. Beware of the possibility of fingerprinting techniques transferring DNA. *Journal of Forensic Sciences* 50(6):1417–1422.

Varetto, L. and O. Curto. 2005. Long persistence of rigor mortis at constant low temperature. *Forensic Science International* 147(1):31–34.

Vass, A.A., W.M. Bass, J.D. Wolt, J.E. Foss and J.T. Ammons. 1992. Time since death determinations of human cadavers using soil solution. *Journal of Forensic Sciences* 37(5):1236–1253.

Vernesi, C., G. Di Benedetto, D. Caramelli, E. Secchieri, L. Simoni, E. Katti, P. Malaspina, A. Novelletto, V.T. Marin and G. Barbujani. 2001. Genetic characterization of the body attributed to the evangelist Luke. *Proceedings of the National Academy of Sciences USA* 98(23):13460–13463.

Vesterby, A., and L.W. Poulsen. 1997. The diagnosis of a murder from skeletal remains: a case report. *International Journal of Legal Medicine* 110(2):97–100.

Vignolo, M., A. Naselli, P. Magliano et al. 1999. Use of the new US90 standards for TW-RUS skeletal maturity scores in youths from the Italian population. *Hormone Research* 51(4):168–172.

von Wurmb-Schwark, N., M. Harbeck, U. Wiesbrock, I. Schroeder, S. Ritz-Timme and M. Oehmichen. 2003. Extraction and amplification of nuclear and mitochondrial DNA from ancient and artificially aged bones. *Legal Medicine (Tokyo)* 5 (Suppl.) 1:S169–S72.

von Wurmb-Schwark, N., A. Ringleb, M. Gebuhr and E. Simeoni. 2005. Genetic analysis of modern and historical burned human remains. *Anthropologischer Anzeiser* 63(1):1–12.

Walker, P.L. 1995. Problems of preservation and sexism in sexing: some lessons from historical collections for paleodemographers. In *Grave Reflections: Portraying the Past Through Skeletal Studies*. A. Herring, S.R. Saunders, eds. Pp. 31–47. Toronto: Canadian Scholars' Press.

Walker, P.L., D.C. Cook and P.M. Lambert. 1997. Skeletal evidence for child abuse: a physical anthropological perspective. *Journal of Forensic Sciences* 42:196–207.

Walker, P.R. 1992. *Great Figures of the Wild West*. New York: Facts on File.

Walsh-Haney, H.A. 1999. Sharp-force trauma analysis and the forensic anthropologist: techniques advocated by William R. Maples, Ph.D. *Journal of Forensic Sciences* 44(4):720–723.

Warren, M.W. 1999. Radiographic determination of developmental age in fetuses and stillborns. *Journal of Forensic Sciences* 44(4):708–712.

Warren, M.W. and J.J. Schultz. 2002. Post-cremation taphonomy and artifact preservation. *Journal of Forensic Sciences* 47(3):656–659.

Weaver, D.S. 1998. Forensic aspects of fetal and neonatal skeletons. In *Forensic Osteology: Advances in the Identification of Human Remains* 2nd edition. K.J. Reichs, ed. Pp. 187–203. Springfield, Il: Charles C. Thomas.

Weddle, J. 1993. *Antrim Is My Stepfather's Name: the Boyhood of Billy the Kid*. Tucson: Arizona Historical Society.

Weedn, V.W. 1998. Postmortem identification of remains. *Clinical Laboratory Medicine* 18(1):115–137.

Weinberger, L.E., S. Sreenivasan, E.A. Gross, E. Markowitz and B.H.Gross. 2000. Psychological factors in the determination of suicide in self-inflicted gunshot head wounds. *Journal of Forensic Sciences* 45(4):815–819.

Wellborn III, O.G. 2005. *Cases and Materials on the Rules of Evidence*, 3rd edition. St. Paul, MN: Thomson/West.

Wessling, R. 2001. Finding without digging?! Basic, controlled experiments in forensic geophysics to investigate the relationship between soil and resistivity. *Proceedings from the American Academy of Forensic Sciences Annual Meeting* 7:249.

White, T.D. and P.A. Folkens. 2005. *The Human Bone Manual*. Burlington, MA: Elsevier-Academic Press.

Whitlow, D. 1992. *Lincoln County Diary*. Santa Fe, NM: Sunstone Press.

Wienker, C.W. and J.S. Rhine. 1989. A professional profile of the physical anthropology section membership, American Academy of Forensic Sciences. *Journal of Forensic Sciences* 34:647–658.

Wilcher G.W. and B. Hulewicz. 2005. Positive identification of a decomposed body using a trilogy of identification criteria: a case report. *Medicine, Science and the Law* 45(3):267–272.

Wilczak, A. and K.A.R. Kennedy. 1998. In *Forensic Osteology: Advances in the Identification of Human Remains* 2nd edition. K. J. Reichs, ed. Pp. 461–490. Springfield, Il: Charles C. Thomas.

Wilkinson, D. 1999. Reframing family ethnicity in America. In *Family Ethnicity, Strength in Diversity*, 2nd edition. H.P. McAdoo, ed. Pp. 15–62. Thousand Oaks, CA: Sage Publications.

Willems, G. 2000. *Forensic Odontology*. Leuven, Belgium: Leuven University Press.

Willey, P. and T. Falsetti. 1991. Inaccuracy of height information on driver's licenses. *Journal of Forensic Sciences* 36(3):813–819.

Willey, P. and A. Heilman. 1987. Estimating time since death using plant roots and stems. *Journal of Forensic Sciences* 32(5):1264–1270.

Willey, P. and L.M. Snyder. 1989. Canid modification of human remains: implications for time-since-death estimations. *Journal of Forensic Sciences* 34(4):894–901.

William, B. 1964. *The Complete and Factual Life of Billy the Kid*. New York: F. Fell.

Williams, B.A. and T.L. Rogers. 2006. Evaluating the accuracy and precision of cranial morphological traits for sex determination. *Journal of Forensic Sciences* 51(4):729–735.

Williams, E.D. and J.D. Crews. 2003. From dust to dust: ethical and practical issues involved in the location, exhumation, and identification of bodies from mass graves. *Croatian Medical Journal* 44(3):251–258.

Williams, F.L.E., R.L. Belcher and G. J. Armelagos. 2005. Forensic misclassification of ancient Nubian crania: implications for assumptions about human variation. *Current Anthropology* 46:340–346.

Wilson, E.F., J.H. Davis, J.D. Bloom, P.J. Batten and S.G. Kamara. 1998. Homicide or suicide: the killing of suicidal persons by law enforcement officers. *Journal of Forensic Sciences* 43(1):46–52.

Wilson, M.R., M. Stoneking, M.M. Holland, J.A. DiZinno and B. Budowle. 1993. Guidelines for the use of mitochondrial DNA sequencing in forensic science. *Crime Laboratory Digest* 20(4):68–77.

Wiredu, E.K., R. Kumoji, R. Seshadri and R.B. Biritwum. 1999. Osteometric analysis of sexual dimorphism in the sternal end of the rib in a West African population. *Journal of Forensic Sciences* 44(5):921–925.

Wittwer-Backhofen, U. and H. Buba. 2002. Age estimation by tooth cementum annulation: perspective of a new validation study. In *Palaeodemography: I. Age Distributions from Skeletal Samples*. R.D. Hoppa, J.W. Vaupel, eds. Pp. 107–128. Cambridge: Cambridge University Press.

Wolf, D.J. 1986. Forensic anthropology scene investigations. In *Forensic Osteology: Advances in the Identification of Human Remains*. K.J. Reichs, ed. Pp. 3–23. Springfield, IL: Charles C. Thomas.

Wood, W.R. and L.A. Stanley. 1989. Recovery and identification of World War II dead: American graves registration activities in Europe. *Journal of Forensic Sciences* 34:1365–1373.

Wright, R.J., C.D. Peters and R.B. Flannery. 1990. Victim identification and family support in mass casualties: the Massachusetts model. *International Journal of Emergency Mental Health* 1(4):237–242.

Wu, L. 1989. Sex determination of Chinese femur by discriminant function. *Journal of Forensic Sciences* 34: 1222–1227.

Wulpi, D.J. 1985. *Understanding How Components Fail*. Metals Park, OH: American Society for Metals.

Wurlitzer, R. 1973. *Pat Garrett and Billy the Kid*. New York: New American Library.

Wyler, D., W. Marty and W. Bar. 1994. Correlation between the postmortem cell content of cerebrospinal fluid and time of death. *International Journal of Legal Medicine* 106(4):194–199.

Yamamoto, T., R. Uchihi, T. Kojima, H. Nozawa, X.L. Huang, K. Tamaki and Y. Katsumata. 1998. Maternal identification from skeletal remains of an infant kept by the alleged mother for 16 years with DNA typing. *Journal of Forensic Sciences* 43(3):701–705.

Yazedjian, L.N., R. Kesetovic and A. Boza-Arlotti. 2005. The importance of using traditional anthropological methods in a DNA-led identification system. *Proceedings from the Annual Meeting of the American Academy of Forensic Sciences* 11:312–313.

Ye, J., A. Ji, E.J. Parra, X. Zheng, C. Jiang, X. Zhao, L. Hu and Z. Tu. 2004. A simple and efficient method for extracting DNA from old and burned bone. *Journal of Forensic Sciences* 49(4):754–759.

Yoder, C., D.H. Ubelaker and J.F. Powell. 2001. Examination of variation sternal rib end morphology relevant to age assessment. *Journal of Forensic Sciences* 46:22

Yoganandan, N. and F.A. Pintar. 2004. Biomechanics of temporo-parietal skull fracture. *Clinical Biomechanics* 19(3):225–239.

Yoshino, M., S. Miyasaka, H. Sato and S. Seta. 1987. Classification system of frontal sinus patterns by radiography: its application to identification of unknown remains. *Forensic Science International* 34(4):289–299.

Yoshino, M., T. Kimijima, S. Miyasaka, H. Sato and S. Seta. 1991. Microscopical study on estimation of time since death in skeletal remains. *Forensic Science International* 49:143–158.

Zagriadskaia, A.P. and N.S. Edelev. 1981. [Forensic medical identification of stabbing and cutting weapons from the marks of the blade's microrelief on rib cartilage: in Russian]. *Sudebno-Meditsinskaia Ekspertiza* 24(1):45–48.

Zharov, V.V. 1996. [Establishing the time of death by putrefactive transformation of the corpse: in Russian] *Sudebuo-Meditsinskaia Ekspertiza* 39(2):5–7.

Zhivotovsky, L.A. 1999. Recognition of the remains of Tsar Nicholas II and his family: a case of premature identification? *Annals of Human Biology* 26(6):569–577.

Zuo, Z. J. and J.Z. Zhu. 1991. Study on the microstructure of skull fracture. *Forensic Science International* 50(1):1–14.

Index